CORROSION AND CORROSION CONTROL

CORROSION AND CORROSION CONTROL

An Introduction to Corrosion Science and Engineering

FOURTH EDITION

R. Winston Revie
Senior Research Scientist
CANMET Materials Technology Laboratory
Natural Resources Canada

Herbert H. Uhlig
Former Professor Emeritus
Department of Materials Science and Engineering
Massachusetts Institute of Technology

A JOHN WILEY & SONS, INC., PUBLICATION

Library of Congress Cataloging-in-Publication Data:

Uhlig, Herbert Henry, 1907–
    Corrosion and corrosion control : an introduction to corrosion science and engineering /
Herbert H. Uhlig, R. Winston Revie.—4th ed.
        p. cm.
    Includes bibliographical references and index.
    ISBN 978-0-471-73279-2 (cloth)
1. Corrosion and anti-corrosives.   I. Revie, R. Winston (Robert Winston), 1944–   II. Title.
    TA462.U39 2008
    620.1′1223–dc22

                                                                                    2007041578

Printed in the United States of America

10 9 8 7 6 5 4 3 2 1

CONTENTS

PREFACE

The three main global challenges for the twenty-first century are energy, water, and air—that is, sufficient energy to ensure a reasonable standard of living, clean water to drink, and clean air to breathe. The ability to manage corrosion is a central part of using materials effectively and efficiently to meet these challenges. For example, oil and natural gas are transmitted across continents using high-pressure steel pipelines that must operate for decades without failure, so that neither the groundwater nor the air is unnecessarily polluted. In design, operation, and maintenance of nuclear power plants, management of corrosion is critical. The reliability of materials used in nuclear waste disposal must be sufficient so that that the safety of future generations is not compromised.

Materials reliability is becoming ever more important in our society, particularly in view of the liability issues that develop when reliability is not assured, safety is compromised, and failure occurs. Notwithstanding the many years over which university, college, and continuing education courses in corrosion have been available, high-profile corrosion failures continue to take place. Although the teaching of corrosion should not be regarded as a dismal failure, it has certainly not been a stellar success providing all engineers and technologists a basic minimum "literacy level" in corrosion that would be sufficient to ensure reliability and prevent failures.

Senior management of some organizations has adopted a policy of "zero failures" or "no failures." In translating this management policy into reality, so that "zero" really does mean "zero" and "no" means "no," engineers and others manage corrosion using a combination of well-established strategies, innovative approaches, and, when necessary, experimental trials.

One objective of preparing the fourth edition of this book is to present to students an updated overview of the essential aspects of corrosion science and engineering that underpin the tools that are available and the technologies that are used for managing corrosion and preventing failures. A second objective is to engage students, so that they are active participants in understanding corrosion and solving problems, rather than passively observing the smorgasbord of information presented. The main emphasis is on quantitative presentation, explanation, and analysis wherever possible; for example, in this new edition, the galvanic series in seawater is presented with the potential range of each material, rather than only as a qualitative list. Considering the potential ranges that can be involved, the student can see how anodic/cathodic effects can develop, not only when different materials form a couple, but also when materials that are nominally the same are coupled. In this edition, some new numerical problems have been added, and the problems are integrated into the book by presenting them at the ends of the chapters.

Since the third edition of this book was published, there have been many advances in corrosion, including advances in knowledge, advances in alloys for application in aggressive environments, and advances of industry in response to public demand. For example, consumer demand for corrosion protection of automobiles has led to a revolution of materials usage in the automotive industry. For this reason, and also because many students have a fascination with cars, numerous examples throughout this book illustrate advances that have been made in corrosion engineering of automobiles. Advances in protecting cars and trucks from corrosion must also be viewed in the context of reducing vehicle weight by using magnesium, aluminum, and other lightweight materials in order to decrease energy usage (increase the miles per gallon, or kilometers per liter, of gasoline) and reduce greenhouse gas emissions.

Although the basic organization of the book is unchanged from the previous edition, there is in this edition a separate chapter on Pourbaix diagrams, very useful tools that indicate the thermodynamic potential–pH domains of corrosion, passivity, and immunity to corrosion. A consideration of the relevant Pourbaix diagrams can be a useful starting point in many corrosion studies and investigations. As always in corrosion, as well as in this book, there is the dual importance of thermodynamics (In which direction does the reaction go? Chapters 3 and 4) and kinetics (How fast does it go? Chapter 5).

After establishing the essential basics of corrosion in the first five chapters, the next 23 chapters expand upon the fundamentals in specific systems and applications and discuss strategies for protection. There are separate chapters on aluminum (Chapter 21), magnesium (Chapter 22), and titanium (Chapter 25) to provide more information on these metals and their alloys than in the previous editions. Throughout this book, environmental concerns and regulations are presented in the context of their impact on corrosion and its control—for example, the EPA Lead and Copper rule enacted in the United States in 1991. The industrial developments in response to the Clean Air Act, enacted in 1970, have reduced air pollution in the United States, with some effect on atmospheric corrosion (Chapter 9). To meet the requirements of environmental regulations and reduce the use of organic solvents, compliant coatings have been developed (Chapter 16).

This is primarily a textbook for students and others who need a basic understanding of corrosion. The book is also a reference and starting point for engineers, researchers, and technologists requiring specific information. The book includes discussion of the main materials that are available, including alloys both old and new. For consistency with current practice in metallurgical and engineering literature, alloys are identified with their UNS numbers as well as with their commonly used identifiers. To answer the question from students about why so many alloys have been developed and are commercially available, the contributions of individual elements to endow alloys with unique properties that are valuable for specific applications are discussed. Throughout the book, there are numerous references to further sources of information, including handbooks, other books, reviews, and papers in journals. At the end of each chapter, there is a list of "General References" pertinent to that chapter, and most of these were published in 2000 and later.

This edition includes introductory discussions of risk (Chapter 1), AC impedance measurements (Chapter 5), Ellingham diagrams (Chapter 11), and, throughout the book, discussions of new alloys that have been developed to meet demands for increasing reliability notwithstanding the increased structural lifetimes that are being required in corrosive environments of ever-increasing severity. Perhaps nowhere are the demands for reliability more challenging than in nuclear reactors, discussed in Chapters 8 and 26. In the discussion of stainless steels (Chapter 19), the concept of critical pitting temperature (CPT) is introduced, as well as the information on critical pitting potential (CPP). The important problem of corrosion of rebar (reinforced steel in concrete) is discussed in Chapter 7 on iron and steel.

In addition to new technologies and new materials for managing corrosion, new tools for presenting books have become available; hence, this book is being published as an electronic book, as well as in the traditional print format. An instructor's manual is also being prepared.

Experience has been invaluable in using the book in a corrosion course in the Department of Mechanical and Aerospace Engineering at Carleton University in Ottawa, which Glenn McRae and I developed along with other members of the Canadian National Capital Section of NACE International.

It would be a delight for me to hear from readers of this book, with their suggestions and ideas for future editions.

I would like to acknowledge my many friends and colleagues at the CANMET Materials Technology Laboratory, with whom it has been my privilege to work for the past nearly 30 years. I would also like to thank the many organizations and individuals who have granted permission to use copyright material; acknowledgments for specific material are provided throughout the book. In addition, I would like to thank Bob Esposito and his staff at John Wiley & Sons, Inc. for their encouragement with this book and also with the Wiley Series in Corrosion.

I would like to thank the Uhlig family for their generosity and hospitality during five decades, beginning when I was a student in the M.I.T. Corrosion Laboratory in the 1960s and 1970s. In particular, I would like to acknowledge Mrs. Greta Uhlig, who continues to encourage initiatives in corrosion education in memory of the late Professor Herbert H. Uhlig (1907–1993).

Lastly, I would like to quote from the Preface of the first edition of this book:

If this book stimulates young minds to accept the challenge of continuing corrosion problems, and to help reduce the huge economic losses and dismaying wastage of natural resources caused by metal deterioration, it will have fulfilled the author's major objective.

Indeed, this remains the main objective today.

Ottawa, Canada   R. WINSTON REVIE
September 2007

1

DEFINITION AND IMPORTANCE OF CORROSION

1.1 DEFINITION OF CORROSION

Corrosion is the destructive attack of a metal by chemical or electrochemical reaction with its environment. Deterioration by physical causes is not called corrosion, but is described as erosion, galling, or wear. In some instances, chemical attack accompanies physical deterioration, as described by the following terms: corrosion–erosion, corrosive wear, or fretting corrosion. Nonmetals are not included in this definition of corrosion. Plastics may swell or crack, wood may split or decay, granite may erode, and Portland cement may leach away, but the term corrosion, in this book, is restricted to chemical attack of metals.

"Rusting" applies to the corrosion of iron or iron-base alloys with formation of corrosion products consisting largely of hydrous ferric oxides. Nonferrous metals, therefore, corrode, but do not rust.

1.1.1 Corrosion Science and Corrosion Engineering

Since corrosion involves chemical change, the student must be familiar with principles of chemistry in order to understand corrosion reactions. Because corrosion processes are mostly electrochemical, an understanding of

Corrosion and Corrosion Control, by R. Winston Revie and Herbert H. Uhlig
Copyright © 2008 John Wiley & Sons, Inc.

electrochemistry is also important. Furthermore, since structure and composition of a metal often determine corrosion behavior, the student should be familiar with the fundamentals of physical metallurgy as well.

The *corrosion scientist* studies corrosion mechanisms to improve (a) the understanding of the causes of corrosion and (b) the ways to prevent or at least minimize damage caused by corrosion. The *corrosion engineer*, on the other hand, applies scientific knowledge to control corrosion. For example, the corrosion engineer uses cathodic protection on a large scale to prevent corrosion of buried pipelines, tests and develops new and better paints, prescribes proper dosage of corrosion inhibitors, or recommends the correct coating. The corrosion scientist, in turn, develops better criteria of cathodic protection, outlines the molecular structure of chemical compounds that behave best as inhibitors, synthesizes corrosion-resistant alloys, and recommends heat treatment and compositional variations of alloys that will improve their performance. Both the scientific and engineering viewpoints supplement each other in the diagnosis of corrosion damage and in the prescription of remedies.

1.2 IMPORTANCE OF CORROSION

The three main reasons for the importance of corrosion are: economics, safety, and conservation. To reduce the economic impact of corrosion, corrosion engineers, with the support of corrosion scientists, aim to reduce material losses, as well as the accompanying economic losses, that result from the corrosion of piping, tanks, metal components of machines, ships, bridges, marine structures, and so on. Corrosion can compromise the safety of operating equipment by causing failure (with catastrophic consequences) of, for example, pressure vessels, boilers, metallic containers for toxic chemicals, turbine blades and rotors, bridges, airplane components, and automotive steering mechanisms. Safety is a critical consideration in the design of equipment for nuclear power plants and for disposal of nuclear wastes. Loss of metal by corrosion is a waste not only of the metal, but also of the energy, the water, and the human effort that was used to produce and fabricate the metal structures in the first place. In addition, rebuilding corroded equipment requires further investment of all these resources— metal, energy, water, and human.

Economic losses are divided into (1) direct losses and (2) indirect losses. Direct losses include the costs of replacing corroded structures and machinery or their components, such as condenser tubes, mufflers, pipelines, and metal roofing, including necessary labor. Other examples are (a) repainting structures where prevention of rusting is the prime objective and (b) the capital costs plus maintenance of cathodic protection systems for underground pipelines. Sizable direct losses are illustrated by the necessity to replace several million domestic hot-water tanks each year because of failure by corrosion and the need for replacement of millions of corroded automobile mufflers. Direct losses include the extra cost of using corrosion-resistant metals and alloys instead of carbon steel where the latter has adequate mechanical properties but not sufficient corrosion resistance; there are also the costs of galvanizing or nickel plating of steel, of adding corrosion inhibitors to water, and of dehumidifying storage rooms for metal equipment.

The economic factor is a very important motivation for much of the current research in corrosion. Losses sustained by industry and by governments amount to many billions of dollars annually, approximately $276 billion in the United States, or 3.1% of the Gross Domestic Product (GDP), according to a recent study [1]. It has been estimated that about 25–30% of this total could be avoided if currently available corrosion technology were effectively applied [1].

Studies of the cost of corrosion to Australia, Great Britain, Japan, and other countries have also been carried out. In each country studied, the cost of corrosion is approximately 3–4 % of the Gross National Product [2].

Indirect losses are more difficult to assess, but a brief survey of typical losses of this kind compels the conclusion that they add several billion dollars to the direct losses already outlined. Examples of indirect losses are as follows:

1. *Shutdown.* The replacement of a corroded tube in an oil refinery may cost a few hundred dollars, but shutdown of the unit while repairs are underway may cost $50,000 or more per hour in lost production. Similarly, replacement of corroded boiler or condenser tubes in a large power plant may require $1,000,000 or more per day for power purchased from interconnected electric systems to supply customers while the boiler is down. Losses of this kind cost the electrical utilities in the United States tens of millions of dollars annually.

2. *Loss of Product.* Losses of oil, gas, or water occur through a corroded-pipe system until repairs are made. Antifreeze may be lost through a corroded auto radiator; or gas leaking from a corroded pipe may enter the basement of a building, causing an explosion.

3. *Loss of Efficiency.* Loss of efficiency may occur because of diminished heat transfer through accumulated corrosion products, or because of the clogging of pipes with rust necessitating increased pumping capacity. It has been estimated that, in the United States, increased pumping capacity, made necessary by partial clogging of water mains with rust, costs many millions of dollars per year. A further example is provided by internal-combustion engines of automobiles where piston rings and cylinder walls are continuously corroded by combustion gases and condensates. Loss of critical dimensions leading to excess gasoline and oil consumption can be caused by corrosion to an extent equal to or greater than that caused by wear. Corrosion processes can impose limits on the efficiencies of energy conversion systems, representing losses that may amount to billions of dollars.

4. *Contamination of Product.* A small amount of copper picked up by slight corrosion of copper piping or of brass equipment that is otherwise durable

may damage an entire batch of soap. Copper salts accelerate rancidity of soaps and shorten the time that they can be stored before use. Traces of metals may similarly alter the color of dyes. Lead equipment, otherwise durable, is not permitted in the preparation of foods and beverages because of the toxic properties imparted by very small quantities of lead salts. The U.S. Bureau of Food and Drugs, for example, permits not more than 1 ppb of lead in bottled drinking water [3].

Similarly, soft waters that pass through lead piping are not safe for drinking purposes. The poisonous effects of small amounts of lead have been known for a long time. In a letter to Benjamin Vaughn dated July 31, 1786, Benjamin Franklin [4] warned against possible ill effects of drinking rain water collected from lead roofs or consuming alcoholic beverages exposed to lead. The symptoms were called in his time "dry bellyache" and were accompanied by paralysis of the limbs. The disease originated because New England rum distillers used lead coil condensers. On recognizing the cause, the Massachusetts Legislature passed an act outlawing use of lead for this purpose.

Another form of contamination is spoilage of food in corroded metal containers. A cannery of fruits and vegetables once lost more than $1 million in one year before the metallurgical factors causing localized corrosion were analyzed and remedied. Another company, using metal caps on glass food jars, lost $0.5 million in one year because the caps perforated by a pitting type of corrosion, thereby allowing bacterial contamination of the contents.

5. *Overdesign.* Overdesign is common in the design of reaction vessels, boilers, condenser tubes, oil-well sucker rods, pipelines transporting oil and gas at high pressure, water tanks, and marine structures. Equipment is often designed many times heavier than normal operating pressures or applied stresses would require in order to ensure reasonable life. With adequate knowledge of corrosion, more reliable estimates of equipment life can be made, and design can be simplified in terms of materials and labor. For example, oil-well sucker rods are normally overdesigned to increase service life before failure occurs by corrosion fatigue. If the corrosion factor were eliminated, losses would be cut at least in half. There would be further savings because less power would be required to operate a lightweight rod, and the expense of recovering a lightweight rod after breakage would be lower.

Indirect losses are a substantial part of the economic tax imposed by corrosion, although it is difficult to arrive at a reasonable estimate of total losses. In the event of loss of health or life through explosion, unpredictable failure of chemical equipment, or wreckage of airplanes, trains, or automobiles through sudden failure by corrosion of critical parts, the indirect losses are still more difficult to assess and are beyond interpretation in terms of dollars.

1.3 RISK MANAGEMENT

In general, risk, R, is defined as the probability, P, of an occurrence multiplied by the consequence, C, of the occurrence; that is,

$$R = P \times C$$

Hence, the risk of a corrosion-related failure equals the probability that such a failure will take place multiplied by the consequence of that failure. Consequence is typically measured in financial terms—that is, the total cost of a corrosion failure, including the cost of replacement, clean-up, repair, downtime, and so on.

Any type of failure that occurs with high consequence must be one that seldom occurs. On the other hand, failures with low consequence may be tolerated more frequently. Figure 1.1 shows a simplified approach to risk management.

Managing risk is an important part of many engineering undertakings today. Managing corrosion is an essential aspect of managing risk. Firstly, risk management must be included in the design stage, and then, after operation starts, maintenance must be carried out so that risk continues to be managed. Engineering design must include corrosion control equipment, such as cathodic protection systems and coatings. Maintenance must be carried out so that corrosion is monitored and significant defects are repaired, so that risk is managed during the operational lifetime.

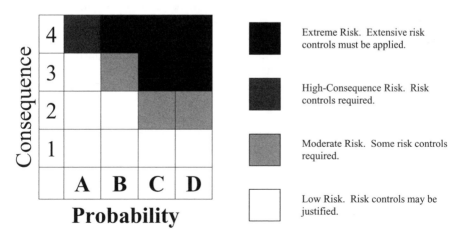

Figure 1.1. A simplified approach to risk management, indicating qualitatively the areas of high risk, where both consequence and probability are high.

1.4 CAUSES OF CORROSION

The many causes of corrosion will be explored in detail in the subsequent chapters of this book. In this introductory chapter, two parameters are mentioned: the change in Gibbs free energy and the Pilling–Bedworth ratio [5].

1.4.1 Change in Gibbs Free Energy

The change in Gibbs free energy, ΔG, for any chemical reaction indicates the tendency of that reaction to go. Reactions occur in the direction that lowers the Gibbs free energy. The more negative the value of ΔG, the greater the tendency for the reaction to go. The role of the change in Gibbs free energy is discussed in detail in Chapter 3.

1.4.2 Pilling–Bedworth Ratio

Although many factors control the oxidation rate of a metal, the Pilling–Bedworth ratio is a parameter that can be used to predict the extent to which oxidation may occur. The Pilling–Bedworth ratio is Md/nmD, where M and D are the molecular weight and density, respectively, of the corrosion product scale that forms on the metal surface during oxidation; m and d are the atomic weight and density, respectively, of the metal, and n is the number of metal atoms in a molecular formula of scale; for example, for Al_2O_3, $n = 2$.

The Pilling–Bedworth ratio indicates whether the volume of the corrosion product is greater or less than the volume of the metal from which the corrosion product formed. If $Md/nmD < 1$, the volume of the corrosion product is less than the volume of the metal from which the product formed. A film of such a corrosion product would be expected to contain cracks and pores and be relatively nonprotective. On the other hand, if $Md/nmD > 1$, the volume of the corrosion product scale is greater than the volume of the metal from which the scale formed, so that the scale is in compression, protective of the underlying metal. A Pilling–Bedworth ratio greater than 1 is not sufficient to predict corrosion resistance. If $Md/nmD \gg 1$, the scale that forms may buckle and detach from the surface because of the higher stresses that develop. For aluminum, which forms a protective oxide and corrodes very slowly in most environments, the Pilling–Bedworth ratio is 1.3, whereas for magnesium, which tends to form a nonprotective oxide, the ratio is 0.8. Nevertheless, there are exceptions and limitations to the predictions of the Pilling–Bedworth ratio, and these are discussed in Chapter 11.

REFERENCES

1. Gerhardus H. Koch, Michiel P. H. Brongers, Neil G. Thompson, Y. Paul Virmani, and J. H. Payer, Corrosion Costs and Preventive Strategies in the United States, Supplement to

Materials Performance, July 2002, Report No. FHWA-RD-01-156, Federal Highway Administration, McLean, VA, 2002.

2. J. Kruger, Cost of metallic corrosion, in *Uhlig's Corrosion Handbook*, 2nd edition, R. W. Revie, editor, Wiley, New York, 2000, pp. 3–10.

3. http://www.fda.gov/fdac/features/1998/198_lead.html

4. Carl Van Doren, editor, *Benjamin Franklin's Autobiographical Writings*, Viking Press, New York, 1945, p. 671.

5. N. Pilling and R. Bedworth, *J. Inst. Metals* **29**, 529 (1923).

GENERAL REFERENCES

R. Bhaskaran, N. Palaniswamy, N. S. Rengaswamy, and M. Jayachandran, Global cost of corrosion—A historical review, in *ASM Handbook*, Vol. 13B, *Corrosion: Materials*, ASM International, Materials Park, Ohio, 2005, pp. 621–628.

M. V. Biezma and J. R. San Cristóbal, Is the cost of corrosion really quantifiable? *Corrosion* **62** (12), 1051 (2006).

Geoff Davies, *Materials for Automobile Bodies*, Elsevier, Oxford, U.K., 2003.

Gerd Gigerenzer, *Reckoning with Risk, Learning to Live with Uncertainty*, Penguin Books, London, 2003.

G. H. Koch, M. P. H. Brongers, N. G. Thompson, Y. P. Virmani, and J. H. Payer, *Corrosion Cost and Preventive Strategies in the United States*, Report No. FHWA-RD-01-156, Federal Highway Administration, U.S. Department of Transportation, McLean VA, March 2002.

G. H. Koch, M. P. H. Brongers, N. G. Thompson, Y. P. Virmani, and J. H. Payer, Direct costs of corrosion in the United States, in *ASM Handbook*, Vol. 13A, *Corrosion: Fundamentals, Testing, and Protection*, ASM International, Materials Park, OH, 2003, pp. 959–967.

W. Kent Muhlbauer, *Pipeline Risk Management Manual: Ideas, Techniques, and Resources*, 3rd edition, Elsevier, Oxford, U.K., 2004.

V. S. Sastri, E. Ghali, and M. Elboujdaini, *Corrosion Prevention and Protection, Practical Solutions*, Wiley, Chichester, England, 2007.

E. D. Verink, Economics of corrosion, in *Uhlig's Corrosion Handbook*, 2nd edition, R. Winston Revie, editor, Wiley, New York, 2000, pp. 11–25.

PROBLEMS

1. A manufacturer provides a warranty against failure of a carbon steel product within the first 30 days after sale. Out of 1000 sold, 10 were found to have failed by corrosion during the warranty period. Total cost of replacement for each failed product is approximately $100,000, including the cost of environmental clean-up, loss of product, downtime, repair, and replacement.

(*a*) Calculate the risk of failure by corrosion, in dollars.

(*b*) If a corrosion-resistant alloy would prevent failure by corrosion, is an incremental cost of $100 to manufacture the product using such an alloy justified? What would be the maximum incremental cost that would be justified in using an alloy that would prevent failures by corrosion?

2. Linings of tanks can fail because of salt contamination of the surface that remains after the surface is prepared for the application of the lining. Between 15% and 80% of coating failures have been attributed to residual salt contamination. The cost of reworking a failed lining of a specific tank has been estimated at $174,000. [Reference: H. Peters, Monetizing the risk of coating failure, *Materials Performance* **45**(5), 30 (2006).]

(*a*) Calculate the risk due to this type of failure assuming that 20% of failures are caused by residual salt contamination.

(*b*) If the cost of testing and removal of contaminating salts is $4100, is this additional cost justified based on the risk calculation in (*a*)?

(*c*) Calculate the minimum percentage of failures caused by residual salt contamination at which the additional cost of $4100 for testing and removal of these salts is justified.

2

ELECTROCHEMICAL MECHANISMS

2.1 THE DRY-CELL ANALOGY AND FARADAY'S LAW

As described in Chapter 1, corrosion processes are most often electrochemical. In aqueous media, the corrosion reactions are similar to those that occur in a flashlight cell consisting of a center carbon electrode and a zinc cup electrode separated by an electrolyte consisting essentially of NH_4Cl solution* (Fig. 2.1). An incandescent light bulb connected to both electrodes glows continuously, with the electrical energy being supplied by chemical reactions at both electrodes. At the carbon electrode (positive pole), chemical reduction occurs, and at the zinc electrode (negative pole) oxidation occurs, with metallic zinc being converted into hydrated zinc ions, $Zn^{2+}\cdot nH_2O$.† The greater the flow of electricity through the cell, the greater the amount of zinc that corrodes. The relationship is

---

*The function of carbon granules for conduction and manganese dioxide as depolarizer, both surrounding the carbon electrode, need not concern us at this point.

†Ions in aqueous solution attach themselves to water molecules, but their number is not well-defined. They differ in this way from gaseous ions, which are not hydrated. It is common practice, however, to omit mention of the appended H_2O molecules and to designate hydrated zinc ions, for example, as Zn^{2+}.

Corrosion and Corrosion Control, by R. Winston Revie and Herbert H. Uhlig
Copyright © 2008 John Wiley & Sons, Inc.

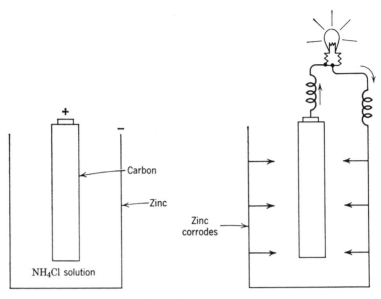

Figure 2.1. Dry cell.

quantitative, as Michael Faraday showed in the early nineteenth century. This is the relationship now known as Faraday's law:

$$\text{Weight of metal reacting} = kIt \tag{2.1}$$

where I is the current in amperes (A), t is in seconds (s), and k is a constant called the *electrochemical equivalent*. The value of k for zinc is 3.39×10^{-4} g/C (gram per coulomb), the coulomb being defined as the amount of electricity represented by 1 A flowing for 1 s. On short-circuiting the cell with a low-resistance metallic connector, the zinc cup perforates by corrosion within a matter of hours; but when the cell is left disconnected (open circuit), the zinc may remain intact for years. The slow consumption of zinc occurring on open circuit is accounted for largely by activity of minute impurities, like iron, embedded in the surface of zinc; these impurities assume the same role as carbon and allow the flow of electricity accompanied by corrosion of zinc. Current of this kind is called *local-action current*, and the corresponding cells are called *local-action cells*. Local-action current, of course, produces no useful energy, but acts only to heat up the surroundings.

Any metal surface, similar to the situation for zinc, is a composite of electrodes electrically short-circuited through the body of the metal itself (Fig. 2.2). So long as the metal remains dry, local-action current and corrosion are not observed. But on exposure of the metal to water or aqueous solutions, local-action cells are able to function and are accompanied by chemical conversion of the metal to corrosion products. Local-action current, in other words, may

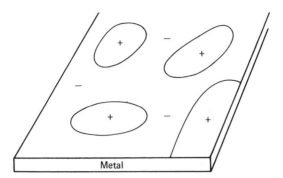

Figure 2.2. Metal surface enlarged, showing schematic arrangement of local-action cells.

account for the corrosion of metals exposed to water, salt solutions, acids, or alkalies.

Whenever impurities in a metal constitute the electrodes of local-action cells, their removal, as might be expected, appreciably improves corrosion resistance. Accordingly, purified aluminum and magnesium are much more resistant to corrosion in seawater or in acids than are the commercial varieties of these metals, and high-purity zinc resists dilute hydrochloric acid much better than does commercial zinc. However, it is not correct to assume, as was done many years ago when the electrochemical theory was first proposed, that pure metals do not corrode at all. As we will see later, local-action cells are also set up when there are variations in the environment or in temperature. With iron or steel in aerated water, for example, the negative electrodes are commonly portions of the iron surface itself covered perhaps by porous rust (iron oxides); and positive electrodes are areas exposed to oxygen, with the positive and negative electrode areas interchanging and shifting from place to place as the corrosion reaction proceeds. Accordingly, high-purity iron in air-saturated water corrodes at essentially the same rate as impure or commercial iron. A difference in rates is observed in acids, however, because impurities now enter predominantly as electrodes of local-action cells. This matter is discussed in Section 7.3.

2.2 DEFINITION OF ANODE AND CATHODE

A combination of two electrical conductors (electrodes) immersed in an electrolyte is called a galvanic cell in honor of Luigi Galvani, a physician in Bologna, Italy, who published his studies of electrochemical action in 1791. A galvanic cell converts chemical energy into electrical energy. On short-circuiting such a cell (attaching a low-resistance wire to connect the two electrodes), positive current flows through the metallic path from positive electrode to negative electrode. This direction of current flow follows an arbitrary convention, established before anything was known about the nature of electricity, and is employed today despite contemporary knowledge that only negative carriers, or electrons, move in a metal. Electrons, of course, go from negative to positive pole, opposite to the imaginary flow of positive carriers. Whenever current is said to flow, however, without designating the sign of the carrier, positive current is always implied.

Within the electrolyte, current is carried by both negative and positive carriers, known as ions (electrically charged atoms or groups of atoms). The current carried by each ion depends on its mobility and electric charge. The total of positive and negative current in the electrolyte of a cell is always exactly equivalent to the total current carried in the metallic path by electrons alone. Ohm's law—that is, $I = E/R$, where I is the current in amperes, E the potential difference in volts, and R the resistance in ohms—applies precisely, under conditions with which we are presently concerned, to current flow in electrolytes as well as in metals.

The electrode at which chemical reduction occurs (or + current enters the electrode from the electrolyte) is called the *cathode*. Examples of cathodic reactions are

$$H^+ \rightarrow \frac{1}{2}H_2 - e^-$$

$$Cu^{2+} \rightarrow Cu - 2e^-$$

$$Fe^{3+} \rightarrow Fe^{2+} - e^-$$

all of which represent reduction in the chemical sense.

The electrode at which chemical oxidation occurs (or + electricity leaves the electrode and enters the electrolyte) is called the *anode*. Examples of anodic reactions are

$$Zn \rightarrow Zn^{2+} + 2e^-$$

$$Al \rightarrow Al^{3+} + 3e^-$$

$$Fe^{2+} \rightarrow Fe^{3+} + e^-$$

These equations represent oxidation in the chemical sense. Corrosion of metals usually occurs at the anode. Nevertheless, alkaline reaction products forming at the cathode can sometimes cause secondary corrosion of amphoteric metals, such as Al, Zn, Pb, and Sn, which corrode rapidly on exposure to either acids or alkalies.

In galvanic cells, the cathode is the positive pole, whereas the anode is the negative pole. However, when current is impressed on a cell from a generator or an external battery—for example, as in electroplating—reduction occurs at the electrode connected to the negative pole of the external current source, and this electrode, consequently, is the cathode. Similarly, the electrode connected to the positive pole of the generator is the anode. It is perhaps best, therefore, not to remember anode and cathode as negative and positive electrodes, or vice versa, but instead to remember the cathode as the electrode at which current enters from the electrolyte and remember the anode as the electrode at which current leaves to return to the electrolyte. This situation is true whether current is impressed on or drawn from the cell.

Cations are ions that migrate toward the cathode when electricity flows through the cell (e.g., H^+, Fe^{2+}) and are always positively charged whether current is drawn from or supplied to the cell. Similarly, anions are always negatively charged (e.g., Cl^-, OH^-, SO_4^{2-}).

2.3 TYPES OF CELLS

There are three main types of cells that take part in corrosion reactions.

1. *Dissimilar Electrode Cells.* Examples of dissimilar electrode cells include: the dry cell (discussed at the beginning of this chapter), a metal containing electrically conducting impurities on the surface as a separate phase, a copper pipe connected to an iron pipe, and a bronze propeller in contact with the steel hull of a ship. Dissimilar electrode cells also include cold-worked metal in contact with the same metal annealed, grain-boundary metal in contact with grains, and a single metal crystal of definite orientation in contact with another crystal of different orientation.*

2. *Concentration Cells.* These are cells with two identical electrodes, each in contact with a solution of different composition. There are two kinds of concentration cells. The first is called a *salt concentration cell.* For example, if one copper electrode is exposed to a concentrated copper sulfate solution, and another to a dilute copper sulfate solution (Fig. 2.3), on short-circuiting the electrodes, copper dissolves (i.e., $Cu \rightarrow Cu^{2+} + 2e^-$) from the electrode in contact with the dilute solution (anode) and plates out (i.e., $Cu^{2+} + 2e^- \rightarrow Cu$) on the other electrode (cathode). These reactions tend to bring the two solutions to the same concentration.

   The second kind of concentration cell, which in practice is the more important, is called a *differential aeration cell.* This may include two iron electrodes in dilute sodium chloride solution, the electrolyte around one electrode being thoroughly aerated (cathode), and the other deaerated

*The various crystal faces of a metal, although initially exhibiting different potentials (tendencies to corrode), tend to achieve the same potential in time when exposed to an environment that reacts with the metal [1]. The most corrodible planes of atoms react first, leaving behind the least corrodible planes; hence, the latter eventually are the only faces exposed regardless of the original orientation. The corrosion rates continue to differ, however, because of differing absolute surface areas of what were previously differing crystal faces. The most corrosion resistant crystal face of any metal is not always the same, but varies with environment. For example, in dilute nitric acid, the (100) face of iron is the least reactive crystallographic plane [2].

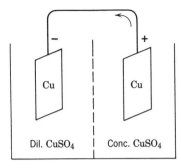

Figure 2.3. Salt concentration cell.

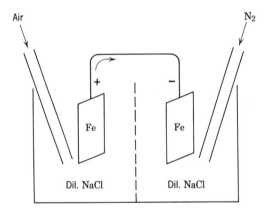

Figure 2.4. Differential aeration cell.

(anode) by, for example, bubbling nitrogen through the solution. The difference in oxygen concentration produces a potential difference and causes current to flow (Fig. 2.4). This type of cell accounts for the pronounced damage at crevices, which is called *crevice corrosion*. Crevices are common in many engineering designs—for example, at the interface of two pipes that are coupled together and at threaded connections. The oxygen concentration is lower within crevices, and the areas of lower oxygen concentration (inside the crevice) are anodic with respect to areas of higher oxygen concentration (outside crevices). Differential aeration cells can also cause pitting damage under rust (Fig. 2.5) and at the water line—that is, at the water–air interface (Fig. 2.6). The amount of oxygen reaching the metal that is covered by rust or other insoluble reaction products is less than the amount that contacts other portions where the permeable coating is thinner or nonexistent.

Differential aeration cells can also lead to localized corrosion at pits (crevice corrosion) in stainless steels, aluminum, nickel, and other passive metals that are exposed to aqueous environments, such as seawater.

3. *Differential Temperature Cells.* Components of these cells are electrodes of the same metal, each of which is at a different temperature, immersed

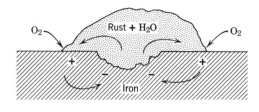

Figure 2.5. Differential aeration cell formed by rust on iron.

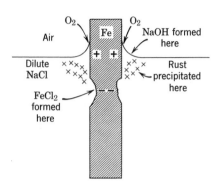

Figure 2.6. Water-line corrosion, showing differential aeration cell.

in an electrolyte of the same initial composition. Less is known about the practical importance and fundamental theory of differential temperature cells than about the cells previously described. These cells are found in heat exchangers, boilers, immersion heaters, and similar equipment.

In copper sulfate solution, the copper electrode at the higher temperature is the cathode, and the copper electrode at the lower temperature is the anode [3]. On short-circuiting the cell, copper deposits on the hot electrode and dissolves from the cold electrode. Lead acts similarly, but for silver the polarity is reversed.

For iron immersed in dilute aerated sodium chloride solutions, the hot electrode is anodic to colder metal of the same composition; but after several hours, depending on aeration, stirring rate, and whether the two metals are short-circuited, the polarity may reverse [4, 5].

In engineering practice, cells responsible for corrosion may be a combination of these three types.

2.4 TYPES OF CORROSION DAMAGE

Corrosion is often thought of only in terms of rusting and tarnishing. However, corrosion damage occurs in other ways as well, resulting, for example, in failure by cracking or in loss of strength or ductility. In general, most types of corrosion, with some exceptions, occur by electrochemical mechanisms, but corrosion products are not necessarily observable and metal weight loss need not be appreciable to result in major damage. The five main types of corrosion classified with respect to outward appearance or altered physical properties are as follows:

1. *General Corrosion, or Uniform Attack.* This type of corrosion includes the commonly recognized rusting of iron or tarnishing of silver. "Fogging" of nickel and high-temperature oxidation of metals are also examples of this type.

Rates of uniform attack are reported in various units, with accepted terminologies being millimeters penetration per year (mm/y) and grams per square meter per day (gmd). Other units that are frequently used include inches penetration per year (ipy), mils (1 mil = 0.001 inch) per year (mpy), and milligrams per square decimeter per day (mdd). These units refer to metal penetration or to weight loss of metal, excluding any adherent or nonadherent corrosion products on the surface. Steel, for example, corrodes at a relatively uniform rate in seawater of about 0.13 mm/y, 2.5 gmd, 25 mdd, or 0.005 ipy. These represent time-averaged values. Generally, for uniform attack, the initial corrosion rate is greater than subsequent rates [6]. Duration of exposure should always be given when corrosion rates are reported because it is often not reliable to extrapolate a reported rate to times of exposure far exceeding the test period.

Conversion of mm/y to gmd or vice versa requires knowledge of the metal density. A given weight loss per unit area for a light metal (e.g., aluminum) represents a greater actual loss of metal thickness than the same weight loss for a heavy metal (e.g., lead). Conversion tables are given in the Appendix, Section 29.8.

For handling chemical media whenever attack is uniform, metals are classified into three groups according to their corrosion rates and intended application. These classifications are as follows:

A. <0.15 mm/y (<0.005 ipy)—Metals in this category have good corrosion resistance to the extent that they are suitable for critical parts, for example, valve seats, pump shafts and impellors, springs.
B. 0.15 to 1.5 mm/y (0.005 to 0.05 ipy)—Metals in this group are satisfactory if a higher rate of corrosion can be tolerated, for example, for tanks, piping, valve bodies, and bolt heads.
C. >1.5 mm/y (>0.05 ipy)—Usually not satisfactory.

2. *Pitting.* This is a localized type of attack, with the rate of corrosion being greater at some areas than at others. If appreciable attack is confined to a relatively small, fixed area of metal, acting as anode, the resultant pits are described as deep. If the area of attack is relatively larger and not so deep, the pits are called shallow. Depth of pitting is sometimes expressed by the *pitting factor*, the ratio of deepest metal penetration to average metal penetration as determined by the weight loss of the specimen. A pitting factor of unity represents uniform attack (Fig. 2.7).

Iron buried in the soil corrodes with formation of shallow pits, whereas stainless steels immersed in seawater characteristically corrode with formation of deep pits. Many metals, when subjected to high-velocity liquids, undergo a pitting

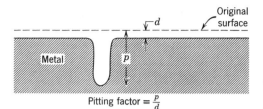

Figure 2.7. Sketch of deepest pit in relation to average metal penetration and the pitting factor.

type of corrosion called *impingement attack*, or sometimes *corrosion-erosion*. Copper and brass condenser tubes, for example, are subject to this type of attack.

Fretting corrosion, which results from slight relative motion (as in vibration) of two substances in contact, one or both being metals, usually leads to a series of pits at the metal interface. Metal-oxide debris usually fills the pits so that only after the corrosion products are removed do the pits become visible.

Cavitation–erosion is the loss of material caused by exposure to cavitation, which is the formation and collapse of vapor bubbles at a dynamic metal–liquid interface—for example, in rotors of pumps or on trailing faces of propellers. This type of corrosion causes a sequence of pits, sometimes appearing as a honeycomb of small relatively deep fissures (see *Uhlig's Corrosion Handbook*, 2nd edition, R. W. Revie, editor, Wiley, New York, 2000, Fig. 12, p. 261).

3. *Dealloying, Dezincification, and Parting.* Dealloying is the selective removal of an element from an alloy by corrosion. One form of dealloying, dezincification, is a type of attack occurring with zinc alloys (e.g., yellow brass) in which zinc corrodes preferentially, leaving a porous residue of copper and corrosion products (Fig. 20.4). The alloy so corroded often retains its original shape, and may appear undamaged except for surface tarnish, but its tensile strength and ductility are seriously reduced. Dezincified brass pipe may retain sufficient strength to resist internal water pressures until an attempt is made to uncouple the pipe, or a water hammer occurs, causing the pipe to split open.

Parting is similar to dezincification in that one or more reactive components of the alloy corrode preferentially, leaving a porous residue that may retain the original shape of the alloy. Parting is usually restricted to such noble metal alloys as gold–copper or gold–silver and is used in gold refining. For example, an alloy of Au–Ag containing more than 65% gold resists concentrated nitric acid as well as does gold itself. However, on addition of silver to form an alloy of approximately 25% Au–75% Ag, reaction with concentrated HNO_3 forms silver nitrate and a porous residue or powder of pure gold.

Copper-base alloys that contain aluminum are subject to a form of corrosion resembling dezincification, with aluminum corroding preferentially.

4. *Intergranular Corrosion.* This is a localized type of attack at the grain boundaries of a metal, resulting in loss of strength and ductility. Grain-boundary

material of limited area, acting as anode, is in contact with large areas of grain acting as cathode. The attack is often rapid, penetrating deeply into the metal and sometimes causing catastrophic failures. Improperly heat-treated 18-8 stainless steels or Duralumin-type alloys (4% Cu–Al) are among the alloys subject to intergranular corrosion. At elevated temperatures, intergranular corrosion can occur because, under some conditions, phases of low melting point form and penetrate along grain boundaries; for example, when nickel-base alloys are exposed to sulfur-bearing gaseous environments, nickel sulfide can form and cause catastrophic failures [7]. This type of attack is usually called *sulfidation*.

5. *Cracking.* If a metal cracks when subjected to repeated or alternate tensile stresses in a corrosive environment, it is said to fail by *corrosion fatigue*. In the absence of a corrosive environment, the metal stressed similarly, but at values below a critical stress, called the *fatigue limit* or *endurance limit*, will not fail by fatigue even after a very large, or infinite, number of cycles. A true endurance limit does not commonly exist in a corrosive environment: The metal fails after a prescribed number of stress cycles no matter how low the stress. The types of environment causing corrosion fatigue are many and are not specific.

If a metal, subject to a constant tensile stress and exposed simultaneously to a specific corrosive environment, cracks immediately or after a given time, the failure is called *stress-corrosion cracking*. Both stress-corrosion cracking and cracking caused by absorption of hydrogen generated by a corrosion reaction follow this definition. Distinguishing differences between the two types of cracking are discussed in Section 8.4. The stress may be residual in the metal, as from cold working or heat treatment, or it may be externally applied. The observed cracks are intergranular or transgranular, depending on the metal and the damaging environment. Failures of this kind differ basically from intergranular corrosion, which proceeds without regard to whether the metal is stressed.

Almost all structural metals (e.g., carbon- and low-alloy steels, brass, stainless steels, Duralumin, magnesium alloys, titanium alloys, nickel alloys, and many others) are subject to stress-corrosion cracking in some environments. Fortunately, either the damaging environments are often restricted to a few chemical species, or the necessary stresses are sufficiently high to limit failures of this kind in engineering practice. As knowledge accumulates regarding the specific media that cause cracking and regarding the limiting stresses necessary to avoid failure within a given time period, it will be possible to design metal structures without incidence of stress-corrosion cracking. Highly stressed metal structures must be designed with adequate assurance that stress-corrosion cracking will not occur.

REFERENCES

1. W. Tragert and W. D. Robertson, *J. Electrochem. Soc.* **102**, 86 (1955).
2. C. S. Barrett and T. B. Massalski, *Structure of Metals: Crystallographic Methods, Principles and Data*, 3rd revised edition, Pergamon Press, Oxford, U.K., 1980, p. 210.

3. N. Berry, *Corrosion* **2**, 261 (1946).
4. H. H. Uhlig and O. Noss, *Corrosion* **6**, 140 (1950).
5. V. Simpson, Jr., S.B. thesis, Department of Chemical Engineering, M.I.T., Cambridge, MA, 1950.
6. R. W. Revie and N. D. Greene, *Corros. Sci.* **9**, 755 (1969).
7. Bopinder Phull, Evaluating intergranular corrosion, in *ASM Handbook*, Vol. 13A, *Corrosion: Fundamentals, Testing, and Protection*, ASM International, Materials Park, OH, 2003, p. 570.

GENERAL REFERENCES

P. Elliott, Gallery of corrosion damage, in *ASM Handbook*, Vol. 13B, *Corrosion: Materials*, ASM International, Materials Park, OH, 2005, pp. 631–646.
E. D. Verink, Designing to prevent corrosion, in *Uhlig's Corrosion Handbook*, 2nd edition, Wiley, New York, 2000, pp. 97–109.

PROBLEMS

1. Derive the general relation between mm/y and gmd.

2. Magnesium corrodes in seawater at a rate of 1.45 gmd. What is the rate in mm/y? If this corrosion rates applies to lead, what is the corresponding rate in mm/y?

3. Laboratory corrosion tests on three alloys in an industrial waste solution show the following results:

Material	Density of Material (g/cm^2)	Weight Loss (gmd)	Pitting Factor
A	2.7	40	1
B	9.0	62	2
C	7.8	5.6	9.2

Calculate maximum penetration in millimeters for each material at the end of one year.

THERMODYNAMICS: CORROSION TENDENCY AND ELECTRODE POTENTIALS

3.1 CHANGE OF GIBBS FREE ENERGY

The tendency for any chemical reaction to go, including the reaction of a metal with its environment, is measured by the Gibbs free-energy change, ΔG. The more negative the value of ΔG, the greater the tendency for the reaction to go. For example, consider the following reaction at 25 °C:

$$Mg + H_2O \ (l) + \frac{1}{2}O_2 \ (g) \rightarrow Mg(OH)_2 \ (s) \qquad \Delta G° = -596{,}600 \, J$$

The large negative value of $\Delta G°$ (reactants and products in standard states) indicates a pronounced tendency for magnesium to react with water and oxygen. On the other hand, we have

$$Cu + H_2O \ (l) + \frac{1}{2}O_2 \ (g) \rightarrow Cu(OH)_2 \ (s) \qquad \Delta G° = -119{,}700 \, J$$

The reaction tendency is less. Or we can say that the corrosion tendency of copper in aerated water is not as pronounced as that of magnesium.

Finally, we have

$$\text{Au} + \frac{3}{2}\text{H}_2\text{O (l)} + \frac{3}{4}\text{O}_2 \text{ (g)} \rightarrow \text{Au(OH)}_3 \text{ (s)} \qquad \Delta G^\circ = +65{,}700\,\text{J}$$

The free energy is positive, indicating that the reaction has no tendency to go at all; and gold, correspondingly, does not corrode in aqueous media to form Au(OH)_3.

It should be emphasized that the tendency to corrode is not a measure of reaction rate. A large negative ΔG may or may not be accompanied by a high corrosion rate, but, when ΔG is positive, it can be stated with certainty that the reaction will not go at all under the particular conditions described. If ΔG is negative, the reaction rate may be rapid or slow, depending on various factors described in detail later in this book.

In view of the electrochemical mechanisms of corrosion, the tendency for a metal to corrode can also be expressed in terms of the electromotive force (emf) of the corrosion cells that are an integral part of the corrosion process. Since electrical energy is expressed as the product of volts by coulombs (joules, J), the relation between ΔG in joules and emf in volts, E, is defined by $\Delta G = -nFE$, where n is the number of electrons (or chemical equivalents) taking part in the reaction, and F is the Faraday (96,500 C/eq). The term ΔG can be converted from calories to joules by using the factor $1\,\text{cal} = 4.184$ absolute joules.

Accordingly, the greater the value of E for any cell, the greater the tendency for the overall reaction of the cell to go. This applies to any of the types of cells described earlier.

3.2 MEASURING THE EMF OF A CELL

The emf of a cell, as set up in the laboratory or in the field, can be measured using a voltmeter of high impedance, $>10^{12}\,\Omega$ (ohms). Alternatively, using a potentiometer, the emf of the cell can be opposed with a known emf until no current flows through a galvanometer in series with the cell. At exact balance (i.e., when the emf of the cell is exactly balanced by the known emf), no current flows through the cell, and the reading of the known emf indicates the exact emf of the cell. In making these measurements, it is essential that any current that flows in the circuit is sufficiently small that the cell is not polarized—that is, that the emf of the cell is not changed because of the current flow. For this reason, sensitive galvanometers of high input impedance, at least $10^{12}\,\Omega$, must be used, so that, if the potentiometer and cell are unbalanced to the extent of 1 V, a current of only $10^{-12}\,\text{A}$ flows. Such a current is not sufficient to polarize (temporarily alter the emf of) the cell.

3.3 CALCULATING THE HALF-CELL POTENTIAL—THE NERNST EQUATION

Based on thermodynamics [1], an equation can be derived to express the emf of a cell in terms of the concentrations of reactants and reaction products. The general reaction for a galvanic cell is

$$lL + mM + \cdots \rightarrow qQ + rR + \cdots$$

meaning that l moles of substance L plus m moles of substance M, and so on, react to form q moles of substance Q, r moles of substance R, and so on. The corresponding change of Gibbs free energy, ΔG, for this reaction is given by the difference in molal free energy of products and reactants, where G_Q represents the molal free energy of substance Q, and so on.

$$\Delta G = (qG_Q + rG_R + \cdots) - (lG_L + mG_M + \cdots) \tag{3.1}$$

A similar expression is obtained for each substance in the standard state or arbitrary reference state, where the symbol $G°$ indicates standard molal free energy:

$$\Delta G° = (qG_Q° + rG_R° + \cdots) - (lG_L° + mG_M° + \cdots) \tag{3.2}$$

If a_L is the corrected concentration or pressure of substance L, called its activity, the difference of free energy for L in any given state and in the standard state is related to a_L by the expression

$$l(G_L - G_L°) = lRT \ln a_L = RT \ln a_L^l \tag{3.3}$$

where R is the gas constant (8.314 J/deg-mole), and T is the absolute temperature (degrees Celsius + 273.16). Subtracting (3.2) from (3.1) and equating to corresponding activities, we have the expression

$$\Delta G - \Delta G° = RT \ln \frac{a_Q^q \cdot a_R^r \cdots}{a_L^l \cdot a_M^m \cdots} \tag{3.4}$$

When the reaction is at equilibrium, there is no tendency for it to go, $\Delta G = 0$, and

$$\frac{a_Q^q \cdot a_R^r \cdots}{a_L^l \cdot a_M^m \cdots} = K$$

where K is the equilibrium constant for the reaction. Hence

$$\Delta G° = -RT \ln K \tag{3.5}$$

On the other hand, when all the activities of reactants and products are equal to unity, the logarithm term becomes zero ($\ln 1 = 0$), and $\Delta G = \Delta G°$.

Since $\Delta G = -nFE$, it follows that $\Delta G° = -nFE°$, where E° is the emf when all reactants and products are in their standard states (activities equal to unity). Corresponding to (3.4), we have

$$E = E^{\circ} - \frac{RT}{nF} \ln \frac{a_Q^q \cdot a_R^r \cdots}{a_L^l \cdot a_M^m \cdots} \tag{3.6}$$

This is the *Nernst equation*, which expresses the exact emf of a cell in terms of activities of products and reactants of the cell. The activity, a_L, of a dissolved substance L is equal to its concentration in *moles per thousand grams of water (molality)* multiplied by a correction factor, γ, called the activity coefficient. The activity coefficient is a function of temperature and concentration and, except for very dilute solutions, must be determined experimentally. If L is a gas, its activity is equal to its fugacity, approximated at ordinary pressures by the pressure in atmospheres. The activity of a pure solid is arbitrarily set equal to unity. Similarly, for water, with concentration essentially constant throughout the reaction, the activity is set equal to unity.

Since the emf of a cell is always the algebraic sum of two electrode potentials or of two half-cell potentials, it is convenient to calculate each electrode potential separately. For example, for the electrode reaction

$$\text{Zn}^{2+} + 2e^{-} \rightarrow \text{Zn} \tag{3.7}$$

we have

$$\phi_{\text{Zn}} = \phi_{\text{Zn}}^{\circ} - \frac{RT}{2F} \ln \frac{(\text{Zn})}{(\text{Zn}^{2+})} \tag{3.8}$$

where (Zn^{2+}) represents the activity of zinc ions (molality $\times$ activity coefficient); (Zn) is the activity of metallic zinc, the latter being a pure solid and equal, therefore, to unity; and ϕ_{Zn}° is the standard potential of zinc (equilibrium potential of zinc in contact with Zn^{2+} at unit activity).

Since it is more convenient to work with logarithms to the base 10, the value of the coefficient RT/F is multiplied by the conversion factor 2.303. Then, from the value of $R = 8.314$ J/deg-mole, $T = 298.2$ K, and $F = 96,500$ C/eq, the coefficient $2.303 RT/F$ at 25 °C becomes 0.0592 V. This coefficient appears frequently in expressions representing potentials or emf.

Measured or calculated values of standard potentials ϕ° at 25 °C are listed in several reference books [2–4]. Some values are listed in Section 3.8, Table 3.2, and in the Appendix, Section 29.9. Values of activity coefficients for various electrolytes are given in the Appendix, Section 29.1, Table 29.1, and definitions and rules applying to the use of activity coefficients are outlined in Section 29.1.

3.4 THE HYDROGEN ELECTRODE AND THE STANDARD HYDROGEN SCALE

Since absolute potentials of electrodes are not known, it is convenient to assume arbitrarily that the standard potential for the reaction

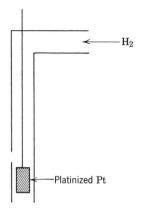

Figure 3.1. Hydrogen electrode.

$$2H^+ + 2e^- \rightarrow H_2 \qquad (3.9)$$

is equal to zero at all temperatures. Hence,

$$\phi_{H_2} = 0 - \frac{RT}{2F} \ln \frac{p_{H_2}}{(H^+)^2} \qquad (3.10)$$

where p_{H_2} is the fugacity of hydrogen in atmospheres and (H^+) is the activity of hydrogen ions. All values of electrode potentials, therefore, are with reference to the hydrogen electrode. By measuring the emf of a cell made up, for example, of a zinc and hydrogen electrode in zinc salt solution of known activity of Zn^{2+} and H^+, the standard potential, ϕ°, for zinc can be calculated; the accepted value is $-0.763\,V$.

The hydrogen electrode potential is measured by immersing a piece of platinized platinum in a solution saturated with hydrogen gas at 1 atm (Fig. 3.1), or, more conveniently, by the glass electrode, for which the potential is also reversible to (in equilibrium with) hydrogen ions. The potential of the electrode equals zero if the hydrogen ion activity and the pressure of hydrogen gas in atmospheres are both unity. This is the standard hydrogen potential. Hence, the half-cell potential for any electrode is equal to the emf of a cell with the standard hydrogen electrode as the other electrode. The half-cell potential for any electrode expressed on this basis is said to be on the *normal hydrogen scale* or on the *standard hydrogen scale*, sometimes expressed as ϕ_H or $\phi(\text{S.H.E.})$.

3.5 CONVENTION OF SIGNS AND CALCULATION OF EMF

In accord with the foregoing discussion, the standard potential of zinc, which is not separately measurable, refers to the emf of a cell with two electrodes—zinc and the standard hydrogen electrode:

$$Pt; H_2, H^+, Zn^{2+}; Zn \qquad (3.11)$$

The corresponding reaction, somewhat simplified, is written arbitrarily subtracting the left-hand reduction reaction from the right-hand reduction reaction, or

$$Zn^{2+} + H_2 \rightarrow Zn + 2H^+ \qquad \text{standard emf} = -0.763\,\text{V} \qquad (3.12)$$

The free-energy change, $\Delta G°$, equals $0.763 \times 2F$ joules; the positive value indicates that the reaction is not thermodynamically possible as written, for products and reactants in their standard states. On the other hand, for the cell

$$Zn; Zn^{2+}, H^+, H_2; Pt \qquad \text{Standard emf} = 0.763\,\text{V} \qquad (3.13)$$

the corresponding reaction is $Zn + 2H^+ \rightarrow Zn^{2+} + H_2$, the standard emf is positive, and $\Delta G°$ is negative, indicating that the reaction is thermodynamically possible.

Clearly, the standard reduction potential for zinc is opposite in sign to the standard oxidation potential for zinc.

It was agreed at the 1953 meeting of the International Union of Pure and Applied Chemistry that the reduction potential for any half-cell electrode reaction would be called *the potential*. This designation of sign has the advantage of conforming to the physicist's concept of potential defined as the work necessary to bring unit positive charge to the point at which the potential is given. It also has the advantage of corresponding in sign to the polarity of a voltmeter or potentiometer to which an electrode may be connected. Thus, zinc has a negative reduction potential and is also the negative pole of a galvanic cell of which the standard hydrogen electrode is the other electrode. It is said to be negative to the hydrogen electrode.

In setting up the emf of a cell, the foregoing conventions of sign dictate the direction regarding spontaneous flow of electricity. If, on short-circuiting a cell, positive current through the electrolyte within the cell flows from left to right, then the emf is positive, and, correspondingly, the left electrode is anode and the right electrode is cathode. If current flows within the cell from right to left, the emf is negative.

To calculate the emf of the cell shown in Fig. 3.2, Cu; Cu^{2+}, Zn^{2+}; Zn, we can first write the reduction reaction of the left electrode, Cu, as if it were cathode (whether it is or not is clarified later):

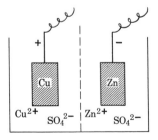

Figure 3.2. Copper–zinc cell.

$$Cu^{2+} + 2e^- \rightarrow Cu \qquad \phi° = 0.337\,V \qquad\qquad (3.14)$$

and

$$\phi_{Cu} = 0.337 - \frac{0.0592}{2}\log\frac{1}{(Cu^{2+})} \qquad\qquad (3.15)$$

The reduction reaction for the right electrode is

$$Zn^{2+} + 2e^- \rightarrow Zn \qquad \phi° = -0.763\,V \qquad\qquad (3.16)$$

and

$$\phi_{Zn} = -0.763 - \frac{0.0592}{2}\log\frac{1}{(Zn^{2+})} \qquad\qquad (3.17)$$

Reaction (3.14) is subtracted from (3.16), multiplying, if necessary, by a numerical factor so that the total electrons are canceled out. This results in the tentative reaction for the cell,

$$Cu + Zn^{2+} \rightarrow Cu^{2+} + Zn \qquad\qquad (3.18)*$$

The emf is obtained by adding, algebraically, the corresponding half-cell potentials, (3.15) and (3.17). Although reversing a reaction changes the sign of potential, multiplying by any factor has no effect on either emf or $\phi°$ values, since the tendency for a reaction to go is independent of the amount of substance reacting (in contrast to the total free-energy change, which does depend on amount of substance reacting):

$$emf = \phi_{Zn} - \phi_{Cu} = -1.100 - \frac{0.0592}{2}\log\frac{(Cu^{2+})}{(Zn^{2+})} \qquad\qquad (3.19)$$

If the activities of Cu^{2+} and Zn^{2+} are chosen to be the same, the emf E is $-1.100\,V$. Since the emf is negative, current flows spontaneously from right to left within the cell. This fixes the true polarity of the cell, with the left electrode, Cu, as positive (cathode) and the right electrode, Zn, as negative (anode). From the relation $\Delta G = -nFE$, the free-energy change of (3.18) is positive, and the reaction is, therefore, not spontaneous as written, but goes instead in the opposite direction. In other words, when current is drawn from the cell, Cu^{2+} plates out on the copper electrode, and the zinc electrode corrodes.

---

*Reaction (3.18) is a simplification of the actual reaction. A more exact approach would include calculation of the liquid-junction potential between $CuSO_4$ and $ZnSO_4$ and elimination of single-ion activities, for which values are not measurable.

Similarly, the emf, polarity, and spontaneous reaction can be determined for any cell for which half-cell reactions and standard potentials are known.

3.6 MEASUREMENT OF pH

Hydrogen ion activity is commonly expressed, for convenience, in terms of pH, defined as

$$pH = -\log(H^+)$$

Hence, for the half-cell reaction $2H^+ + 2e^- \to H_2$, with the pressure of hydrogen equal to 1 atm, we have

$$\phi_{H_2} = -0.0592 \, pH$$

Since pure water contains equal concentrations of H^+ and OH^- in equilibrium with undissociated water, $H_2O \to H^+ + OH^-$, it is possible to calculate the activity of either the hydrogen ion or the hydroxyl ion from the ionization constant, the value of which at 25 °C is 1.01×10^{-14}. Therefore, the pH of pure water at 25 °C is

$$-\log \sqrt{1.01 \times 10^{-14}} = 7.0$$

If (H^+) exceeds (OH^-), as in acids, the pH is less than 7. If the pH is greater than 7, the solution is alkaline. The pH of strong acids can be negative, and the pH of strong alkalies can be greater than 14.

At temperatures above 25 °C, the ionization constant of H_2O is greater than at 25 °C; therefore, above 25 °C, the pH of pure water is less than 7 (Table 3.1).

3.7 THE OXYGEN ELECTRODE AND DIFFERENTIAL AERATION CELL

The oxygen electrode can be represented by platinized platinum immersed in an electrolyte saturated with oxygen. This electrode is particularly important to cor-

TABLE 3.1. Ionization Constant K_W and pH of Pure Water at Various Temperatures

Temperature (°C)	K_W	pH
0	0.115×10^{-14}	7.47
10	0.293	7.27
25	1.008	7.00
40	2.916	6.77
60	9.614	6.51

rosion studies because of its role in differential aeration cells in the mechanism of crevice corrosion and pitting.

The equilibrium for such an electrode is expressed as

$$\frac{1}{2}H_2O + \frac{1}{4}O_2 + e^- \rightarrow OH^- \qquad \phi^\circ = 0.401\,V \qquad (3.20)$$

and

$$\phi_{O_2} = 0.401 - 0.0592\log\frac{(OH^-)}{p_{O_2}^{1/4}} \qquad (3.21)$$

This reaction, however, unlike the hydrogen electrode reaction, is not strictly reversible under practical conditions of measurement, and hence the measured potential may vary with time and is not reproducible. The observed potential tends to be less noble than the calculated reversible value. Nevertheless, it is useful to know the direction of expected potential change as, for example, when oxygen pressure is altered. For illustration, consider two oxygen electrodes in an aqueous environment: one in contact with O_2 at 1 atm (left) and the other with O_2 at 0.2 atm (right). The potentials of the left- and right-hand electrodes, respectively, are as follows:

$$\frac{1}{2}H_2O + \frac{1}{4}O_2(1\,atm) + e^- \rightarrow OH^- \qquad (3.22)$$

$$\phi_1 = 0.401 - 0.0592\log\frac{(OH^-)}{1^{1/4}} \qquad (3.23)$$

$$\frac{1}{2}H_2O + \frac{1}{4}O_2(0.2\,atm) + e^- \rightarrow OH^- \qquad (3.24)$$

$$\phi_2 = 0.401 - 0.0592\log\frac{(OH^-)}{0.2^{1/4}} \qquad (3.25)$$

Subtracting, Eqs. (3.24)–(3.22),

$$\frac{1}{4}O_2(0.2\,atm) \rightarrow \frac{1}{4}O_2(1\,atm) \qquad (3.26)$$

Subtacting, Eqs. (3.25)–(3.23),

$$\phi_2 - \phi_1 = -0.0592\log\frac{1^{1/4}}{0.2^{1/4}} = \frac{0.0592}{4}\log 0.2 = -0.0103\,V \qquad (3.27)$$

The negative value of emf indicates that ΔG for (3.26) is positive, and hence the reaction is not spontaneous as written. Instead, positive electricity flows spontaneously within the cell from right to left. Therefore, the left-hand electrode, (3.22), is positive (cathode) and the right-hand electrode, (3.24), is negative (anode). This expresses the fact formulated earlier that, in any differential aeration cell, the electrode in contact with lower-pressure oxygen tends to be the anode, and the electrode in contact with higher-pressure oxygen tends to be the cathode.

When a similar cell is made of iron electrodes instead of platinum, an adherent, electrically conducting oxide of iron forms at cathodic areas; when in contact with aerated solution, this oxide acts as an oxygen electrode. But at anodic areas, Fe^{2+} forms, and the electrode acts as an iron electrode ($\phi° = -0.440\,V$). The operating emf of such a cell is much larger than that of the cell made up of platinum electrodes, with the value being given by

$$E = -0.440 - 0.401 - \frac{0.0592}{2} \log \frac{p_{O_2}^{1/2}}{(Fe^{2+})(OH^-)^2} \tag{3.28}$$

If the ferrous ion activity at the anode is assumed equal to 0.1, the pH of water at the cathode equal to 7.0, and the partial pressure of oxygen at the cathode equal to that of air (0.2 atm), the operating emf of the corresponding cell is 1.27 V. This is an appreciable value for resultant flow of current and accompanying corrosion at the anode. In practice, the emf is less than this because of the irreversible nature of the oxygen electrode, especially as approximated by an iron oxide film on iron, but the emf would, in general, be larger than the small value calculated for two platinum electrodes.

3.8 THE EMF AND GALVANIC SERIES

The Emf Series is an orderly arrangement of the standard potentials for all metals. The more negative values correspond to the more reactive metals (Table 3.2). Position in the Emf Series is determined by the *equilibrium* potential of a metal in contact with its ions at a concentration equal to unit activity. Of two metals composing a cell, the anode is the more active metal in the Emf Series, provided that the ion activities in equilibrium are both unity. Since unit activity corresponds in some cases to impossible concentrations of metal ions because of restricted solubility of metal salts, the Emf Series has only limited use for predicting which metal is anodic to another.

Tin, for example, is noble to iron according to the Emf Series. This is also the normal galvanic relationship of tin to iron in tinplate exposed to aerated aqueous media. But on the inside of tin-plated iron containers ("tin cans") in contact with food, certain constituents of foods combine chemically with Sn^{2+} ions to form soluble tin complexes. Reactions of this kind greatly lower the activity of Sn^{2+} ions with which the tin is in equilibrium, causing the potential of tin to become much more active, perhaps even more active than iron. The polarity of the iron–tin couple under these conditions reverses sign. The ratio of Sn^{2+} to Fe^{2+} within the can must be very small for the reversal of polarity to occur, as can be calculated from $\phi°$ values for iron and tin in accord with the following reaction:

TABLE 3.2. Electromotive Force (Emf) Series

Electrode Reaction	Standard Potential, $\phi°$, in volts at 25 °C
$Au^{3+} + 3e^- = Au$	1.50
$Pt^{2+} + 2e^- = Pt$	~1.2
$Pd^{2+} + 2e^- = Pd$	0.987
$Hg^{2+} + 2e^- = Hg$	0.854
$Ag^+ + e^- = Ag$	0.800
$Hg_2^{2+} + 2e^- = 2Hg$	0.789
$Cu^+ + e^- = Cu$	0.521
$Cu^{2+} + 2e^- = Cu$	0.342
$2H^+ + 2e^- = H_2$	0.000
$Pb^{2+} + 2e^- = Pb$	−0.126
$Sn^{2+} + 2e^- = Sn$	−0.136
$Mo^{3+} + 3e^- = Mo$	~−0.2
$Ni^{2+} + 2e^- = Ni$	−0.250
$Co^{2+} + 2e^- = Co$	−0.277
$Tl^+ + e^- = Tl$	−0.336
$In^{3+} + 3e^- = In$	−0.342
$Cd^{2+} + 2e^- = Cd$	−0.403
$Fe^{2+} + 2e^- = Fe$	−0.440
$Ga^{3+} + 3e^- = Ga$	−0.53
$Cr^{3+} + 3e^- = Cr$	−0.74
$Zn^{2+} + 2e^- = Zn$	−0.763
$Cr^{2+} + 2e^- = Cr$	−0.91
$Nb^{3+} + 3e^- = Nb$	~−1.1
$Mn^{2+} + 2e^- = Mn$	−1.18
$Zr^{4+} + 4e^- = Zr$	−1.53
$Ti^{2+} + 2e^- = Ti$	−1.63
$Al^{3+} + 3e^- = Al$	−1.66
$Hf^{4+} + 4e^- = Hf$	−1.70
$U^{3+} + 3e^- = U$	−1.80
$Be^{2+} + 2e^- = Be$	−1.85
$Mg^{2+} + 2e^- = Mg$	−2.37
$Na^+ + e^- = Na$	−2.71
$Ca^{2+} + 2e^- = Ca$	−2.87
$K^+ + e^- = K$	−2.93
$Li^+ + e^- = Li$	−3.05

$$Fe^{2+} + Sn \rightarrow Sn^{2+} + Fe \tag{3.29}$$

and

$$E = 0.136 - 0.440 - \frac{0.0592}{2} \log \frac{(Sn^{2+})}{(Fe^{2+})} \tag{3.30}$$

The cell reverses polarity when $E = 0$. Hence,

$$\log \frac{(Sn^{2+})}{(Fe^{2+})} = -0.304 \times \frac{2}{0.0592} = -10.27$$

or, the ratio $(Sn^{2+})/(Fe^{2+})$ must be less than 5×10^{-11} for tin to become active to iron. This small ratio can occur only through formation of tin complexes resulting from certain organic substances in foods. Complexing agents in general, such as EDTA, cyanides, and strong alkalies, tend to increase the corrosion rates of many metals by reducing the metal-ion activity, thereby shifting metal potentials markedly in the active direction.

Another factor that alters the galvanic position of some metals is the tendency, especially in oxidizing environments, to form specific surface films. These films shift the measured potential in the noble direction. In this state, the metal is said to be passive (see Chapter 6). Hence, chromium, although normally near zinc in the EMF Series, behaves galvanically more like silver in many air-saturated aqueous solutions because of a passive film that forms over its surface. The metal acts like an oxygen electrode instead of like chromium; hence, when coupled with iron, chromium becomes the cathode and current flow accelerates the corrosion of iron. In the active state (e.g., in hydrochloric acid), the reverse polarity occurs; that is, chromium becomes anodic to iron. Many metals, especially the transition metals of the periodic table, commonly exhibit passivity in aerated aqueous solutions.

Because of the limitations of the Emf Series for predicting galvanic relations, and also because alloys are not included, the *Galvanic Series* has been developed. The Galvanic Series is an arrangement of metals and alloys in accord with their actual measured potentials in a given environment. The potentials that determine the position of a metal in the Galvanic Series may include steady-state values in addition to truly reversible values; hence, alloys and passive metals are included. The Galvanic Series for metals in seawater is given in Fig. 3.3. Some metals occupy two positions in the Galvanic Series, depending on whether they are active or passive, whereas in the Emf Series only the active positions are possible, since only in this state is true equilibrium attained. The passive state, on the contrary, represents a nonequilibrium state in which the metal, because of surface films, is no longer in normal equilibrium with its ions. Although only one Emf Series exists, there can be several Galvanic Series because of differing complexing tendencies of various environments and differences in tendency to form surface films. In general, therefore, a specific Galvanic Series exists for each environment, and the relative positions of metals in such Series may vary from one environment to another.

The damage incurred by coupling two metals depends not only on how far apart they are in the Galvanic Series (potential difference), but also on their relative areas and the extent to which they are polarized (see Chapter 5). The potential difference of the polarized electrodes and the conductivity of the corrosive environment determine how much current flows between them.

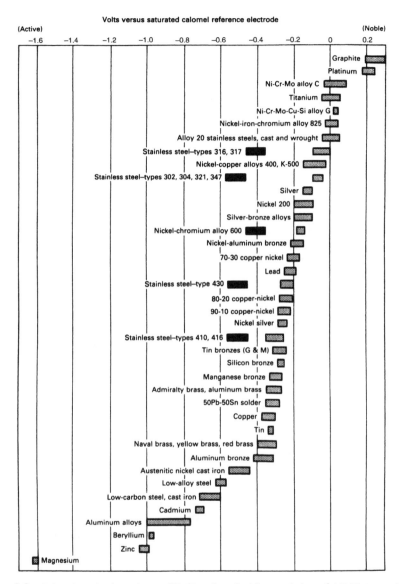

Figure 3.3. Galvanic series in seawater [5]. (Reprinted with permission of ASM International®. All rights reserved. www.asminternational.org.)

3.9 LIQUID JUNCTION POTENTIALS

In addition to potential differences between two metals in an electrolyte, potential differences also arise whenever two solutions of different composition or concentration come into contact. The potential difference is called the liquid

TABLE 3.3. Characteristic Liquid Junction Potentials of
Salt Solutions [6]

Electrolyte	Concentration	
	0.1 N	0.01 N
(−)HCl	35.65 mV	33.87 mV
KCl	8.87	8.20
NH$_4$Cl	6.92	6.89
NaCl	2.57	2.63
(+)LiCl	0.00	0.00

junction potential, and its sign and magnitude are determined by the relative mobility of ions and their concentration differences across the liquid junction. For example, in a junction formed between dilute and concentrated hydrochloric acid, H$^+$ ions move with greater velocity than Cl$^-$ ions; mobility at infinite dilution is 36×10^{-4} and 7.9×10^{-4} cm/s, respectively. Hence, the dilute aqueous solution acquires a positive charge with respect to the concentrated solution. For potassium chloride, the mobilities of K$^+$ and Cl$^-$ are similar; hence, liquid junction potentials between dilute and concentrated KCl are small in comparison with HCl junctions. If the HCl solutions discussed previously are saturated with KCl, so that most of the current across the boundary is carried by K$^+$ and Cl$^-$ ions, the liquid junction potential is decreased greatly. Use of a saturated KCl solution whenever liquid junctions are formed is one practical approach to minimizing liquid junction potentials.

Calculations of liquid junction potentials can be made on the basis of certain assumptions, but the derivation of such calculations is relatively complex even for simple junctions. Potentials of junctions formed between salt solutions of the same concentration having a common ion, such as the chloride ion, are additive. Characteristic values are given in Table 3.3, assigning zero arbitrarily to LiCl. For example, the potential of the junction HCl(0.1 N) : KCl(0.1 N) is equal to 35.65 − 8.87 = 26.78 mV with KCl solution positive and HCl solution negative. On the other hand, for the junction LiCl(0.1 N) : NH$_4$Cl(0.1 N), the value is 0.00 − 6.92 = −6.92 mV; the NH$_4$Cl solution is negative, and the LiCl solution is positive. Values at 0.01 N and 0.1 N are nearly alike and, except perhaps for strongly acidic or basic solutions, the values are small and are not of great concern for most corrosion measurements.

3.10 REFERENCE ELECTRODES

In the measurement of emf, the observed value represents a tendency for simultaneous reactions to occur at both electrodes of the cell. Interest is usually focused on the reaction that occurs at one electrode only. The criterion of complete cathodic protection based on potential measurements is an example. Measurements of this kind are made by using an electrode, called a *reference electrode*, that has a fixed value of potential regardless of the environment in which it is used. Any change in emf is the result of a change in potential of the electrode under study and not of the reference electrode. Reference electrodes use stable reversible electrode systems, as discussed in the following paragraphs. Details on procedures for preparing reference electrodes are available elsewhere [7].

3.10.1 Calomel Reference Electrode

The calomel reference electrode has long been a standard reference electrode used in the laboratory. It consists of mercury in equilibrium with Hg_2^{2+}, the activity of which is determined by the solubility of Hg_2Cl_2 (mercurous chloride, or calomel). The half-cell reaction is

$$Hg_2Cl_2 + 2e^- \rightarrow 2Hg + 2Cl^- \qquad \phi^\circ = 0.268 \, V$$

One design of electrode is illustrated in Fig. 3.4. Pure mercury covers a platinum wire sealed through the bottom of a glass tube. The mercury is covered with powdered mercurous chloride, which is only slightly soluble in potassium chloride solution, the latter filling the cell. The activity of Hg_2^{2+} depends on the concentration of KCl since the solubility product $(Hg_2^{2+})(Cl^-)^2$ is a constant. Potentials on the standard hydrogen scale for various KCl concentrations are listed in Table 3.4.

The saturated KCl electrode is convenient to prepare, but the potential is somewhat more sluggish in responding to temperature changes than are the unsaturated KCl electrodes. The $0.1\,N$ electrode has the lowest temperature coefficient.

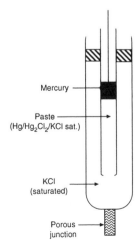

Mercury

Paste
(Hg/Hg₂Cl₂/KCl sat.)

KCl
(saturated)

Porous
junction

Figure 3.4. A type of calomel reference electrode. (B. Eggins, *Biosensors: an Introduction*, p. 56. Copyright Wiley and Teubner, 1996.)

TABLE 3.4. Potentials of Calomel Reference Electrodes

KCl Concentration	ϕ (V)	Temperature Coefficient (V/°C)
0.1 N	0.3338	-0.88×10^{-4}
1.0 N	0.2801	-2.75×10^{-4}
Saturated KCl	0.2416	-6.6×10^{-4}

The values cited neglect the liquid-junction potential at the KCl boundary, which in the case of strong acids, for example, increases the absolute value an average of several millivolts.

3.10.2 Silver–Silver Chloride Reference Electrode

The silver–silver chloride reference electrode, can be prepared by electroplating with silver a platinum wire sealed into a glass tube. The silver coating is then converted partly into silver chloride by making it anode in dilute hydrochloric acid. Alternatively, a silver wire can be chloridized as just described. More details on electrode preparation are available elsewhere [7]. When the silver–silver chloride electrode is immersed in a chloride solution, the following equilibrium is established:

$$AgCl + e^- \rightarrow Ag + Cl^- \qquad \phi° = 0.222 \text{ V}$$

Like the calomel electrode, the potential is more active the higher the KCl concentration. In 0.1 N KCl, the value is 0.288 V, and the temperature coefficient is -4.3×10^{-4} V/°C. Potentials for other concentrations of KCl can be obtained by substituting the corresponding mean ion activity of Cl$^-$ into the Nernst equation.

A schematic of a silver–silver chloride reference electrode is shown in Fig. 3.5.

3.10.3 Saturated Copper–Copper Sulfate Reference Electrode

The saturated copper–copper sulfate reference electrode consists of metallic copper immersed in saturated copper sulfate, as shown in Fig. 3.6. It is used primarily in field measurements where the electrode must be resistant to shock and where its usual large size minimizes polarization errors. The accuracy of this electrode is adequate for most corrosion investigations, even though it falls somewhat below the precision obtainable with the calomel or silver chloride electrodes. The half-cell reaction is

$$Cu^{2+} + 2e^- \rightarrow Cu \qquad \phi° = 0.342 \text{ V}$$

For saturated copper sulfate, the potential is 0.316 V, and the temperature coefficient is 7×10^{-4} V/°C [8].

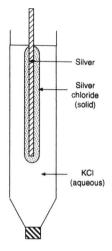

Figure 3.5. Schematic of a silver–silver chloride reference electrode. (B. Eggins, *Biosensors: an Introduction*, p. 57. Wiley and Teubner, 1996.)

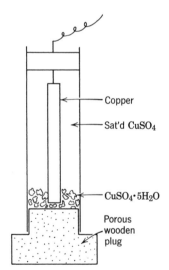

Figure 3.6. Copper–saturated copper sulfate reference electrode.

REFERENCES

1. See, for example, G. Lewis, M. Randall, K. Pitzer, and L. Brewer, *Thermodynamics*, McGraw-Hill, New York, 1961.

2. W. M. Latimer, *The Oxidation States of the Elements and Their Potentials in Aqueous Solutions*, 2nd edition, Prentice-Hall, Englewood Cliffs, NJ, 1952.

3. G. Milazzo and S. Caroli, *Tables of Standard Electrode Potentials*, Wiley, Chichester, 1978.

4. M. S. Antelman, *The Encyclopedia of Chemical Electrode Potentials*, Plenum Press, New York, 1982.

5. H. P. Hack, Evaluating galvanic corrosion, in *ASM Handbook*, Vol. 13A, *Corrosion: Fundamentals, Testing, and Protection*, ASM International, Materials Park, OH, 2003, p. 563; ASTM G82–98(2003), Standard Guide for Development and Use of a Galvanic Series for Predicting Galvanic Corrosion Performance, ASTM International, West Conshohocken, Pennsylvania; E. D. Verink, Designing to prevent corrosion, in Uhlig's Corrosion Handbook, 2nd edition, Wiley, New York, 2000, p. 99.

6. D. A. MacInnes, *The Principles of Electrochemistry*, Reinhold, New York, 1939; Dover, New York, 1961.

7. D. J. G. Ives and G. J. Janz, *Reference Electrodes: Theory and Practice*, Academic Press, New York, 1961, pp. 198–213.

8. S. Ewing, The Copper–Copper Sulfate Half-Cell for Measuring Potentials in the Earth, Committee Report, American Gas Association, 1939, p. 10.

GENERAL REFERENCES

B. R. Eggins, *Chemical Sensors and Biosensors*, Wiley, Chichester, U.K., 2002.

D. R. Gaskell, *Introduction to the Thermodynamics of Materials*, 3rd edition, Taylor & Francis, Washington, D.C., 1995.

E. Protopopoff, Potential measurements with reference electrodes, in *ASM Handbook*, Vol. 13A, *Corrosion: Fundamentals, Testing, and Protection*, ASM International, Materials Park, OH 2003, pp. 13–16.

PROBLEMS

In these problems, temperature is 25 °C unless otherwise stated.

1. Calculate the value for $2.303\,RT/F$ at 50 °C.

2. Calculate the exact half-cell potential for the Ag–AgCl electrode in $1\,M$ NaCl.

3. Calculate the exact half-cell potential of zinc in $0.01\,M$ $ZnCl_2$.

4. Calculate the half-cell potential of the hydrogen electrode in a solution of pH = 7 and partial pressure of H_2 = 0.5 atm at 40 °C.

5. The emf of a cell made up of zinc (anode) and hydrogen electrode (cathode) immersed in $0.5\,M$ $ZnCl_2$ is +0.590 V. What is the pH of the solution?

6. Calculate the theoretical tendency of zinc to corrode (in volts) with evolution of hydrogen when immersed in $0.01\,M$ $ZnCl_2$ acidified to pH = 2.

7. Calculate the theoretical tendency for nickel to corrode (in volts) in deaerated water of pH = 7. Assume that the corrosion products are H_2 and $Ni(OH)_2$, the solubility product of which is 1.6×10^{-16}.

8. (a) What is the emf of a cell made up of a copper electrode and a hydrogen electrode ($p_{H_2} = 2\,atm$) in copper sulfate (activity $Cu^{2+} = 1$) of pH = 1?

   (b) What is the polarity of the cell and which electrode is anode?

9. (a) Calculate whether copper will corrode in deaerated $CuSO_4$, pH = 0, to form Cu^{2+} (activity = 0.1) and H_2 (1 atm). What is the corrosion tendency in volts?

   (b) Similarly, calculate whether copper will corrode in deaerated KCN (activity $CN^- = 0.5$) of pH = 10, assuming that $Cu(CN)_2^-$ is formed, the activity of which is 10^{-4}.

$$Cu(CN)_2^- + e^- \rightarrow 2CN^- + Cu \qquad \phi° = -0.446\,V$$

10. (a) Calculate the emf of the following cell:

$$Pt;\ Fe^{3+}\ (activity = 0.1),\ Fe^{2+}\ (activity = 0.001),\ Ag^+\ (activity = 0.01);\ Ag$$

   (b) Write the spontaneous reaction for the cell. What is the polarity and which electrode is anode?

11. (a) Calculate the emf of a concentration cell made up of copper electrodes in $0.1\,M$ $CuSO_4$ and $0.5\,M$ $CuSO_4$, neglecting the liquid-junction potential.

   (b) Write the spontaneous reaction of the cell and indicate which electrode is anode.

12. (a) Calculate the emf of the following cell at 40°C:

$$O_2\ (1\,atm),\ Pt;\ H_2O;\ O_2\ (0.1\,atm),\ Pt$$

   (b) What is the polarity and which electrode is anode?

13. (a) Calculate the emf of a cell made up of a hydrogen electrode ($p_{H_2} = 1\,atm$) and an oxygen electrode ($p_{O_2} = 0.5\,atm$) in $0.5\,M$ NaOH.

   (b) What is the polarity and which electrode is anode? (Assume that the oxygen electrode is reversible.)

14. Calculate whether silver will corrode if immersed in $0.1M$ $CuCl_2$ to form solid AgCl. What is the corrosion tendency in volts?

15. Calculate whether silver will corrode with hydrogen evolution in deaerated KCN solution, pH = 9, when CN^- activity = 1.0 and $Ag(CN)_2^-$ activity is 0.001.

16. Zinc is immersed in a solution of $CuCl_2$. What is the reaction and at what activity ratio Zn^{2+}/Cu^{2+} will the reaction stop?

17. Calculate the emf of a cell made up of iron and lead electrodes in a solution of Fe^{2+} and Pb^{2+} of equal activity. Which electrode tends to corrode when the cell is short-circuited?

18. Calculate the emf as in Problem 17 in an air-saturated alkaline solution of pH = 10. Which electrode corrodes on short-circuiting the cell? (Assume that $HPbO_2^-$ forms as corrosion product and that its activity is 0.1; also assume that iron is passive and that its potential approximates the oxygen electrode.)

$$HPbO_2^- + H_2O + 2e^- \rightarrow Pb + 3OH^- \qquad \phi° = -0.54\,V$$

19. Given $Fe^{2+} + 2e^- \rightarrow Fe$, $\phi° = -0.440\,V$, and $Fe^{3+} + e^- \rightarrow Fe^{2+}$, $\phi° = 0.771\,V$, calculate $\phi°$ for $Fe^{3+} + 3e^- \rightarrow Fe$.

20. Given $O_2 + 2H_2O + 4e^- \rightarrow 4OH^-$, $\phi° = 0.401\,V$, calculate $\phi°$ for $O_2 + 4H^+ + 4e^- \rightarrow 2H_2O$.

21. Calculate the pressure (fugacity) of hydrogen required to stop corrosion of iron immersed in $0.1\,M$ $FeCl_2$, pH = 3.

22. Calculate the pressure of hydrogen, as in Problem 21, in deaerated water with $Fe(OH)_2$ as the corrosion product. [Solubility product $Fe(OH)_2$ = 1.8×10^{-15}.]

23. Calculate the pressure of hydrogen required to stop corrosion of cadmium at $25\,°C$ in deaerated water, with $Cd(OH)_2$ as the corrosion product. [Solubility product $Cd(OH)_2 = 2.0 \times 10^{-14}$.]

24. A copper storage tank containing dilute H_2SO_4 at pH = 0.1 is blanketed with hydrogen at $1\,atm$. Calculate the maximum Cu^{2+} contamination of the acid in moles Cu^{2+} per liter. What is the corresponding contamination if the hydrogen partial pressure is reduced to $10^{-4}\,atm$?

Answers to Problems

1. 0.0641 V.
2. 0.233 V.
3. –0.827 V.
4. –0.426 V.
5. 3.28.
6. 0.709 V.
7. –0.111 V.
8. –0.405 V; H_2 electrode is negative and is the anode.
9. (a) No; –0.307 V.
   (b) Yes; 0.056 V.
10. (a) 0.208 V.
    (b) Ag is negative and is the anode.
11. (a) 0.0096 V.
    (b) $0.1M$ $CuSO_4$ electrode is anode.

12. (*a*) 0.0155 V.
   (*b*) 0.1 atm O_2 electrode is negative and is the anode.
13. 1.225 V; H_2 electrode is negative and is the anode.
14. Yes; 0.019 V.
15. No; emf = −0.045 V.
16. 1.45×10^{37}.
17. 0.314 V; Fe.
18. −0.84 V; Pb.
19. −0.036 V.
20. 1.23 V.
21. 1.26×10^{10} atm.
22. 42 atm.
23. 0.42 atm.
24. 2.6×10^{-12}; 2.6×10^{-8} mole/liter.

4

THERMODYNAMICS: POURBAIX DIAGRAMS

4.1 BASIS OF POURBAIX DIAGRAMS

M. Pourbaix devised a compact summary of thermodynamic data in the form of potential–pH diagrams, which relate to the electrochemical and corrosion behavior of any metal in water. These diagrams, known as Pourbaix diagrams, are now available for most of the common metals [1]. They have the advantage of showing at a glance specific conditions of potential and pH under which the metal either does not react (immunity) or can react to form specific oxides or complex ions; that is, Pourbaix diagrams indicate the potential–pH domain in which each species is stable.

From a corrosion engineering perspective, the value of Pourbaix diagrams is their usefulness in identifying potential–pH domains where corrosion does not occur—that is, where the metal itself is the stable phase. By controlling potential (e.g., by cathodic protection) and/or by adjusting the pH in specific domains identified using Pourbaix diagrams, it may be possible to prevent corrosion from taking place. For example, in the potential–pH domain labeled "Fe (immunity)," iron is stable, and no corrosion is predicted.

In general, thermodynamics is an excellent starting point for many corrosion studies, and potential–pH diagrams are essential tools. For this reason, Pourbaix diagrams continue to be developed for interpreting corrosion studies in specific systems of engineering importance [2].

As every good engineer knows, however, predictions must be tested experimentally and validated before using them, and predictions arrived at using Pourbaix diagrams are no exception to this general rule.

4.2 POURBAIX DIAGRAM FOR WATER

Each line of a Pourbaix diagram represents conditions of thermodynamic equilibrium for some reaction. The Pourbaix diagram for water is presented in Fig. 4.1. Above line b, oxygen is evolved in accord with the reaction $H_2O \rightarrow \frac{1}{2}O_2 + 2H^+ + 2e^-$. For this equilibrium, the relationship between potential and pH is, from the Nernst equation,

$$\phi = \phi^\circ - 2.303 \frac{RT}{2F} \log \frac{1}{(H^+)^2}$$

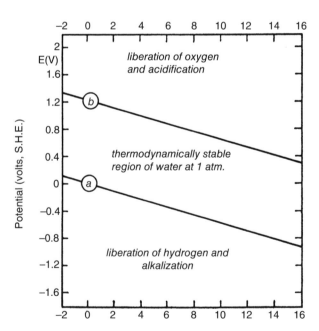

Figure 4.1. Pourbaix diagram for water at 25°C, showing the oxygen line, b, above which oxygen is evolved, and the hydrogen line, a, below which hydrogen is evolved, from the surface of an immersed electrode. Between these two lines, water is stable. (M. Pourbaix, *Atlas of Electrochemical Equilibria in Aqueous Solutions*, 2nd English edition, p. 100, copyright NACE International 1974 and CEBELCOR.)

With $\phi° = 1.229$ (Appendix, Section 29.9), $T = 298.2\,K$, $R = 8.314\,J/deg\text{-}mole$, and $F = 96,500\,C/eq$, we have

$$\phi = 1.229 - 0.0592\,pH$$

Above line b, defined by this equation, oxygen is evolved at the surface of an immersed electrode. Below this line, water is stable.

Below line a, hydrogen is evolved in accord with the reaction $2H^+ + 2e^- \rightarrow H_2$. Using the Nernst equation for this equilibrium, the relationship between potential and pH is

$$\phi = \phi° - 2.303\frac{RT}{2F}\log\frac{1}{(H^+)^2} = -0.0592\,pH$$

With $\phi° = 0$ (Section 3.8, Table 3.2), we obtain

$$\phi = -0.0592\,pH$$

Below line a, represented by this equation, hydrogen gas is evolved from the surface of an immersed electrode.

Between lines a and b, water is stable.

4.3 POURBAIX DIAGRAM FOR IRON

The Pourbaix diagram for iron is presented in Fig. 4.2.

A horizontal line represents a reaction that does not involve pH; that is, neither H^+ nor OH^- is involved, as in the reaction, $Fe^{2+} + 2e^- \rightarrow Fe$. For this equilibrium, using the Nernst equation, we obtain

$$\phi = \phi° - 2.303\frac{RT}{nF}\log\frac{1}{(Fe^{2+})}$$

$$\phi = -0.440 + 0.0296\log(Fe^{2+})$$

If (Fe^{2+}) is taken as 10^{-6}, then $\phi = -0.617\,V$, a horizontal line on the Pourbaix diagram.

A vertical line involves H^+ or OH^-, but not electrons; for example, $2Fe^{3+} + 3H_2O \rightarrow Fe_2O_3 + 6H^+$. In Figure 4.2, the vertical line separating Fe^{3+} from Fe_2O_3 corresponds to this reaction. For this equilibrium, we have

$$K = \frac{(H^+)^6}{(Fe^{3+})^2}$$

$$\log K = 6\log(H^+) - 2\log(Fe^{3+})$$

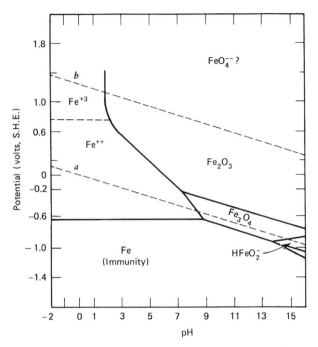

Figure 4.2. Pourbaix diagram for the iron–water system at 25°C, considering Fe, Fe_3O_4, and Fe_2O_3 as the only solid substances.

$$\log K = -6\,pH - 2\log(Fe^{3+})$$

Since $\Delta G° = -RT \ln K$ and $\Delta G° = -8240\,J/mole$, we obtain

$$\log K = 1.43$$
$$\log(Fe^{3+}) = -0.72 - 3\,pH$$

Taking $(Fe^{3+}) = 10^{-6}$, we have $pH = 1.76$.

In the Pourbaix diagram for iron, Fig. 4.2, the vertical line at pH 1.76 represents the equilibrium reaction, $2Fe^{3+} + 3H_2O \rightarrow Fe_2O_3 + 6H^+$. To the right of this line (i.e., at pH > 1.76), Fe_2O_3 is the stable phase; and this oxide, as a protective film, would be expected to provide some protection against corrosion. To the left of this line (i.e., at pH < 1.76), ferric ions in solution are stable, and corrosion is expected to take place without any protection afforded by a surface oxide film.

A sloping line involves H^+, OH^-, and electrons. For example, the sloping line separating Fe^{2+} from Fe_2O_3 represents the reaction $Fe_2O_3 + 6H^+ + 2e^- \rightarrow 2Fe^{2+} + 3H_2O$. For this reaction, we have

$$\phi = \phi° - 2.303\frac{RT}{nF}\log\frac{(Fe^{2+})^2}{(H^+)^6}$$

Since $\phi° = 0.728\,V$ and $n = 2$, we get

$$\phi = 0.728 - 0.0296\,(Fe^{2+})$$

$$\phi = 0.728 - \frac{0.0592}{2}\log(Fe^{2+})^2 + \frac{0.0592}{2}\log(H^+)^6$$

$$\phi = 0.728 - 0.0592\log(Fe^{2+}) - 0.1776\,pH$$

Taking $(Fe^{2+}) = 10^{-6}$, we obtain

$$\phi = 1.082 - 0.1776\,pH$$

This line in Fig. 4.2 represents the equilibrium, $Fe_2O_3 + 6H^+ + 2e^- \rightarrow 2Fe^{2+} + 3H_2O$. To the right of this line, Fe_2O_3 is a stable phase that is expected to form a surface oxide film that protects the underlying metal from corrosion. To the left of this line, Fe^{2+} is a stable species in solution.

The pH values in Pourbaix diagrams are those of solution in immediate contact with the metal surface. This value in some instances (e.g., Fe in aerated H_2O) differs from that of the bulk solution.

Soluble hypoferrites ($HFeO_2^-$) can form in very alkaline solutions within a restricted active potential range. Soluble ferrates (FeO_4^{2-}) can form in alkaline solutions at very noble potentials, but the stable field is not well-defined.

When any reaction involves ions other than H^+ or OH^-, it is assumed, in general, that the activity equals 10^{-6}. Thus the horizontal line at $-0.62\,V$ means that iron will not corrode below this value to form a solution of concentration $> 10^{-6}\,M$ Fe^{2+} in accord with $Fe \rightarrow Fe^{2+} + 2e^-$, $\phi = -0.44 + (0.059/2)\log(10^{-6}) = -0.62\,V$. If a value other that 10^{-6} is used, the lines separating the phases are shifted (as can be seen in Figs. 4.3 and 4.4).

The fields marked Fe_2O_3 and Fe_3O_4 are sometimes labeled "passivation" on the assumption that iron reacts in these regions to form protective oxide films. This is correct only insofar as passivity is accounted for by a diffusion-barrier oxide layer (Definition 2, Section 6.1). Actually, the Flade potential, above which passivity of iron is observed in media such as sulfuric or nitric acid, parallels line a and b, intersecting $0.6\,V$ at pH = 0. For this reason, the passive film (Definition 1, Section 6.1) may not be any of the equilibrium stoichiometric iron oxides, as is further discussed in Chapter 6.

4.4 POURBAIX DIAGRAM OF ALUMINUM

The Pourbaix diagram of aluminum, presented in Fig. 4.3, indicates that hydrargillite, $Al_2O_3 \cdot 3H_2O$, is the stable phase between about pH 4 and 9 [3]. Indeed, this film is considered to be responsible for the successful use of aluminum in many structural applications. This diagram also predicts the amphoteric nature

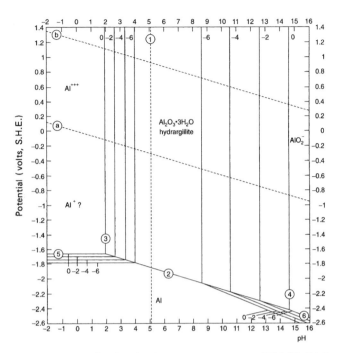

Figure 4.3. Pourbaix diagram for the aluminum–water system at 25°C [2]. (M. Pourbaix, *Atlas of Electrochemical Equilibria in Aqueous Solutions*, 2nd English edition, p. 171, copyright NACE International 1974 and CEBELCOR.)

of aluminum, with corrosion under highly acidic and highly alkaline conditions and protection by an oxide film between about pH 4 and 9. Behavior of aluminum is discussed further in Chapter 21.

4.5 POURBAIX DIAGRAM OF MAGNESIUM

The Pourbaix diagram of magnesium (Fig. 4.4) contrasts with that of aluminum in that Mg^{2+} is the stable species in solution over most of the potential–pH range [4]. The potential–pH domain in which magnesium is stable is well below the domain in which water is stable; therefore, magnesium reduces water, and hydrogen evolution is the cathodic reaction. Between pH about 8.5 and 11.5, a protective film of oxide or hydroxide may provide some protection. At pH greater than about 11.5, a stable film of $Mg(OH)_2$ protects magnesium from corrosion. Because of the very active value of the equilibrium potential of the reaction, $Mg^{2+} + 2e^- \rightarrow Mg$, magnesium cannot be prepared by electrolysis of its aqueous solutions, which would evolve hydrogen without forming magnesium.

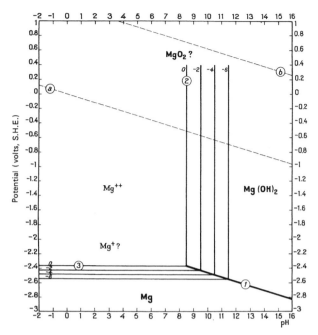

Figure 4.4. Pourbaix diagram for the magnesium–water system at 25°C [3]. (M. Pourbaix, *Atlas of Electrochemical Equilibria in Aqueous Solutions*, 2nd English edition, p. 141, copyright NACE International 1974 and CEBELCOR.)

4.6 LIMITATIONS OF POURBAIX DIAGRAMS

Since Pourbaix diagrams are based on thermodynamic data, they convey no information on rates of reaction—that is, whether any possible reaction proceeds rapidly or slowly when the energy changes are favorable. The data indicate the conditions for which diffusion-barrier films may form on an electrode surface, but they provide no measure of how effective such barrier films may be in the presence of specific anions, such as SO_4^{2-} or Cl^-. Similarly, they do not indicate the detailed conditions under which nonstoichiometric metal compound films are possible; some of these films are important in determining observed corrosion rates (see Chapter 6). However, the diagrams do clearly outline the nature of the stoichiometric compounds into which any less stable compounds may transform whenever equilibrium is achieved. Conventional Pourbaix diagrams do not include alloys; for example, the corrosion of iron–nickel alloys cannot be accurately predicted by superimposing conventional Pourbaix diagrams for iron and nickel [5]. With these limitations, the diagrams usefully describe the equilibrium status of a metal either as immersed in acids or alkalies or when a given potential is impressed on it.

REFERENCES

1. Marcel Pourbaix, *Atlas of Electrochemical Equilibria in Aqueous Solutions*, translated from the French by James A. Franklin, 2nd English edition, National Association of Corrosion Engineers, Houston, TX, and Centre Belge d'Etude de la Corrosion (CEBEL-COR), Brussels, Belgium, 1974.
2. M. J. Muñoz-Portero, J. García-Antón, J. L. Guiñón, and V. Pérez-Herranz, *Corrosion* **63** (7), 625 (2007).
3. Ref. 1., p. 171.
4. Ibid., p. 141.
5. W. T. Thompson, M. H. Kaye, C. W. Bale, and A. D. Pelton, Pourbaix diagrams for multi-element systems, in *Uhlig's Corrosion Handbook*, 2nd edition, R. W. Revie, editor, Wiley, New York, 2000, pp. 125–136.

GENERAL REFERENCES

Marcel Pourbaix, *Atlas of Electrochemical Equilibria in Aqueous Solutions*, translated from the French by James A. Franklin, 2nd English edition, National Association of Corrosion Engineers, Houston, Texas, and Centre Belge d'Etude de la Corrosion (CEBEL-COR), Brussels, Belgium, 1974.

W. T. Thompson, M. H. Kaye, C. W. Bale, and A. D. Pelton, Pourbaix diagrams for multielement systems, in *Uhlig's Corrosion Handbook*, 2nd edition, R. W. Revie, editor, Wiley, New York, 2000, pp. 125–136.

E. D. Verink, Jr., Simplified procedure for constructing Pourbaix diagrams, in *Uhlig's Corrosion Handbook*, 2nd edition, R. W. Revie, editor, Wiley, New York, 2000, pp. 111–124.

PROBLEMS

1. Referring to the Pourbaix diagram for iron (Fig. 4.2), list equilibrium reactions corresponding to lines separating (*a*) Fe and Fe_3O_4, (*b*) Fe_3O_4 and Fe_2O_3, and (*c*) Fe^{2+} and Fe_2O_3. Calculate slopes $\partial\phi/\partial pH$ in each case and compare with the diagram.

2. Using the Pourbaix diagram for iron (Fig. 4.2), determine the direction of each of the following four reactions in solution of pH 7 containing 0.06 ppm iron in each of the following potential ranges.

   Potential ranges:
   (*a*) $\phi < -0.7\,V$.
   (*b*) $-0.6 < \phi < -0.4\,V$.
   (*c*) $-0.4\,V < \phi < -0.2\,V$.
   (*d*) $-0.2\,V < \phi < +0.8\,V$.
   (*e*) $+0.80\,V < \phi$

Reactions:

$$Fe^{2+} + 2e^- \rightarrow Fe$$

$$2H^+ + 2e^- \rightarrow H_2$$

$$Fe_2O_3 + 6H^+ + 2e^- \rightarrow Fe^{2+} + 3H_2O$$

$$O_2 + 4H^+ + 4e^- \rightarrow 2H_2O$$

Answers to Problem

1. (a) $Fe_3O_4 + 8H^+ + 8e^- \rightarrow 3Fe + 4H_2O$
   $\partial\phi/\partial pH = -0.059$
   (b) $\partial\phi/\partial pH = -0.18$

5

KINETICS: POLARIZATION AND CORROSION RATES

5.1 POLARIZATION

Thermodynamics—the equilibrium between metals and their environments—has been discussed in the previous chapters. The concept of corrosion tendency is based on thermodynamics. In practice, however, we are concerned primarily with rates of corrosion—that is, with kinetics. Some metals with pronounced tendency to react (e.g., aluminum) nevertheless react so slowly that they meet the requirements of a structural metal and may actually be more resistant in some media than other metals that have inherently less tendency to react.

It must not be concluded, however, that considerations of equilibria are irrelevant to the study of corrosion. Instead, a fundamental approach to nonequilibrium states, along with calculation of corrosion rates, begins with the primary consideration that equilibrium has been disturbed. In general, therefore, we must know the equilibrium state of the system before we can appreciate the various factors that control the rate at which the system tends toward equilibrium, that is, the rate of corrosion.

An electrode is not at equilibrium when a net current flows to or from its surface. The measured potential of such an electrode is altered to an extent that depends on the magnitude of the external current and its direction. The direction

Corrosion and Corrosion Control, by R. Winston Revie and Herbert H. Uhlig
Copyright © 2008 John Wiley & Sons, Inc.

of potential change always opposes the shift from equilibrium and, hence, opposes the flow of current, whether the current is impressed externally or is of galvanic origin. When current flows in a galvanic cell, for example, the anode is always more cathodic in potential and the cathode always becomes more anodic, with the difference of potential between the anode and cathode becoming smaller as current is increased. The potential change caused by net current to or from an electrode, measured in volts, is called polarization.

Electrode kinetics is the study of reaction rates at the interface between an electrode and a liquid. The science of electrode kinetics has made possible many advances in the understanding of corrosion and the practical measurement of corrosion rates. The interpretation of corrosion processes by superimposing electrochemical partial processes was developed by Wagner and Traud [1]. Important concepts of electrode kinetics that will be introduced in this chapter are the *corrosion potential* (also called the *mixed potential* and the *rest potential*), *corrosion current density, exchange current density*, and *Tafel slope*. The treatment of electrode kinetics in this book is, of necessity, elementary and directed toward application of corrosion science. For more detailed discussion of electrode kinetics, the reader should refer to specialized texts listed at the end of the chapter.

5.2 THE POLARIZED CELL

Consider a cell made up of zinc in $ZnSO_4$ solution and copper in $CuSO_4$ solution (the Daniell cell), the electrodes of which are connected to a variable resistance R, voltmeter V, and ammeter A, as shown in Fig. 5.1. The potential difference (emf) of zinc and copper electrodes of the cell without current flow is about 1 V. If a small current is allowed to flow through the external resistance, the measured potential difference falls below 1 V because both electrodes polarize. The voltage continues to fall as the current increases. On complete short-circuiting (very small external resistance), maximum current flows and the potential difference of copper and zinc electrodes becomes almost zero.

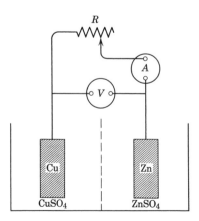

Figure 5.1. Polarized copper–zinc cell.

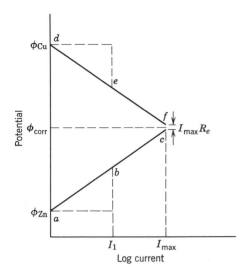

Figure 5.2. Polarization diagram for copper–zinc cell.

The effect of net current flow on voltage of the Daniell cell can be represented by plotting a *polarization diagram*—that is, a graph showing potential, ϕ, of copper and zinc electrodes (as described in Section 5.3, "How Polarization Is Measured") with total current, I, as shown in Fig. 5.2. The *thermodynamic potentials* (no current flow through the cell) are given by ϕ_{Zn} and ϕ_{Cu}.

In Fig. 5.2, the zinc electrode polarizes along curve abc, and the copper electrode polarizes along curve def. At a value of current through the ammeter equal to I_1, the polarization of zinc in volts is given by the difference between the actual potential of zinc at b and the thermodynamic value a or ϕ_{Zn}. Similarly, the polarization of copper is given by the difference of potential $e - d$. The potential difference of the polarized electrodes, $b - e$, is equal to the current i_1 multiplied by the total resistance of both the external metallic resistance R_m and the internal electrolytic resistance, R_e, in series, or $I_1(R_e + R_m)$. On short-circuiting, the current becomes maximum, I_{max}. Then R_m can be neglected, and the potential difference of both electrodes decreases to a minimum equal to $I_{max}R_e$. The maximum current is equivalent to $(65.38/2)I_{max}/F$ grams zinc corroding per second, where I_{max} is in amperes, F is equal to 96,500 C/eq, and 65.38/2 is the equivalent weight of zinc.

The cathodic reaction corresponds to the identical chemical equivalents of copper depositing per second on the cathode. The corrosion rate of zinc can exceed the indicated equivalent corrosion rate, I_{max}, only if a method is introduced for reducing the polarization of zinc or copper, or both, thereby reducing the slopes of abc or def, causing an approach to intersection at a larger value of I. Similarly, any factor tending to increase polarization will decrease current through the cell and decrease the corresponding corrosion rate of zinc. Obviously, the polarization curves can never actually intersect, although they can approach each other very closely if anodes and cathodes are closely spaced in media of moderate to good conductivity. There will always be a finite potential difference accompanying an observed flow of current.

Electrolytic cells that account for the corrosion of metals are analogous to the short-circuited cell. The measured potential of a corroding metal, the mixed potential of both polarized anodes and cathodes, is also referred to as the corrosion potential, ϕ_{corr}. The value, I_{max}, is known as the corrosion current, I_{corr}. By Faraday's law, the corrosion rate of anodic areas on a metal surface is proportional to I_{corr}, and hence the corrosion rate per unit area can always be expressed as a current density. For zinc, a corrosion rate of 1 gmd is equivalent to 0.0342 A/m^2. For Fe corroding to Fe^{2+}, the corresponding value is 0.0400 A/m^2. Values for other metals are listed in the Appendix (Section 29.8.2).

Referring to Fig. 5.2, we can calculate the corrosion rate of a metal if data are available for the corrosion potential and for the polarization behavior and thermodynamic potential of either anode or cathode. In general, the relative anode–cathode area ratio for the corroding metal must also be known, since polarization data are usually obtained under conditions where the electrode surface is all anode or all cathode.

5.3 HOW POLARIZATION IS MEASURED

Referring to Fig. 5.3(a), showing a two-compartment cell separated by a sintered glass disk, G, assume electrode B to be polarized by current from electrode D. In order to achieve uniform current density at B, a three-compartment cell can be used, with B in the center and two auxiliary electrodes in the outer compartments. The probe L, sometimes called the *Luggin capillary* or *salt bridge*, of reference cell R (or of a salt bridge between R and B) is placed close to the surface of B, thereby minimizing extraneous potentials caused by IR drop through the electrolyte. The emf of cell B–R is recorded for each value of current as read on ammeter A, allowing sufficient time for steady-state conditions. Polarization of B, whether anode or cathode, is recorded in volts with reference to half-cell electrode R for various values of current density. The potentials are often converted to the standard hydrogen scale.

Figure 5.3(b) shows the type of cell that is commonly used in corrosion studies to measure polarization, in which the working electrode (the electrode being studied), two (for uniformity of current flow) counter electrodes, gas inlet and outlet, Luggin capillary, and thermometer are all included [2].

The Luggin capillary can disturb the distribution of the current applied to electrode (B) in Fig. 5.3(a). In addition, particularly in low conductivity solutions, the distance of the tip of the Luggin capillary from the electrode being studied [the distance between L and B in Fig. 5.3(a)] can have a significant effect on IR drop that arises because of the current flow through the electrolyte.

Potentials can be measured with probe L adjusted at various distances from B, with subsequent extrapolation to zero distance. This correction, as shown in

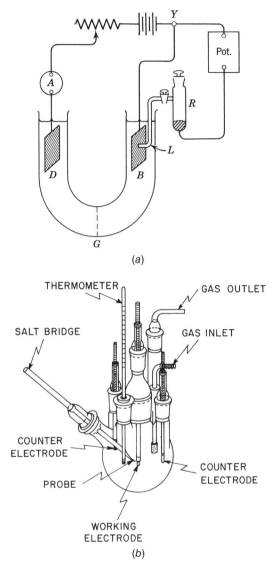

(a)

(b)

Figure 5.3. (a) Cell for measuring polarization. (b) Schematic diagram of commonly used polarization cell [2]. (Copyright ASTM INTERNATIONAL. Reprinted with permission.)

the next paragraph, is needed only when the measurements require highest accuracy, when the current densities are unusually high, or when the conductivity of the electrolyte is unusually low, as in distilled water. This correction, however, *does not include* any high-resistance reaction product film on the surface of the electrode.

5.3.1 Calculation of *IR* Drop in an Electrolyte

The resistance of an electrolyte solution measuring l cm long and S cm^2 in cross section is equal to $l/\kappa S$ ohms, where κ is the specific conductivity. Hence, the *IR* drop in volts equals il/κ, where i is the current density. For seawater, $\kappa = 0.05\,\Omega^{-1}\,cm^{-1}$; therefore, a current density of 1×10^{-5} A/cm^2 (0.1 A/m^2)(the magnitude of current density that might be applied for cathodically protecting steel) produces an *IR* drop correction for a 1-cm separation of probe from cathode equal to $(1 \times 10^{-5}\,V)/0.05 = 0.2$ mV. This value is negligible in establishing the critical minimum current density for adequate cathodic protection. In some soft waters, however, where κ may be $10^{-5}\,\Omega^{-1}\,cm^{-1}$, the corresponding *IR* drop equals 1 V/cm.

5.4 CAUSES OF POLARIZATION

There are three types of electrode polarization: concentration polarization, activation polarization, and *IR* drop.

1. *Concentration Polarization (Diffusion Overpotential).* If copper is made cathode in a solution of dilute CuSO$_4$ in which the activity of cupric ion is represented by (Cu^{+2}), then the potential ϕ_1, in absence of external current, is given by the Nernst equation

$$\phi_1 = 0.337 + \frac{0.0592}{2}\log(Cu^{2+}) \tag{5.1}$$

When current flows, copper is deposited on the electrode, thereby decreasing surface concentration of copper ions to an activity (Cu^{2+})$_s$. The potential ϕ_2 of the electrode becomes

$$\phi_2 = 0.337 + \frac{0.0592}{2}\log(Cu^{2+})_s \tag{5.2}$$

Since (Cu^{2+})$_s$ is less than (Cu^{2+}), the potential of the polarized cathode is less noble, or more active, than in the absence of external current. The difference of potential, $\phi_2 - \phi_1$, is the *concentration polarization*, equal to

$$\phi_2 - \phi_1 = -\frac{0.0592}{2}\log\frac{(Cu^{2+})}{(Cu^{2+})_s} \tag{5.3}$$

The larger the current, the smaller the surface concentration of copper ion, or the smaller the value of (Cu^{2+})$_s$, thus the larger the corresponding polarization. Infinite concentration polarization is approached when (Cu^{2+})$_s$ approaches zero at the electrode surface; the corresponding current density that results in this limiting lower value of (Cu^{2+})$_s$ is called the *limiting current density*. In practice, polarization can never reach infinity; instead, another electrode reaction establishes itself at a more active potential than corresponds to the first reaction. In the case of copper deposition, for example, the potential moves to that for hydrogen evolution, $2H^+ + 2e^- \rightarrow H_2$, and hydrogen gas is liberated while copper is simultaneously plated out.

If i_L is the limiting current density for a cathodic process and i is the applied current density, it can be shown [3] that

$$\phi_2 - \phi_1 = -\frac{RT}{nF} \ln \frac{i}{i_L - i} \qquad (5.4)$$

As i approaches i_L, $\phi_2 - \phi_1$ approaches $-\infty$, that is, minus infinity. This is shown by the plot of $\phi_2 - \phi_1$ versus i in Fig. 5.4.

The limiting current density (A/cm^2) can be evaluated from the expression

$$i_L = \frac{DnF}{\delta t} c \times 10^{-3} \qquad (5.5)$$

where D is the diffusion coefficient for the ion being reduced, n and F have their usual significance, δ is the thickness of the stagnant layer of electrolyte next to the electrode surface (about 0.05 cm in an unstirred solution), t is the transference number of all ions in solution except the ion being reduced (equal to unity if

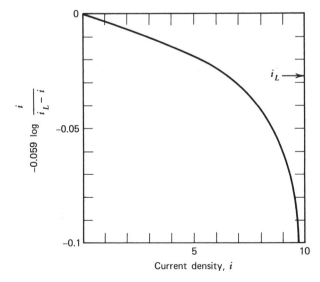

Figure 5.4. Dependence of concentration polarization at a cathode on applied current density.

many other ions are present), and c is the concentration of diffusing ion in moles/liter. Since D for all ions at 25°C in dilute solution, except for H^+ and OH^-, averages about $1 \times 10^{-5}\,cm^2/s$, the limiting current density is approximated by

$$i_L = 0.02\,nc \qquad (5.6)$$

For H^+ and OH^-, D equals 9.3×10^{-5} and $5.2 \times 10^{-5}\,cm^2/s$, respectively (infinite dilution), so that the corresponding values of i_L are higher.

Should the copper electrode be polarized anodically, concentration of copper ion at the surface is larger than that in the body of solution. The ratio $(Cu^{2+})/(Cu^{2+})_s$ then becomes less than unity and $\phi_2 - \phi_1$ of (5.3) changes sign. In other words, concentration polarization at an anode polarizes the electrode in the cathodic or noble direction, opposite to the direction of potential change when the electrode is polarized as cathode. For a copper anode, the limiting upper value for concentration polarization corresponds to formation of saturated copper salts at the electrode surface. This limiting value is not as large as for cathodic polarization where the Cu^{2+} activity approaches zero.

2. *Activation Polarization.* Activation polarization is caused by a slow electrode reaction. The reaction at the electrode requires an activation energy in order to proceed. The most important example is that of hydrogen ion reduction at a cathode, $H^+ + e^- \rightarrow \frac{1}{2}H_2$. For this reaction, the polarization is called *hydrogen overpotential.*

Overpotential is defined as the polarization (potential change) of an equilibrium electrode that results from current flow across the electrode–solution interface. Overpotential, η, is the difference between the measured potential and the thermodynamic, or reversible, potential;* that is,

$$\eta = \phi_{meansured} - \phi_{reversible}$$

At a platinum cathode, for example, the following reactions are thought to occur in sequence:

$$H^+ + e^- \rightarrow H_{ads}$$

where H_{ads} represents hydrogen atoms adsorbed on the metal surface. This relatively rapid reaction is followed by another reaction, namely, adsorbed hydrogen atoms combining to form hydrogen molecules and bubbles of gaseous hydrogen:

$$2H_{ads} \rightarrow H_2$$

Overvoltage, on the other hand, is defined as polarization (potential change) of a corroding electrode caused by flow of an applied current; that is, overvoltage, ε, is the difference between the measured potential and the corrosion (or mixed) potential; that is, $\varepsilon = \phi_{measured} - \phi_{corr}$.

This reaction is relatively slow, and its rate determines the value of hydrogen overpotential on platinum. The controlling slow step of H^+ discharge is not always the same, but varies with metal, current density, and environment.

Pronounced activation polarization also occurs with discharge of OH^- at an anode accompanied by oxygen evolution:

$$2OH^- \rightarrow \frac{1}{2}O_2 + H_2O + 2e^-$$

The polarization corresponding to this reaction is called *oxygen overpotential*. Overpotential may also occur with Cl^- or Br^- discharge, but the values at a given current density are much smaller than those for O_2 or H_2 evolution.

Activation polarization is also characteristic of metal-ion deposition or dissolution. The value may be small for nontransition metals, such as silver, copper, and zinc, but it is larger for the transition metals, such as iron, cobalt, nickel, and chromium (see Table 5.1). The anion associated with the metal ion influences metal overpotential values more than in the case of hydrogen overpotential. The controlling step in the reaction is not known precisely, but, in some cases, it is probably a slow rate of hydration of the metal ion as it leaves the metal lattice, or dehydration of the hydrated ion as it enters the lattice.

Activation polarization, η, increases with current density, i, in accord with the Tafel equation*:

$$\eta = \beta \log \frac{i}{i_0}$$

where β and i_0 are constants for a given metal and environment and are both dependent on temperature. The *exchange current density, i_0*, represents the current density equivalent to the equal forward and reverse reactions at the electrode at equilibrium. The larger the value of i_0 and the smaller the value of β, the smaller the corresponding overpotential.

A typical plot of activation polarization or overpotential for H^+ discharge is shown in Fig. 5.5. At the equilibrium potential for the hydrogen electrode (-0.059pH), for example, overpotential is zero. At applied current density, i_1, it is given by η, the difference between measured and equilibrium potentials. Although usually listed as positive, hydrogen overpotential values are negative and, correspondingly, oxygen overpotential values are positive on the ϕ scale.

3. *IR Drop.* Polarization measurements include a so-called ohmic potential drop through a portion of the electrolyte surrounding the electrode, through a metal-reaction product film on the surface, or both. An ohmic potential drop always occurs between the working electrode and the capillary tip of the reference electrode. This contribution to polarization is equal to iR, where I is the

---

*Named after J. Tafel [*Z. Phys. Chem.* **50**, 641 (1904)], who first proposed a similar equation to express hydrogen overpotential as a function of current density.

TABLE 5.1. Overpotential Values

$$\eta = \beta \log \frac{i}{i_0}$$

Metal	Temperature (°C)	Solution	β (volts)	i_0 (A/m^2)	η at 1 mA/cm^2 (V)[a]
Hydrogen Overpotential					
Pt (smooth)	20	1 N HCl	0.03	10	0.00
	25	0.1 N NaOH	0.11	0.68	0.13
Pd	20	0.6 N HCl	0.03	2	0.02
Mo	20	1 N HCl	0.04	10^{-2}	0.12
Au	20	1 N HCl	0.05	10^{-2}	0.15
Ta	20	1 N HCl	0.08	10^{-1}	0.16
W	20	5 N HCl	0.11	10^{-1}	0.22
Ag	20	0.1 N HCl	0.09	5×10^{-3}	0.30
Ni	20	0.1 N HCl	0.10	8×10^{-3}	0.31
	20	0.12 N NaOH	0.10	4×10^{-3}	0.34
Bi	20	1 N HCl	0.10	10^{-3}	0.40
Nb	20	1 N HCl	0.10	10^{-3}	0.40
Fe	16	1 N HCl	0.15	10^{-2}	0.45
	25	4% NaCl pH 1–4	0.10	10^{-3}	0.40 (Stern)
Cu	20	0.1 N HCl	0.12	2×10^{-3}	0.44
	20	0.15 N NaOH	0.12	1×10^{-2}	0.36
Sb	20	2 N H$_2$SO$_4$	0.10	10^{-5}	0.60
Al	20	2 N H$_2$SO$_4$	0.10	10^{-6}	0.70
Be	20	1 N HCl	0.12	10^{-5}	0.72
Sn	20	1 N HCl	0.15	10^{-4}	0.75
Cd	16	1 N HCl	0.20	10^{-3}	0.80
Zn	20	1 N H$_2$SO$_4$	0.12	1.6×10^{-7}	0.94
Hg	20	0.1 N HCl	0.12	7×10^{-9}	1.10
	20	0.1 N H$_2$SO$_4$	0.12	2×10^{-9}	1.16
	20	0.1 N NaOH	0.10	3×10^{-11}	1.15
Pb	20	0.01–8 N HCl	0.12	2×10^{-9}	1.16
Oxygen Overpotential					
Pt (smooth)	20	0.1 N H$_2$SO$_4$	0.10	9×10^{-8}	0.81
	20	0.1 N NaOH	0.05	4×10^{-9}	0.47
Au	20	0.1 N NaOH	0.05	5×10^{-9}	0.47
Metal Overpotential (Deposition)					
Zn	25	1 M ZnSO$_4$	0.12	0.2	0.20(Bockris)
Cu	25	1 M CuSO$_4$	0.12	0.2	0.20(Bockris)
Fe	25	1 M FeSO$_4$	0.12	10^{-4}	0.60(Bockris)
Ni	25	1 M NiSO$_4$	0.12	2×10^{-5}	0.68(Bockris)

[a] 1 mA/cm^2 = 10 A/m^2.

Source: Data from B. E. Conway, *Electrochemical Data*, Elsevier, New York, 1952; *Modern Aspects of Electrochemistry*, J. Bockris, editor, Academic Press, New York, 1954; M. Stern, *J. Electrochem. Soc.* **102**, 609 (1955); H. Kita, *J. Electrochem. Soc.* **113**, 1095 (1966).

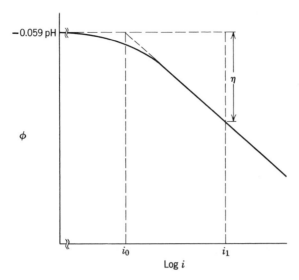

Figure 5.5. Hydrogen overvoltage as a function of current density.

current density, and R, equal to l/κ, represents the value in ohms of the resistance path of length l cm and specific conductivity κ. The product, iR, decays simultaneously with shutting off the current, whereas concentration polarization and activation polarization usually decay at measurable rates. The iR contribution can be calculated as described in Section 5.3.1.

Note: Concentration polarization decreases with stirring, whereas activation polarization and iR drop are not affected significantly.

5.5 HYDROGEN OVERPOTENTIAL

The polarization term that controls the corrosion rate of many metals in deaerated water and in nonoxidizing acids is hydrogen overpotential. In accord with the previously discussed definition of polarization, hydrogen overpotential is the difference of potential between a cathode at which hydrogen is being evolved, $\phi_{measured}$, and a hydrogen electrode at equilibrium in the same solution; that is,

$$H_2 \text{ overpotential} = \phi_{measured} - (-0.059\text{pH})$$

Hydrogen overpotential, therefore, is measured in the same way as polarization is measured. Ideally, hydrogen overpotential includes only the activation polarization term corresponding to the reaction $2H^+ + 2e^- \rightarrow H_2$, but reported values often include iR drop and sometimes concentration polarization as well.

The absolute values of hydrogen overpotential for a given metal *decrease* with:

1. Increasing temperature (i_0 increases). For metals that corrode by hydrogen evolution, decreasing hydrogen overpotential is one factor accounting for increase of corrosion as the temperature is raised.
2. Roughening of the surface. A sandblasted surface has a lower hydrogen overpotential than a polished surface. Greater area and improved catalytic activity of a rough surface account for the observed effect.
3. Decreasing current density. The Tafel equation,

$$\eta = \beta \log \frac{i}{i_0}$$

is obeyed within the region of applied current density, i, below the limiting current density for concentration polarization, and above the exchange current density, i_0. The term β is equal to $2.3RT/\alpha F$, where α is a constant and the other terms have their usual significance. The term α is approximately 2 for platinum and palladium and is approximately 0.4–0.6 for Fe, Ni, Cu, Hg, and several other metals. Although hydrogen overpotential values usually differ in acid compared with alkaline media, values are not greatly sensitive to pH within either range.

Stern [4] showed that for a corroding metal we have

$$\eta = \beta \log \frac{i + i_r + i_{corr}}{i_0}$$

where i_r is the reverse current for the reaction $H_2 \rightarrow 2H^+ + 2e^-$, which varies with potential and which, at equilibrium, is equal to i_0. This equation describes the small observed slope of $d\eta/d \log i$ for small values of impressed current density, i, with the slope increasing as i approaches $i_{corr} + i_r$ and reaching the true Tafel slope, β, at $i \gg i_r + i_{corr}$. Similarly, the overpotential for a noncorroding metal can be represented by a modified Tafel equation,

$$\eta = \beta \log \frac{i + i_r}{i_0}$$

This equation holds from zero to finite values of i (Fig. 5.5). Values of i_r determined from measured values of η also follow the Tafel equation, but with a slope of opposite sign.

The slow step in the discharge of hydrogen ions on platinum, or palladium, as described earlier, appears to be recombination of adsorbed hydrogen atoms. This assumption is consistent, from kinetic considerations, with an observed value of $\alpha = 2$. For iron, the value of α is approximately 0.5 and, correspondingly, $\beta = 0.1$. To account for these values, the proposal has been made that the slow step in the hydrogen evolution reaction on iron is probably

$$H^+ + H_{ads} \rightarrow H_2 - e^-$$

The same slow step may apply to various metals having intermediate values of hydrogen overpotential (e.g., iron, nickel, copper).

For metals of high hydrogen overpotential, such as mercury and lead, the slow discharge of the hydrated hydrogen ion is apparently the slow step:

$$H^+ \rightarrow H_{ads} - e^-$$

For many metals at high current densities, this slow discharge step is also the controlling reaction. The slow discharge step may, instead, be the reduction of H_2O,

$$H_2O \rightarrow OH^- + \frac{1}{2}H_2 - e^-$$

The reduction of water as the controlling reaction applies generally to metals in alkaline solutions at high and low current densities.

The rapidity with which H_{ads} combines to form H_2, either by combination with itself or with H^+, is affected by the catalytic properties of the electrode surface. A good catalyst, such as platinum or iron, leads to a low value of hydrogen overpotential, whereas a poor catalyst, such as mercury or lead, accounts for a high value of overpotential. If a catalyst poison, like hydrogen sulfide or certain arsenic or phosphorus compounds, is added to the electrolyte, it retards the rate of formation of molecular H_2 and increases the accumulation of adsorbed hydrogen atoms on the electrode surface.* The increased concentration of surface hydrogen favors entrance of hydrogen atoms into the metal lattice, causing *hydrogen embrittlement* (loss of ductility), and, in some stressed high-strength ferrous alloys, may induce spontaneous cracking, called hydrogen cracking (see Section 8.4). Catalyst poisons increase absorption of hydrogen whether the metal

*Increase of hydrogen overpotential normally decreases the corrosion rate of steel in acids, but presence of sulfur or phosphorus in steels is observed instead to increase the rate. This increase probably results from the low hydrogen overpotential of ferrous sulfide or phosphide, either existing in the steel as separate phases or formed as a surface compound by reaction of iron with H_2S or phosphorus compounds in solution. It is also possible [4] that the latter compounds, in addition, stimulate the anodic dissolution reaction, $Fe \rightarrow Fe^{2+} + 2e^-$ (reduce activation polarization), or alter the anode–cathode area ratio.

Similarly, arsenious oxide in small amount accelerates corrosion of steel in acids (e.g., H_2SO_4), perhaps forming arsenides, but when present in larger amount (e.g., 0.05% As_2O_3 in 72% H_2SO_4), it is an effective corrosion inhibitor, probably because elementary arsenic, having a high hydrogen overpotential, deposits out at cathodic areas. Tin salts have the same inhibiting effect and have been used to protect steels from attack by pickling acids during descaling operations.

is polarized by externally applied current or by a corrosion reaction with accompanying hydrogen evolution. For this reason, some oil-well brines containing H_2S are difficult to handle in low-alloy steel tubing subject to the usual high stresses of a structure extending several thousand feet underground. Slight general corrosion of the tubing produces hydrogen, a portion of which enters the stressed steel to cause hydrogen cracking. In the absence of hydrogen sulfide, general corrosion occurs, but without hydrogen cracking. High-strength steels, because of their limited ductility, are more susceptible to hydrogen cracking than are low-strength steels, but hydrogen enters the lattice in either case, tending to blister low-strength steels instead.

Values of β, i_0, and of η at $1\,mA/cm^2$ for H^+ discharge are given in Table 5.1. As can be seen from the table, values of hydrogen overpotential vary greatly with the metal. Values also change with concentration of electrolyte, but the effect, comparatively, is not large.

5.6 POLARIZATION DIAGRAMS OF CORRODING METALS

Polarization diagrams of corroding metals, sometimes called Evans diagrams, are graphs of potential versus log current or log current density. They were originally developed by U. R. Evans at the University of Cambridge in England, who recognized the usefulness of such diagrams for predicting corrosion behavior [5]. To establish a polarization diagram, the usual electrodes that are employed, in addition to the electrode being studied (the "working" electrode), are the reference electrode and the inert counter (or auxiliary) electrode that is usually made of platinum. The design of a 1-liter electrochemical cell used in many corrosion laboratories [2] is shown in Fig. 5.3(b). A gas bubbler in used for atmospheric control—for example, to deaerate the solution or to saturate the solution with a specific gas.

The measurements are usually made using a potentiostat—an instrument that automatically maintains the desired potential between the working and reference electrodes by passing the appropriate current between the working and counter electrodes. Various electronic circuits for potentiostat design have been presented in the corrosion literature and their applications to corrosion studies have been discussed [6].

Alternatively, depending on the type of measurements to be made, a galvanostatic circuit, consisting of a power supply, resistor, ammeter, and potentiometer, can be used. The current between working and counter electrodes is controlled, and the potential of the working electrode with respect to the reference electrode is measured.

In experimentally establishing a polarization diagram, the first measurement is usually that of the corrosion potential, ϕ_{corr}, when the applied current, I_{appl}, is zero. The working electrode is then polarized either anodically or cathodically to establish one of the dashed lines in Fig. 5.6. The polarization procedure is then repeated, but with I_{appl} reversed, to obtain the second dashed line. Using the

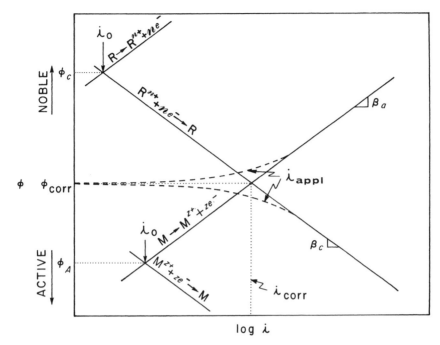

Figure 5.6. Polarization diagram.

potentiostat, polarization may be carried out either in potential steps (i.e., potentiostatically) or continuously (i.e., potentiodynamically). Having established ϕ versus $\log I_{appl}$ on the more noble and the more active sides of the corrosion potential, the complete polarization diagram is then constructed, as shown in Fig. 5.6 for metal M. In this system, the oxidation reaction may be the dissolution of metal, $M \rightarrow M^{z+} + ze^-$, and the reduction reaction may be symbolized as $R^{n+} + ne^- \rightarrow R$. In an aerated neutral or basic aqueous solution, the reduction reaction could be $O_2 + 2H_2O + 4e^- \rightarrow 4OH^-$, whereas in a deaerated acid, the reduction reaction could be $2H^+ + 2e^- \rightarrow H_2$.

For any corroding metal, the chemical equivalents of metal going into solution at the anodic sites are equal to the chemical equivalents of reduction products produced at cathodic sites. In terms of corrosion current, for a given area of metal surface, $I_a = I_c$; that is, the anodic and cathodic currents are equal in magnitude. The corresponding current density, i_a, at anodic areas depends on the fraction of metal surface, A_a, that is anode, and, similarly, i_c depends on A_c. In general, $I_a/A_a = i_a$ and $I_c/A_c = i_c$; but $i_a = i_c$ only if $A_a = A_c$—that is, if 50% of the surface is anodic and 50% is cathodic. Under the latter condition, the general relation for any anode–cathode area ratio $I_{appl} = I_a - I_c$ can be translated to corresponding current densities $i_{appl} = i_a - i_c$.

When the electrode is polarized at sufficiently high current densities to shift the potential more than approximately 100 mV from the corrosion potential, the reverse, or "back" reactions, are usually negligible [7]; depending on the direction of I_{appl}, the metal surface acts either as all anode or all cathode. Accordingly, for anodic polarization, $i_{appl} \approx i_a$, and, similarly, for cathodic polarization, $i_{appl} \approx i_c$. Tafel slopes can then be determined (see paragraph "Activation Polarization," Section 5.4). By extrapolating from the anodic Tafel region to the reversible (equilibrium) anode potential, ϕ_A, the exchange current density, i_{0a}, for the reaction $M^{z+} + ze^-$ $\rightarrow$ M is determined; that is, the equal rates of oxidation and reduction reactions are expressed as a current density. Similarly, by extrapolating from the Tafel region to the reversible potential ϕ_c, the exchange current density i_{0c} for the cathodic reaction is determined. By extrapolating from either the anodic or cathodic Tafel region to the corrosion potential ϕ_{corr}, where $i_c = i_a$, the corrosion rate i_{corr} can be determined for the condition that $A_a = A_c$ (anode–cathode area ratio = 1). Although the latter condition often closely approximates the true situation, a more exact approximation to the corrosion rate would require information about the actual anode–cathode area ratio.

5.7 INFLUENCE OF POLARIZATION ON CORROSION RATE

Both resistance of the electrolyte and polarization of the electrodes limit the magnitude of current produced by a galvanic cell. For local-action cells on the surface of a metal, electrodes are in close proximity to each other; consequently, resistance of the electrolyte is usually a secondary factor compared to the more important factor of polarization. When polarization occurs mostly at the anodes, the corrosion reaction is said to be *anodically controlled* (see Fig. 5.7). Under anodic control, the corrosion potential is close to the thermodynamic potential of the cathode. A practical example is impure lead immersed in sulfuric acid, where a lead sulfate film covers the anodic areas and exposes cathodic impurities, such as copper. Other examples are magnesium exposed to natural waters and iron immersed in a chromate solution.

When polarization occurs mostly at the cathode, the corrosion rate is said to be *cathodically controlled*. The corrosion potential is then near the thermodynamic anode potential. Examples are zinc corroding in sulfuric acid and iron exposed to natural waters.

Resistance control occurs when the electrolyte resistance is so high that the resultant current is not sufficient to appreciably polarize anodes or cathodes. An example occurs with a porous insulating coating covering a metal surface. The corrosion current is then controlled by the *IR* drop through the electrolyte in pores of the coating.

It is common for polarization to occur in some degree at both anodes and cathodes. This situation is described as *mixed control*.

The extent of polarization depends not only on the nature of the metal and electrolyte, but also on the actual exposed area of the electrode. If the anodic area of a corroding metal is very small, caused, for example, by porous surface films, there may be considerable anodic polarization accompanying corrosion, even

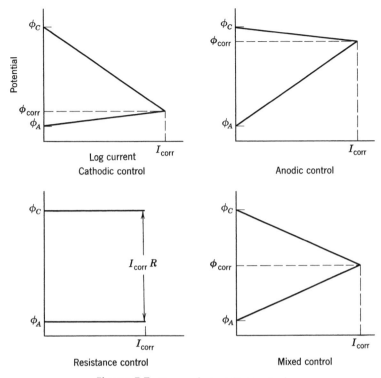

Figure 5.7. Types of corrosion control.

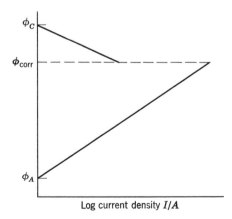

Figure 5.8. Polarization diagram for corroding metal when anode area equals one-half of cathode area.

though measurements show that unit area of the bare anode polarizes only slightly at a given current density. Consequently, the anode–cathode area ratio is also an important factor in determining the corrosion rate. If current density is plotted instead of total corrosion current, as, for example, when the anode area is half the cathode area, corresponding polarization curves are shown in Fig. 5.8.

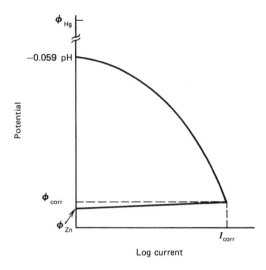

Figure 5.9. Polarization diagram for zinc amalgam in deaerated HCl.

An important experiment that illustrates the electrochemical mechanism of corrosion was performed by Wagner and Traud [1]. They measured the corrosion rate of a dilute zinc amalgam in an acid calcium chloride mixture and the cathodic polarization of mercury in the same electrolyte. The current density equivalent to the corrosion rate was found to correspond to the current density necessary to polarize mercury to the corrosion potentiual of the zinc amalgam (Fig. 5.9). In other words, mercury atoms of the amalgam composing the majority of the surface apparently act as cathodes, or hydrogen electrodes,* and zinc atoms act as anodes of corrosion cells. The amalgam polarizes anodically very little, and the corrosion reaction is controlled almost entirely by the rate of hydrogen evolution at cathodic areas. Consequently, it is the high hydrogen overpotential of mercury that limits the corrosion rate of amalgams in nonoxidizing acids. A piece of platinum coupled to the amalgam considerably increases the rate of corrosion because hydrogen is more readily evolved from a low-overpotential cathode at the operating emf of the zinc–hydrogen electrode cell.

The very low corrosion rate and the absence of appreciable anodic polarization explain why amalgams in corresponding metal salt solutions exhibit corrosion potentials closely approaching the reversible potential of the alloyed component. For example, the corrosion potential of cadmium amalgam in $CdSO_4$ solution is closer to the thermodynamic value for $Cd \rightarrow Cd^{2+} + 2e^-$ than is observed for pure cadmium in the same solution. The steady-state corrosion rate of pure cadmium is appreciably higher than that of cadmium amalgam, leading

---

*Mercury in aqueous solutions acts first as a mercury electrode, but, when cathodically polarized, all mercury ions in solution are deposited before H^+ is discharged. Any conducting surface on which H^+ ions are discharged acts as a polarized hydrogen electrode and can be so considered in evaluating a corrosion cell.

to greater deviation of the measured corrosion potential from the corresponding thermodynamic value. In general, the steady-state potential of any metal more active than hydrogen (e.g., iron, nickel, zinc, cadmium) in an aqueous solution of its ions deviates from the true thermodynamic value by an amount that depends on the prevailing corrosion rate accompanied by H^+ discharge. The measured value tends to be more noble than the true value. This situation also holds for the steady-state potentials of more noble metals (e.g., copper, mercury) which undergo corrosion in the presence of dissolved oxygen.

5.8 CALCULATION OF CORROSION RATES FROM POLARIZATION DATA

The corrosion current can be calculated from the corrosion potential and the thermodynamic potential if the equation expressing polarization of the anode or cathode is known, and if the anode–cathode area ratio can be estimated. For corrosion of active metals in deaerated acids, for example, the surface of the metal is probably covered largely with adsorbed H atoms and can be assumed, therefore, to be mostly cathode. The thermodynamic potential is -0.059 pH, and if i_{corr} is sufficiently larger than i_0 for $H^+ \rightarrow \frac{1}{2}H_2 - e^-$, the Tafel equation expresses cathodic polarization behavior. Then,

$$\phi_{corr} = -\left(0.059\text{pH} + \beta \log \frac{i_{corr}}{i_0}\right) \tag{5.7}$$

from which i_{corr} and the equivalent corrosion rate can be calculated. Stern [4] showed that calculated corrosion rates for iron, using (5.7) and employing empirical values for β and i_0, were in excellent agreement with observed rates. Typical values are given in Table 5.2.

TABLE 5.2. Comparison of Calculated and Observed Corrosion Currents for Pure Iron in Various Deaerated Acids [4]

Solution	$\phi_{Hcorr} + 0.059\text{pH}$ (volts, S.H.E.)	β (V)	i_0 ($\mu A/cm^2$)	i_{corr} ($\mu A/cm^2$) Calculated	i_{corr} ($\mu A/cm^2$) Observed
0.1 M Citric acid pH = 2.06	−0.172	0.084	0.093	10.4	11.5
0.1 M Malic acid pH = 2.24	−0.158	0.083	0.015	1.2	1.2
4% NaCl + HCl					
pH = 1	−0.203	0.100	0.10	10.5	11.1
pH = 2	−0.201	0.100	0.11	11.0	11.3

Subsequently, Stern and Geary [8] derived the very important and useful equation, now known as the Stern–Geary equation,

$$I_{corr} = \frac{I_{appl}}{2.3\Delta\phi}\left(\frac{\beta_c\beta_a}{\beta_c+\beta_a}\right) \tag{5.8}$$

or

$$I_{corr} = \frac{1}{2.3R}\left(\frac{\beta_c\beta_a}{\beta_c+\beta_a}\right)$$

where β_c and β_a refer to Tafel constants for the cathodic and anodic reactions, respectively, and $I_{appl}/\Delta\phi$ is the polarization slope (the reciprocal of the polarization resistance, $R = \Delta\phi/I_{appl}$) in the region near the corrosion potential, for which the change of potential, $\Delta\phi$, with I_{appl} is essentially linear (for derivation, see the Appendix, Section 29.2). Under conditions of slight polarization, for which $\Delta\phi$ is not more than about 10 mV, the anode–cathode area ratio, which need not be known, remains essentially constant and conditions otherwise at the surface of the corroding metal are largely undisturbed.

If corrosion is controlled by concentration polarization at the cathode, as when oxygen depolarization is controlling, Equation (5.8) simplifies to

$$I_{corr} = \frac{\beta_a}{2.3}\frac{I_{appl}}{\Delta\phi} \tag{5.9}$$

Although values of β are relatively well known for H^+ discharge, they are not as generally available for other electrode reactions. Stern showed, however, that the majority of reported β values are between 0.06 and 0.12 V. If β_c is known to be 0.06 V, for example, and β_a is between 0.06 and 0.12 V, the calculated corrosion current is within at least 20% of the correct value. Under other assumptions, the corrosion rate can be calculated to at least a factor of 2.

The general validity of Eqs. (5.8) and (5.9) is shown by data summarized in Fig. 5.10. The observed corrosion current, corresponding to data on corrosion of nickel in HCl and on corrosion of steels and cast iron in acids and in natural waters, extends over six orders of magnitude. Some exchange current densities for $Fe^{3+} \rightarrow Fe^{2+} - e^-$ on passive surfaces are included because the same principle applies in calculating i_0 for a noncorroding electrode as in calculating I_{corr} for a corroding electrode. Also, straight lines are shown representing values calculated on the basis of several assumed β values within which most of the empirical data lie. Equations (5.7), (5.8), and (5.9) have been widely used with considerable success for determining the corrosion rates of various metals in aqueous environments at ambient and elevated temperatures.

The linear polarization method has obvious advantages in calculating instantaneous corrosion rates for many metals in a wide variety of environments and under various conditions of velocity and temperature. It can also be used to

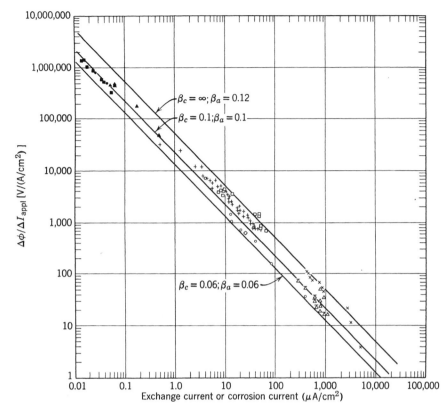

Figure 5.10. Relation between polarization slope at low applied current densities and observed corrosion or exchange current densities [9]. (Copyright ASTM INTERNATIONAL. Reprinted with permission.)

evaluate inhibitors and protective coatings, as well as for detecting changes of corrosion rate with time. A correction is required if an *IR* drop is involved in the measurement.

The Stern–Geary equation has been modified to minimize errors under particular conditions. Errors in Eq. (5.8) are especially significant in systems in which the corrosion potential is close to one of the reversible potentials—that is, is outside the Tafel region. Mansfeld and Oldham [10] have developed a set of equations that provides less error than Eq. (5.8) when this situation applies (see Appendix, Section 29.2).

5.9 ANODE–CATHODE AREA RATIO

The reduction of H^+ at a platinum cathode at a given rate is accompanied by the simultaneous reverse reaction at a lower rate for H_2 oxidizing to H^+. The oxidation and reduction reactions are assumed to take place on the same surface sites. At equilibrium, the forward and reverse reactions are equal and the equivalent current density is called the exchange current density. For a corroding metal, the anode and cathode reactions are distinct and different; one is not merely the reverse reaction of the other. Hence, the oxidation reaction can take place only on sites of the metal surface that are different from those on which reduction takes place. For this situation, the distance between anode and cathode may vary from atomic dimensions to a separation of many meters.

Correspondingly, an anode–cathode area ratio exists, and since the observed polarization of anodic or cathodic sites depends, in part, on the area over which oxidation or reduction occurs, the anode–cathode area ratio is an important factor in the observed corrosion rate.

Stern [11] developed the general relation for which anodic and cathodic polarization follows Tafel behavior and IR drop is negligible. Using a typical polarization diagram, as illustrated in the Appendix, Section 29.2, Fig. 29.1, we obtain

$$\phi_c' = \phi_c - \beta_c \log \frac{I_c}{A_c i_{0c}} \tag{5.10}$$

and

$$\phi_a' = \phi_a + \beta_a \log \frac{I_a}{A_a i_{0a}} \tag{5.11}$$

where subscripts a and c refer to anode and cathode, respectively, ϕ' is the polarized potential, ϕ is the thermodynamic potential, β is the Tafel slope, I is the current per unit metal surface area, A_a is the fraction of surface that is anode such that $A_a + A_c = 1$, and i_{0a} and i_{0c} are the exchange current densities for the anode and cathode reactions, respectively. At steady state, $\phi_c' = \phi_a' = \phi_{corr}$ and $I_a = I_c = I_{corr}$. Then

$$\beta_c \log I_{corr} = \phi_c - \phi_{corr} + \beta_c \log A_c i_{0c} \tag{5.12}$$

and

$$\phi_{corr} = \phi_a + \beta_a \log I_{corr} - \beta_a \log A_a i_{0a} \tag{5.13}$$

Substituting (5.13) into (5.12), we obtain

$$\log I_{corr} = \frac{\phi_c - \phi_a}{\beta_c + \beta_a} + \frac{\beta_a}{\beta_c + \beta_a} \log A_a i_{0a} + \frac{\beta_c}{\beta_c + \beta_a} \log A_c i_{0c} \tag{5.14}$$

Since $A_a = 1 - A_c$, we get

$$\frac{d \log I_{corr}}{dA_c} = -\frac{\beta_a}{2.3(\beta_c + \beta_a)}\frac{1}{(1 - A_c)} + \frac{\beta_c}{2.3(\beta_c + \beta_a)}\frac{1}{A_c} \tag{5.15}$$

$$= \frac{\beta_c(1 - A_c) - \beta_a A_c}{2.3(\beta_c + \beta_a)A_c A_a} \tag{5.16}$$

Maximum I_{corr} occurs at $d \log I_{corr}/dA_c = 0$—that is, when the numerator of (5.16) equals zero. This occurs when $A_c = \beta_c/(\beta_c + \beta_a)$. If, as is frequently observed, $\beta_c = \beta_a$, then the maximum corrosion rate occurs at $A_c = \frac{1}{2}$—that is, at an anode–cathode area ratio of unity. At any other anode–cathode area ratio, the corrosion rate is less, reaching zero at a ratio equal to either zero or infinity.

5.10 ELECTROCHEMICAL IMPEDANCE SPECTROSCOPY

Electrochemical methods based on alternating currents can be used to obtain insights into corrosion mechanisms and to establish the effectiveness of corrosion control methods, such as inhibition and coatings. In an alternating-current circuit, impedance determines the amplitude of current for a given voltage. Impedance is the proportionality factor between voltage and current. In electrochemical impedance spectroscopy (EIS), the response of an electrode to alternating potential signals of varying frequency is interpreted on the basis of circuit models of the electrode/electrolyte interface. Figure 5.11 shows two circuit models that can

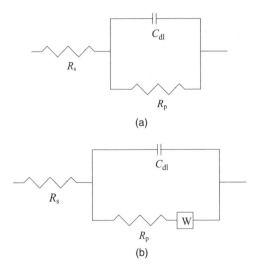

Figure 5.11. (a) Electrical equivalent circuit model used to represent an electrochemical interface undergoing corrosion in the absence of diffusion control. R_p is the polarization resistance, C_{dl} is the double layer capacitance, R_p is the polarization resistance, and R_s is the solution resistance [15]. (b) Electrical equivalent circuit model when diffusion control applies; W is the Warburg impedance [13].

be used for analyzing EIS spectra. The simplest model for characterizing the metal–solution interface, Fig. 5.11(a), includes the three essential parameters, R_s (the solution resistance), C_{dl} (the capacitance of the double layer), and R_p (the polarization resistance). When direct-current measurements are carried out (i.e., frequency is zero), the impedance of the capacitor approaches infinity. In parallel electrical circuits, the circuit with the smallest impedance dominates, with the result that, under these conditions, the sum of R_s and R_p is measured. If R_s if significant, the corrosion rate is underestimated.

When diffusion control is important, another element, Z_D, sometimes called the Warburg impedance, is added in series with R_p, as shown in Fig. 5.11(b).

In electrochemical impedance spectroscopy, the impedance of the corroding metal is analyzed as a function of frequency. A sinusoidal potential change is applied to the corroding electrode at a number of frequencies, ω. At each frequency, the resulting sinusoidal current waveform is out of phase with the applied potential signal by an amount, the phase angle, θ, that depends on the circuit parameters. The current amplitude is inversely proportional to the impedance of the interface. The electrochemical impedance, $Z(\omega)$, is the frequency-dependent proportionality factor in the relationship between the voltage signal and the current response,

$$Z(\omega) = \frac{E(\omega)}{i(\omega)} \tag{5.17}$$

where E is the voltage signal, $E = E_0 \sin(\omega t)$; i is the current density, $i = i_0 \sin(\omega t + \theta)$; Z is the impedance (ohm-cm^2); and t is the time (seconds) [12].

Impedance is a complex number that is described by the frequency-dependent modulus, $|Z|$, and the phase angle, θ, or, alternatively, by the real component, Z', and the imaginary component, Z''. The mathematical convention for separating the real and imaginary components is to multiply the magnitude of the imaginary component by j $[=\sqrt{-1}]$ and report the real and imaginary values as a complex number.

The equations for electrochemical impedance are [13]

$$E = E_{real} + E_{imaginary} = E' + jE'' \tag{5.18}$$

$$I = I_{real} + I_{imaginary} = I' + jI'' \tag{5.19}$$

$$Z = Z' + jZ'' = \frac{E' + jE''}{I' + jI''} \tag{5.20}$$

$$\tan\theta = \frac{Z''}{Z'} \tag{5.21}$$

$$|Z|^2 = (Z')^2 + (Z'')^2 \tag{5.22}$$

In electrochemical impedance analysis, three different types of plots are commonly used: Nyquist plots (complex plane, showing $-Z''$ versus Z') and two dif-

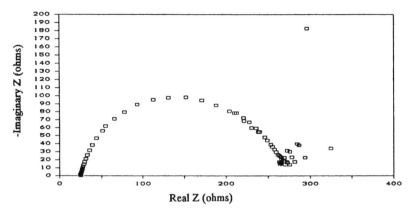

Figure 5.12. Electrochemical impedance data for cast 99.9% magnesium in pH 9.2 sodium borate presented in the Nyquist format [14]. (Reproduced by permission of ECS, The Electrochemical Society.)

ferent types of Bode plots, showing the impedance magnitude versus log frequency and showing phase angle versus log frequency. One example of each type of plot is shown in Figs. 5.12 and 5.13. Figure 5.12 illustrates electrochemical impedance data for cast 99.9% magnesium in pH 9.2 sodium borate in the Nyquist format. This figure shows a single capacitive semicircle, indicating that the only process occurring is charge transfer.

Figure 5.13 shows Bode magnitude and phase angle plots for a system with the equivalent circuit shown in Fig. 5.11(a) [15]. From Fig. 5.13, it can be seen that, at very low frequencies,

$$Z = R_s + R_p \tag{5.23}$$

At very high frequencies,

$$Z = R_s \tag{5.24}$$

5.11 THEORY OF CATHODIC PROTECTION

Observing the polarization diagram for the copper–zinc cell in Fig. 5.2, it is clear that, if polarization of the cathode is continued, using external current, beyond the corrosion potential to the thermodynamic potential of the anode, both electrodes attain the same potential and no corrosion of zinc can occur. This is the basis for cathodic protection of metals. Cathodic protection, discussed further in Chapter 13, is one of the most effective engineering means for reducing the corrosion rate to zero. Cathodic protection is accomplished by supplying an external current to the corroding metal that is to be protected, as shown in Fig. 5.14. Current leaves the auxiliary anode (composed of any metallic or nonmetallic

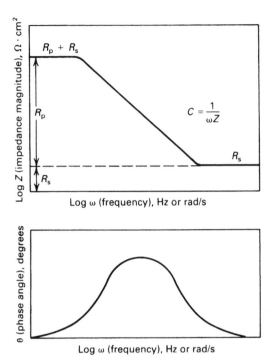

Figure 5.13. Bode magnitude and phase angle plots showing the frequency dependence of electrochemical impedance for the equivalent circuit model in Fig. 5.11(a) [15]. (Reprinted with permission of ASM International®. All rights reserved. www.asminternational.org.)

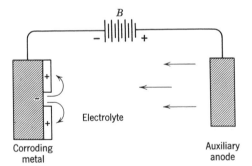

Figure 5.14. Cathodic protection by impressed current.

conductor), passes through the electrolyte, enters the metal that is to be protected, and returns to the source of direct current, B. The corroding metal is fully protected—and the corrosion rate is zero—when it is polarized to the open-circuit anode potential, also known as the reversible potential, the thermodynamic potential, and the equilibrium potential. The metal cannot corrode as long as the current is applied to maintain this potential. The polarization diagram is shown in Fig. 5.15, where I_{appl} is the applied current necessary for complete protection.

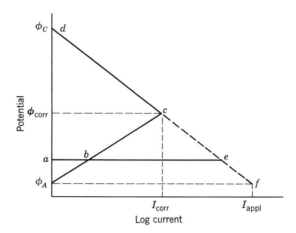

Figure 5.15. Polarization diagram illustrating principle of cathodic protection.

If the metal is polarized slightly beyond the open-circuit potential, ϕ_A, of the anode, the corrosion rate remains zero. Net current flows from the electrolyte to the metal; hence, metal ions cannot enter the solution. Current in excess of the required does no good, however, and may damage amphoteric metals and coatings. In practice, therefore, the impressed current is kept close to the theoretical minimum. Should the applied current fall below that required for complete protection, some degree of protection nevertheless occurs. For example, if the corrosion potential is moved from ϕ_{corr} to a in Fig. 5.15, by applied current e–b, the corrosion current decreases from I_{corr} to b. As the applied current e–b is increased, the potential a moves to more active values, and the corrosion current b becomes smaller. When a coincides with ϕ_A, the corrosion current becomes zero, and the applied current for complete cathodic protection equals I_{appl}.

REFERENCES

1. C. Wagner and W. Traud, Z. *Elektrochem.* **44**, 391 (1938).

2. N. D. Greene, Experimental Electrode Kinetics, Rensselaer Polytechnic Institute, Troy, NY, 1965; ASTM Standard G5-94 (Reapproved 1999), ASTM, West Conshohocken, PA.

3. J. O'M. Bockris, A. K. N. Reddy, and M. Gamboa-Aldeco, *Modern Electrochemistry*, 2nd edition, Vol. 2A, *Fundamentals of Electrodics*, Kluwer Academic/Plenum Publishers, New York, 2000, pp. 1246–1247.

4. M. Stern, *J. Electrochem. Soc.* **102**, 609; ibid., 663 (1955).

5. U. R. Evans, *J. Franklin Inst.* **208**, 45 (1929).

6. F. Mansfeld and R. V. Inman, *Corrosion* **35** (1), 21 (1979).

7. R. G. Kelly, in *Electrochemical Techniques in Corrosion Science and Engineering*, R. G. Kelly, J. R. Scully, D. W. Shoesmith, and R. G. Buchheit, editor, Marcel Dekker, New York, 2003, pp. 43–45; see also Ref. 3, pp. 1054–1055.

8. M. Stern and A. Geary, *J. Electrochem. Soc.* **104**, 56 (1957).

9. M. Stern and E. Weisert, *Am. Soc. Testing Mater. Proc.* **59**, 1280 (1959).

10. F. Mansfeld and K. Oldham, *Corros. Sci.* **11**, 787 (1971); F. Mansfeld, in *Advances in Corrosion Science and Technology*, edited by M. Fontana and R. Staehle, Vol. 6, Plenum Press, New York, 1976, p. 163.

11. M. Stern, *Corrosion* **14**, 329t (1958).

12. J. R. Scully, in *Electrochemical Techniques in Corrosion Science and Engineering*, Marcel Dekker, New York, 2003, p. 125.

13. ASTM G 106-89 (Reapproved 1999), Standard Practice for Verification of Algorithm and Equipment for Electrochemical Impedance Measurements, ASTM, West Conshohocken, PA.

14. G. L. Makar and J. Kruger, Corrosion studies of rapidly solidified magnesium alloys, *J. Electrochem. Soc.* **137**, 414 (1990).

15. J. R. Scully and R. G. Kelly, Methods for DeterminIng Aqueous Corrosion Reaction Rates, in *ASM Handbook*, Vol. 13A, *Corrosion: Fundamentals, Testing, and Protection*, ASM International, Materials Park, OH, 2003, p. 73.

GENERAL REFERENCES

E. Barsoukov and J. R. Macdonald, editors, *Impedance Spectroscopy, Theory, Experiment, and Applications*, 2nd edition, Wiley, New York, 2005.

J. O'M. Bockris and A. K. N. Reddy, *Modern Electrochemistry*, 2nd edition, Kluwer Academic/Plenum Publishers, New York, 2000.

A Century of Tafel's Equation: A Commemorative Issue of *Corrosion Science*, *Corrosion Science* **47**(12), December 2005.

M. Orazem and B. Tribollet, *Electrochemical Impedance Spectroscopy*, Wiley, New York, 2008.

D. C. Silverman, Practical corrosion prediction using electrochemical techniques, in *Uhlig's Corrosion Handbook*, 2nd edition, R. W. Revie, editor, Wiley, New York, 2000, p. 1179.

W. H. Smyrl, in *Comprehensive Treatise of Electrochemistry*, Vol. 4, *Electrochemical Materials Science*, J. O'M. Bockris et al., editors, Plenum Press, New York, 1981, p. 97.

PROBLEMS

1. The potential of an iron electrode when polarized as cathode at $0.001 \, \text{A/cm}^2$ is $-0.916 \, \text{V}$ versus $1 \, N$ calomel half-cell. The pH of the electrolyte is 4.0. What is the value of the hydrogen overpotential?

2. The potential of a cathode at which H^+ discharges at $0.001\,A/cm^2$ is $-0.92\,V$ versus Ag–AgCl in $0.01\,N$ KCl at $25°C$.

   (a) What is the cathode potential on the standard hydrogen scale?

   (b) If the pH of the electrolyte is 1, what is the value of hydrogen overpotential?

3. The potential of a platinum anode at which oxygen evolves in an electrolyte of pH 10 is $1.30\,V$ with respect to the saturated calomel electrode. What is the value of oxygen overpotential?

4. The potential of a copper electrode on which Cu^{2+} deposits from $0.2\,M$ $CuSO_4$ versus $1\,N$ calomel electrode is $-0.180\,V$. What is the polarization of the electrode in volts? Is the electrode polarized in the active or noble direction?

5. Closely separated short-circuited zinc and mercury electrodes are immersed in deaerated HCl of pH = 3.5. What is the current through the cell if the total exposed area of each electrode is $10\,cm^2$? What is the corresponding corrosion rate of zinc in gmd? (Corrosion potential of zinc versus $1\,N$ calomel half-cell is $-1.03\,V$.)

6. The corrosion potential of mild steel in a deaerated solution of pH = 2 is $-0.64\,V$ versus saturated Cu–$CuSO_4$ half-cell. The hydrogen overpotential (volts) for the same steel follows the relation $0.7 + 0.1\log i$, where $i = A/cm^2$. Assuming that approximately all the steel surface acts as cathode, calculate the corrosion rate in mm/y.

7. Derive an expression for the slope of corrosion rate versus pH for a dilute cadmium amalgam in a deaerated cadmium ion solution. Neglect concentration polarization and assume that the amalgam is approximately all cathode.

8. The potential of platinum versus saturated calomel electrode when polarized cathodically in deaerated H_2SO_4, pH = 1.0 at $0.01\,A/cm^2$ is $-0.334\,V$ and at $0.1\,A/cm^2$ it is $-0.364\,V$. Calculate β and i_0 for discharge of H^+ on platinum in this solution.

9. The corrosion rate of iron in deaerated HCl of pH = 3 is $3.0\,gmd$. Calculate the corrosion potential of iron in this acid versus $0.1\,N$ calomel electrode. Assume that the entire iron surface acts as a cathode.

10. The linear polarization slope $d\phi/dI$ at low current densities for iron in a corrosive solution equals $2\,mV/\mu A\text{-}cm^2$. Using Eq. (5.8), calculate the corrosion rate in gmd. Assume $\beta_a = \beta_c = 0.1\,V$.

11. Anodic activation overpotential η for small applied current density i follows the relation $\eta = ki$. Derive the value of k in terms of the exchange current density i_0, assuming $\beta_a = \beta_c = 0.1\,V$.

12. Determine the corrosion potential and corrosion rate of an iron pipe carrying $1N$ sulfuric acid at $0.2\,m/s$ at $25°C$. Assume that the entire iron surface acts as cathode, that Tafel slopes are $\pm 0.100\,V$, and that the exchange current densities for Fe/Fe^{2+} and for hydrogen on iron are 10^{-3} and $10^{-2}\,A/m^2$, respectively.

13. Determine the corrosion potential and corrosion rate of zinc in $1N$ hydrochloric acid. Assume that the entire zinc surface acts as cathode, that Tafel slopes are $\pm 0.100\,V$, and that the exchange current densities for zinc and for hydrogen on zinc are 0.1 and $10^{-4}\,A/m^2$, respectively.

14. Show the complex-plane impedance spectra for the following electrochemical cells:
(*a*) The cell is represented by a single electrical resistance of value, $R\ (\Omega)$.
(*b*) The cell is represented by a capacitance of value, $C\ (F)$.
(*c*) The cell is represented by an equivalent circuit with a resistance, $R_s\ (\Omega)$, in series with a capacitance, $C_s\ (F)$.
(*d*) The cell is represented by an equivalent circuit with a resistance, $R_p\ (\Omega)$, in parallel with a capacitance, $C_p\ (F)$.
(*e*) The cell is represented by the circuit shown in Fig. 5.11(*a*).

Answers to Problems

1. $-0.40\,V$.
2. (*a*) $-0.58\,V$; (*b*) $-0.52\,V$.
3. $0.90\,V$.
4. $-0.188\,V$; active direction.
5. $2.3 \times 10^{-7}\,A$; $0.0067\,gmd$.
6. $0.134\,mm/y$.
7. $di_{corr}/dpH = -2.3(0.059i_{corr})/\beta$.
8. $\beta = 0.03\,V$; $i_0 = 8 \times 10^{-4}\,A/cm^2$.
9. $-0.719\,V$.
10. $2.73\,gmd$.
11. $k = 0.0217/i_0$.
12. $-0.17\,V$ (S.H.E.); $0.58\,mm/y$.
13. $-0.53\,V$ (S.H.E.); $29.9\,mm/y$.

6

PASSIVITY

6.1 DEFINITION

A passive metal is one that is active in the Emf Series, but that corrodes never-theless at a very low rate. Passivity is the property underlying the useful natural corrosion resistance of many structural metals, including aluminum, nickel, and the stainless steels. Some metals and alloys can be made passive by exposure to passivating environments (e.g., iron in chromate or nitrite solutions) or by anodic polarization at sufficiently high current densities (e.g., iron in H_2SO_4).

As early as the eighteenth century, it was observed that iron reacts rapidly in dilute HNO_3, but is visibly unattacked by concentrated HNO_3 [1]. Upon removing iron from the concentrated acid and immersing it into the dilute acid, a temporary state of corrosion resistance persists. In 1836, Schönbein [2] defined iron in the corrosion-resistant state as "passive." He also showed that iron could be made passive by anodic polarization. Around the same time, Faraday [3] performed several experiments showing, among other things, that a cell made up of passive iron coupled to platinum in concentrated nitric acid produced little or no current, in contrast to the high current produced by amalgamated zinc coupled to platinum in dilute sulfuric acid. Although passive iron corrodes only slightly in concentrated HNO_3, and similarly amalgamated zinc corrodes only slightly in

Corrosion and Corrosion Control, by R. Winston Revie and Herbert H. Uhlig
Copyright © 2008 John Wiley & Sons, Inc.

dilute H_2SO_4, Faraday emphasized that a low corrosion rate is not alone a measure of passivity. Instead, he stated, the magnitude of current produced in the cell versus platinum is a better criterion, for on this basis iron is passive, but not zinc. This, in essence, defined a passive metal, as we do today, as one that is appreciably polarized by a small anodic current. Later investigators deviated from this definition, however, and also called metals passive if they corrode only slightly despite their pronounced tendency to react in a given environment. This usage brought about two definitions of passivity:

> **Definition 1.** A metal is passive if it substantially resists corrosion in a given environment resulting from marked anodic polarization.
> **Definition 2.** A metal is passive if it substantially resists corrosion in a given environment despite a marked thermodynamic tendency to react.

C. Wagner [4] offered an extension of Definition 1, the essence of which is the following: A metal is passive if, on increasing the electrode potential toward more noble values, the rate of anodic dissolution in a given environment under steady-state conditions becomes less than the rate at some less noble potential. Alternatively, a metal is passive if, on increasing the concentration of an oxidizing agent in an adjacent solution or gas phase, the rate of oxidation, in absence of external current, is less than the rate at some lower concentration of the oxidizing agent. These alternative definitions are equivalent under conditions where the electrochemical theory of corrosion applies.

Thus, lead immersed in sulfuric acid, or magnesium in water, or iron in inhibited pickling acid, would be called passive by Def. 2 based on low corrosion rates, despite pronounced corrosion tendencies; but these metals are not passive by Def. 1. Their corrosion potentials are relatively active, and polarization is not pronounced when they are made the anode of a cell.

Examples of metals that are passive under Definition 1, on the other hand, include chromium, nickel, molybdenum, titanium, zirconium, the stainless steels, 70%Ni–30%Cu alloys (Monel), and several other metals and alloys. Also included are metals that become passive in passivator solutions, such as iron in dissolved chromates. Metals and alloys in this category show a marked tendency to polarize anodically. Pronounced anodic polarization reduces observed reaction rates, so that metals passive under Definition 1 usually conform as well to Definition 2 based on low corrosion rates. The corrosion potentials of metals passive by Definition 1 approach the open-circuit cathode potentials (e.g., the oxygen electrode); hence, as components of galvanic cells, they exhibit potentials near those of the noble metals.

6.2 CHARACTERISTICS OF PASSIVATION AND THE FLADE POTENTIAL

Suppose iron is made anode in $1N$ H_2SO_4 and is arranged so that, as the potential is gradually increased, the corresponding polarizing current reaches the value required to maintain the prevailing potential with respect to a reference electrode. In the laboratory, this is usually accomplished using a potentiostat, which controls the potential between the "working" electrode and the reference electrode by automatically adjusting the current between the "working" electrode and an inert electrode, usually platinum, that is called the "auxiliary" or "counter" electrode. The resulting polarization curve, shown in Fig. 6.1 [5], is called a potentiostatic polarization curve.

Galvanostatic polarization is an alternative to potentiostatic polarization. In galvanostatic polarization measurements, the current between working and counter electrodes is controlled, and the potential between working and reference electrodes is automatically adjusted to the value required to maintain the current. Galvanostatic measurements can be made using, for example, the circuit shown in Fig. 5.3(a) in Section 5.3. The galvanostatic polarization curve for iron in $1N$ H_2SO_4 is shown in Fig. 6.2.

The potentiostatic polarization curve provides considerably more information about passivity than does the galvanostatic curve.

Referring to Fig. 6.1, iron is active at small current densities and corrodes anodically as Fe^{2+} in accord with Faraday's law. As current increases, a partially insulating film forms over the electrode surface, composed probably of $FeSO_4$. At a critical current density, $i_{critical}$, of about 0.2 A/cm^2 (higher on stirring or on lowering pH of the environment), the current suddenly drops to a value orders of magnitude lower, called the passive current density, $i_{passive}$. At this point, the thick insulating film dissolves, being replaced by a much thinner film, and iron becomes passive. The value of $i_{passive}$ decreases with time, diminishing to about

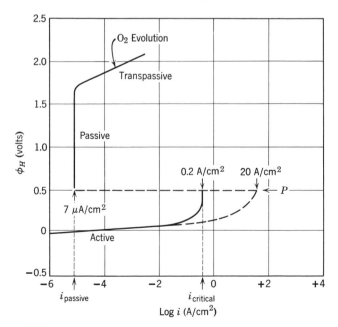

Figure 6.1. Potentiostatic anodic polarization curve for iron in $1N$ H_2SO_4.

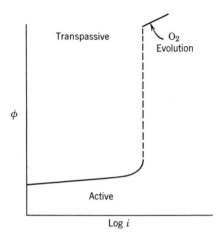

Figure 6.2. Galvanostatic anodic polarization curve for iron in $1N$ H_2SO_4.

$7 \mu A/cm^2$ in $1N$ H_2SO_4. In other electrolytes, $i_{passive}$ may be higher or lower than in $1N$ H_2SO_4. The true critical current density for achieving passivity of iron in absence of an insulating reaction-product layer is estimated to be about 10–20 A/cm^2, as shown by short-time current–pulse measurements.

On further gradual change of the potential, the current density remains at the above low value, and the corrosion product is now Fe^{3+}. At about 1.2 V, the equilibrium oxygen electrode potential is reached, but oxygen is not evolved appreciably until the potential exceeds the equilibrium value by several tenths of a volt (oxygen overpotential). The increased current densities in the region labeled "transpassive" represent O_2 evolution plus Fe^{3+} formation.

When the applied potential is removed, passivity decays within a short time in the manner shown in Fig. 6.3. The potential first changes quickly to a value still noble on the hydrogen scale, and then it changes slowly for a matter of seconds to several minutes. Finally, it decays rapidly to the normal active potential of iron. The noble potential arrived at just before rapid decay to the active value was found by Flade [6] to be more noble the more acid the solution in which passivity decayed. This characteristic potential, ϕ_F, was later called the Flade potential, and Franck established it to be a linear function of pH [7]. His measurements in acid media, combined with later data by others, provide the relation at 25°C:

$$\phi_F(\text{volts, S.H.E.}) = +0.63 - 0.059\text{pH} \tag{6.1}$$

This reproducible Flade potential and its 0.059 pH dependence is characteristic of the passive film on iron. A similar potential–pH relation is found for the passive film on chromium, for Cr–Fe alloys,* and for nickel, for which the stan-

---

*When activated cathodically, the Flade potential of chromium and stainless steels follows the relation $n(0.059\,\text{pH})$, where n may be as high as 2. For self-activation, n is 1 [8].

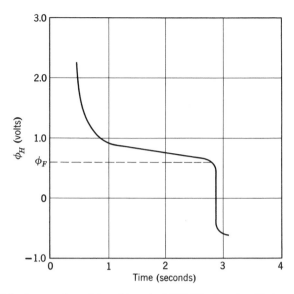

Figure 6.3. Decay of passivity of iron in 1N H$_2$SO$_4$ showing Flade potential, φ_F.

dard Flade potentials (pH = 0) are less noble than for iron, in accord with more stable passivity.

Stability of passivity is related to the Flade potential, assuming the following schematic reaction to take place during anodic passivation:

$$M + H_2O \rightarrow O \cdot M + 2H^+ + 2e^- \tag{6.2}$$

where $-\phi_F$ is the potential for the reaction, and O·M refers to oxygen in the passive film on metal M whatever the passive film composition and structure may be. The amount of oxygen assumed to be combined with M has no effect on present considerations. It follows, as observed, that

$$\phi_F = \phi_F^\circ + \frac{0.059}{2}\log(H^+)^2 = \phi_F^\circ - 0.059\text{pH} \tag{6.3}$$

The positive value of ϕ_F° for iron (0.63 V) indicates considerable tendency for the passive film to decay [reverse reaction of (6.2)], whereas an observed negative value of $\phi_F^\circ = -0.2\,\text{V}$ for chromium indicates conditions more favorable to passive-film formation and, hence, greater stability of passivity. The value of ϕ_F° for nickel is 0.2 V. For chromium–iron alloys, values range from 0.63 V for pure iron to increasingly negative values as chromium is alloyed, changing most rapidly in the range 10–15% chromium and reaching about –0.1 V at 25% chromium. The corresponding ϕ_F° values are shown in Fig. 6.4.

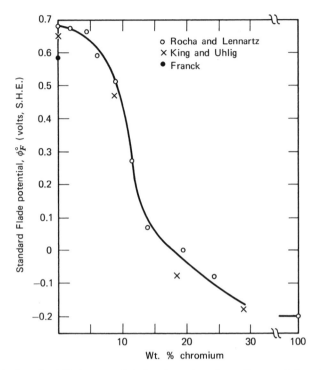

Figure 6.4. Standard Flade potentials for chromium–iron alloys and chromium [7–9].

The potential (P in Fig. 6.1) at which passivity of iron initiates (passivating potential) approximates, but is not the same as, the Flade potential because of IR drop through the insulating layer first formed and because the pH of the electrolyte at the base of pores in this layer differs from that in the bulk of solution (concentration polarization). These effects are absent on decay of passivity.

6.3 BEHAVIOR OF PASSIVATORS*

The same Flade potential is obtained whether iron is passivated by concentrated nitric acid or is anodically polarized in sulfuric acid, indicating that the passive film is essentially the same in both instances. In fact, when iron is passivated by immersion in solutions of passivators, whether chromates (CrO_4^{2-}), nitrites (NO_2^-), molybdates (MoO_4^{2-}), tungstates (WO_4^{2-}), ferrates (FeO_4^{2-}), or pertechnetates (TcO_4^-), the corresponding Flade potentials are close to values obtained

---

*See also Chapter 17.

otherwise [10, 11]. Hence, it can be concluded that the passive film on iron is essentially the same regardless of the passivation process. The amount of passive-film substance, as determined by coulometric and other measurements, is approximately $0.01\,C/cm^2$ in all cases. Passivation proceeds by an electrochemical mechanism, with some exceptions discussed later.

The standard Flade potential of iron passivated by chromates ($\phi_F^\circ = 0.54\,V$) is less noble than for iron passivated by HNO_3 ($\phi_F^\circ = 0.63\,V$). One explanation [11] is that chromates adsorb on the passive film more strongly than do nitrates, thereby reducing the overall free energy of the system and increasing the stability of the passive film. Other passivators presumably adsorb similarly, but with differing energies of adsorption.

Passivators are reduced at cathodic areas at a current density equivalent to a true current density at anodic areas equaling or exceeding $i_{critical}$ ($10–20\,A/cm^2$) for passivation of iron. The passivator is reduced over a large cathodic area of the metal surface to an extent not less than that necessary to form a chemically equivalent passive film at small residual anodic areas. The small passive areas, in turn, adsorb passivator, thereby becoming noble to adjoining passive or non-passive areas, causing passivity to spread. When the passive film is complete, it acts as cathode over its entirety, and further reduction of passivator proceeds at a much slower rate, equivalent to the rate of continuous passive film break-down, or to $i_{passive}$. Since breakdown is accelerated by presence of chlorides and by elevated temperatures, consumption of passivator is also increased correspondingly.

Passivators as a group are inorganic oxidizing agents that have the unique property of reacting only slowly when in direct contact with iron, but they are reduced more rapidly by cathodic currents. For this reason, they can adsorb first on the metal surface, with each site of adsorption adding to the cathodic area. The higher the concentration of passivator, the more readily it adsorbs, and the smaller the residual anodic areas become; this situation obviously favors increased anodic polarization and ultimately passivation. It requires about 0.5–2 h for the passive film to form completely when iron is immersed in 0.1% K_2CrO_4, with the shorter time being characteristic of aerated solution while the longer time is typical of deaerated solution [12, 13].

6.3.1 Passivation of Iron by HNO_3

In nitric acid, the cathodic depolarizer (passivator) is nitrous acid, HNO_2. This must form first in sufficient quantity by an initial rapid reaction of iron with HNO_3. As nitrous acid accumulates, anodic current densities increase, eventually reaching $i_{critical}$. Passivity is then achieved, and the corrosion rate falls to the comparatively low value of about 2 gmd [14].

If urea is added to concentrated HNO_3, passivation is interrupted because urea reacts with nitrous acid in accord with

$$(NH_2)_2CO + 2HNO_2 \rightarrow 2N_2 + CO_2 + 3H_2O \qquad (6.4)$$

thereby decreasing the nitrous acid concentration. The rate of reaction is, nevertheless, sufficiently below that of HNO_2 production so that passivity can still occur, although periodic breakdown and formation of the passive film usually result.

Hydrogen peroxide added to concentrated nitric acid also causes periodic breakdown and formation of passivity, probably by oxidizing HNO_2 to HNO_3 [14]. The peroxide, of itself, is not as efficient a cathodic depolarizer as HNO_2, and hence the passive film in its presence can repair itself only when the momentary surface concentration of HNO_2 formed by reaction of iron with HNO_3 is sufficiently high. After passivity is achieved, the surface concentration of HNO_2 is diminished by reaction with peroxide below that needed to maintain passivity, whereupon the cycle is repeated.

If iron is first immersed in dilute chromate solution for several minutes, it remains passive in concentrated nitric acid without the initial reaction to form HNO_2. The passive film is preformed by chromate, and nitrous acid is no longer necessary as depolarizer to reach $i_{critical}$, but is needed only in concentration sufficient to maintain the film already present.

6.4 ANODIC PROTECTION AND TRANSPASSIVITY

The electrochemical nature of the passivation process is the basis on which anodic polarization leads to passivation. The anodic polarization that is required for passivation can be achieved by applying current or by increasing either the cathodic area or the cathodic reaction rate. High-carbon steels, for example— which contain areas of cementite (Fe_3C) that act as cathodes—are more readily passivated by concentrated nitric acid than is pure iron. For this reason, nitrating mixtures of sulfuric and nitric acids are best stored or shipped in drums of steel of the highest carbon content consistent with required mechanical properties. Similarly, stainless steels that may lose passivity in dilute sulfuric acid retain their corrosion resistance if alloyed with small amounts of more noble constituents of low hydrogen overpotential, or of low overpotential for cathodic reduction of dissolved oxygen, such as Pd, Pt, or Cu [15]. The polarization diagram corresponding to increased passivity by low overpotential cathodes is shown in Fig. 6.5, and the corresponding improved corrosion resistance to sulfuric acid is shown in Fig. 6.6.

Because titanium, unlike 18–8 stainless steel, has a low critical current density for passivity in chlorides as well as in sulfates, passivity in boiling 10% HCl is made possible by alloying titanium with 0.1% Pd or Pt [16]. Pure titanium, on the other hand, corrodes in the same acid at very high rates (see Fig. 25.2, Section 25.3).

Based on the same principle, the corrosion resistance of metals and alloys with polarization curves having active–passive transitions can be greatly improved by an impressed anodic current initially equal to or greater than the critical current for passivity. The potential of the metal moves into the passive region

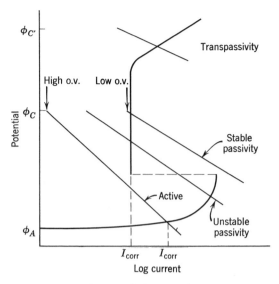

Figure 6.5. Polarization diagram for metal that is either active or passive, depending on overvoltage of cathodic areas (differing cathodic reaction rates).

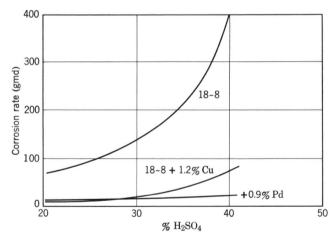

Figure 6.6. Corrosion rates in sulfuric acid of 18–8 stainless steel alloyed with copper or palladium, 360-h test, 20°C [15].

(Fig. 6.1), so that the current density and accompanying corrosion rate correspond to the low value of $i_{passive}$. This process is called *anodic protection* (see Section 13.9) because the current flow is in the direction opposite to that which is used in cathodic protection. Whereas cathodic protection can, in principle, be applied to both passive and nonpassive metals, anodic protection is applicable only to metals that can be passivated when anodically polarized (see Definition 1, Section 6.1).

If the cathodic polarization curves of Fig. 6.5 intersect the anodic curve at a still more noble potential, within the transpassive region, the corrosion rate of, for example, stainless steel, is greatly increased over the corrosion rate at less noble potentials within the passive region, and the corrosion products become $Cr_2O_7^{2-}$ and Fe^{3+}. Transpassivity occurs not only with stainless steels, but also with chromium, for which the potential for the reaction

$$Cr_2O_7^{2-} + 14H^+ + 12e^- \rightarrow 2Cr + 7H_2O \qquad \phi° = 0.30\,V \qquad (6.5)$$

is less noble than the potential for the oxygen evolution reaction

$$O_2 + 4H^+ + 4e^- \rightarrow 2H_2O \qquad \phi° = 1.23\,V \qquad (6.6)$$

Appreciable corrosion in the transpassive region does not occur for iron in sulfuric acid (oxygen evolution is the primary reaction), but increased corrosion does occur in alkaline solutions that favor formation of ferrate, FeO_4^{2-}. Transpassivity accounts for an observed increase of corrosion rate with time for 18–8 stainless steels in boiling concentrated nitric acid, in which corrosion products accumulate, in particular $Cr_2O_7^{2-}$, and move the corrosion potential into the transpassive region.

6.5 THEORIES OF PASSIVITY

There are two commonly expressed points of view regarding the composition and structure of the passive film. The first holds that the passive film (Definition 1 or 2) is always a diffusion-barrier layer of reaction products—for example, metal oxide or other compound that separates metal from its environment and that decreases the reaction rate. This theory is sometimes referred to as the oxide-film theory.

The second holds that metals passive by Definition 1 are covered by a chemisorbed film—for example, of oxygen. Such a layer displaces the normally adsorbed H_2O molecules and decreases the anodic dissolution rate involving hydration of metal ions. Expressed another way, adsorbed oxygen decreases the exchange current density (increases anodic overvoltage) corresponding to the overall reaction $M \rightarrow M^{z+} + ze^-$. Even less than a monolayer on the surface is observed to have a passivating effect [17]; hence, it is suggested that the film cannot act primarily as a diffusion-barrier layer. This second point of view is called the adsorption theory of passivity.

There is no question on either viewpoint that a diffusion-barrier film is the basis of passivity of many metals that are passive only by Definition 2. Examples of protective films that isolate the metal from its environment are (a) a visible lead sulfate film on lead immersed in H_2SO_4 and (b) an iron fluoride film on steel immersed in aqueous HF.

For metals that are passive by Definition 1, based on marked anodic polarization, the films are usually invisible, about 2 to 3 nm thick. Metals and alloys in this category have been the source of extended debate and discussion on the mechanism of passivity over the past 150 years. If the surface is abraded, local high temperatures generated at the surface produce a detectable oxide, but this is not the passive film.

Low-energy electron-diffraction (LEED) techniques are used to detect adsorbed films, including those responsible for passivity. Surface analytical techniques that can be used to study films on metals include Auger electron spectroscopy (AES), X-ray photoelectron spectroscopy (XPS), secondary ion mass spectrometry (SIMS), and others [18, 19]. Using these advanced techniques, evidence has been developed to show that the passive film formed on iron in borate solution is an anhydrous, crystalline spinel with a "γ-Fe_2O_3/Fe_3O_4-like" structure [19].

The adsorption theory derives support from the fact that most of the metals that are passive by Definition 1 are transition metals of the periodic table; that is, they contain electron vacancies or uncoupled electrons in the d shells of the atom. The uncoupled electrons account for strong bond formation with components of the environment, especially O_2, which also contains uncoupled electrons (hence its slight paramagnetic susceptibility) resulting in electron-pair or covalent bonding supplementary to ionic bonding. Furthermore, transition metals have high heats of sublimation compared to nontransition metals, a property that favors adsorption of the environment because metal atoms tend to remain in their lattice, whereas oxide formation requires metal atoms to leave their lattice. The characteristic high energies for adsorption of oxygen on transition metals correspond to chemical-bond formation; hence, such films are called *chemisorbed* in contrast to lower-energy films, which are called *physically adsorbed*. On the nontransition metals, such as copper and zinc, oxides tend to form immediately, and any chemisorbed films on the metal surface are short-lived. On transition metals, the initially chemisorbed oxygen is generally more stable thermodynamically than is the metal oxide [20]. This situation reverses for multilayer adsorbed oxygen films (lower energy of bonding to the metal) that revert in time to metal oxides. But such oxides are less important in accounting for passivity than the chemisorbed films that form initially and continue to form on metal exposed at pores in the oxide.

The adsorption theory emphasizes that the observed Flade potential of passive iron is too noble by about 0.6 V to be explained by any of the known iron oxides in equilibrium with iron. Observed values of the Flade potential are consistent with a chemisorbed film of oxygen on the surface of iron, the corresponding potential of which is calculated (see Problem 2 in Chapter 6) using the observed heat and estimated entropy of adsorption of oxygen on iron in accord with the reaction [21]

$$O_2 \cdot O_{\text{ads on Fe}} + 6H^+ + 6e^- \rightarrow 3H_2O + \text{Fe surface} \qquad \phi^{\circ}_{\text{calc}} = 0.57 \text{ V} \qquad (6.7)$$

The measured chemical equivalents of passive-film substance (about $0.01 \, C/cm^2$) correspond (roughness factor = 4) to one atomic layer of oxygen atoms ($r = 0.07 \, nm$) over which one layer of oxygen molecules ($r = 0.12 \, nm$) is chemisorbed; hence, the adsorbed passive film may be represented above by $O_2 \cdot O_{ads \, on \, Fe}$. The value of the Flade potential, as cited in Eq. (6.1), corresponds to $\phi^\circ_{obs} = 0.63 \, V$, which is in reasonable agreement with the calculated value, $0.57 \, V$. The agreement is still better if account is taken of adsorbed water displaced from the metal surface by the passive film during the passivation reaction, this displacement presumably involving a greater free energy change than the adsorption of water on the passive film itself [8].

Further confirming evidence that the passive film on iron contains oxygen in a higher-energy state than corresponds to any iron oxide is obtained from the ability of the passive film to oxidize chromite, CrO_2^-, to chromate, CrO_4^{2-}, in NaOH solution [14]. This oxidation does not occur with active iron. The maximum oxidizing capacity corresponds to $0.012 \, C/cm^2$ of passive-film substance in accord with the reaction

$$O_2 \cdot O_{ads \, on \, Fe} + Fe + OH^- + CrO_2^- + H_2O \rightarrow Fe(OH)_3 + CrO_4^{2-}$$

$$\Delta G^\circ = -138{,}000 \, J \tag{6.8}$$

Calculations show that the same reaction will not go for chromium or the stainless steels, on which the passive film is more stable.

On breakdown of passivity, $0.01 \, C/cm^2$ of adsorbed oxygen on iron reacts with underlying metal in accord with

$$O_2 \cdot O_{ads \, on \, Fe} + 2Fe + 3H_2O \rightarrow 2Fe(OH)_3 \quad \text{or} \quad Fe_2O_3 \cdot 3H_2O \tag{6.9}$$

A layer of Fe_2O_3 (mol. wt. = 159.7; $d = 5.12 \, g/cm^3$) is formed equal to a minimum of $(0.01 \times 159.7)/(6 \times 96{,}500 \times 5.12) = 5.4 \, nm$ thick (based on apparent area). The hydrated oxide would be thicker. This value compares in magnitude with measured values for the thickness of the decomposed passive film (2.5–10 nm). It is the decomposed passive film that is presumably isolated in experiments designed to remove the passive film from iron.

According to the adsorption theory, passivity of chromium and the stainless steels, because of their pronounced affinity for oxygen, can occur by direct chemisorption of oxygen from the air or from aqueous solutions, and the equivalents of oxygen so adsorbed were found [22] to be of the same order of magnitude as the equivalents of passive film formed on iron when passivated either anodically, by concentrated nitric acid, or by exposure to chromates. Similarly, oxygen in air can adsorb directly on iron and passivate it in aerated alkaline solutions, or also in near-neutral solutions if the partial pressure of oxygen is increased sufficiently.

6.5.1 More Stable Passive Films with Time

Flade observed that the longer iron remains in concentrated nitric acid, the longer the passive film remains stable when the iron is subsequently immersed in sulfuric acid [6]. In other words, the film is stabilized by continued exposure to the passivating environment. Similarly, Frankenthal [17] noted that the film on 24% Cr–Fe thickened and became more resistant to cathodic reduction when the alloy was passivated for longer times in $1N$ H_2SO_4 at potentials noble to the passivating potential (P in Fig. 6.1), although less than monolayer quantities of oxygen (measured coulometrically) suffice to passivate this alloy in this solution. The observed stabilizing effect is likely the result of positively charged metal ions entering the adsorbed layers of negatively charged oxygen ion and molecules, with the coexisting opposite charges tending to stabilize the adsorbed film. Low-energy electron-diffraction data for nickel single crystals [23]. indicate that the first-formed adsorbed film consists of a regular array of oxygen and nickel ions located in the same approximate plane of the surface. This initial adsorbed layer is found to be more stable thermally than the oxide, NiO. At increasing oxygen pressures, several adsorbed layers, probably consisting of O_2, form on top of the first layer and result in an amorphous film. Additional metal ions in time succeed in entering such a film, particularly in the noble potential range, becoming a mobile species within the adsorbed oxygen layer. Protons from the aqueous environment are also incorporated. Okamoto and Shibata [24], for example, showed that the passive film on 18–8 stainless steel contains H_2O. Eventually, the stoichiometric oxide is nucleated at favorable sites on the metal surface, and such nuclei then grow laterally to form a uniform oxide film; but the adsorbed (passive) film remains intact at pores in the oxide. A schematic structure of the first-formed adsorbed passive film is shown in Fig. 6.7 [20]. The initial passive film grows into a multilayer adsorbed M·O·H structure, which can

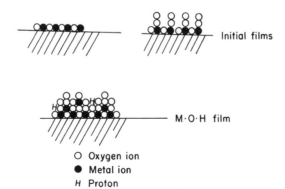

O Oxygen ion
● Metal ion
H Proton

Figure 6.7. Schematic structure of initial passive films containing less or more than monolayer amounts of adsorbed oxygen, along with schematic structure of a thicker passive film containing additional metal ions and protons in nonstoichiometric amounts.

be considered to be an amorphous nonstoichiometric oxide. It differs markedly in protective properties from the stoichiometric oxide into which it may eventually convert.

6.5.2 Action of Chloride Ions and Passive–Active Cells

Chloride ions—and, to a lesser extent, other halogen ions—break down passivity or prevent its formation in iron, chromium, nickel, cobalt, and the stainless steels. From the perspective of the oxide-film theory, Cl^- penetrates the oxide film through pores or defects easier than do other ions, such as SO_4^{2-}. Alternatively, Cl^- may colloidally disperse the oxide film and increase its permeability.

On the other hand, according to the adsorption theory, Cl^- adsorbs on the metal surface in competition with dissolved O_2 or OH^-. Once in contact with the metal surface, Cl^- favors hydration of metal ions and increases the ease with which metal ions enter into solution, opposite to the effect of adsorbed oxygen, which decreases the rate of metal dissolution. In other words, adsorbed chloride ions increase the exchange current (decrease overvoltage) for anodic dissolution of the above-mentioned metals over the value prevailing when oxygen covers the surface. The effect is so pronounced that iron and the stainless steels are not readily passivated anodically in solutions containing an appreciable concentration of Cl^-. Instead, the metal continues to dissolve at high rates in both the active and passive potential ranges.

Breakdown of passivity by Cl^- occurs locally rather than generally, with the preferred surface sites being determined perhaps by small variations in the passive-film structure and thickness. Minute anodes of active metal are formed surrounded by large cathodic areas of passive metal. The potential difference between such areas is 0.5 V or more, and the resulting cell is called a *passive–active cell*. High current densities at the anode cause high rates of metal penetration, accompanied by cathodic protection of the metal area immediately surrounding the anode. This fixes the anode in place and results in pitting corrosion. Also, the greater the current flow and cathodic protection at any pit, the less likely it is that another pit will initiate nearby; hence, the observed number of deep pits per unit area is usually less than that of smaller shallow pits. It is evident, because of the possibility of passive–active cell formation, that deep pitting is much more common with passive metals than with nonpassive metals.

Halogen ions have less effect on the anodic behavior of titanium, tantalum, molybdenum, tungsten, and zirconium. Passivity of these metals may continue in media of high chloride concentration, in contrast to the behavior of iron, chromium, and Fe–Cr alloys, which lose passivity. This behavior is sometimes explained by formation of insoluble protective Ti, Ta, Mo, etc., basic chloride films. However, the true situation is probably related to the high affinity of these metals for oxygen, making it more difficult for Cl^- to displace oxygen of the passive film, in accord with the noble critical potentials of these metals above which pitting is initiated, if pitting occurs at all.

6.6 CRITICAL PITTING POTENTIAL

When iron is anodically polarized potentiostatically in $1N$ H_2SO_4 to which sodium chloride is added ($>3 \times 10^{-4}$ moles/liter), the apparent transpassive region is shifted to lower potentials. However, instead of oxygen being evolved, the metal corrodes locally with formation of visible pits. Similarly, if 18–8 stainless steel is anodically polarized in $0.1N$ NaCl within the lower passive region, the alloy remains passive, analogous to its behavior in Na_2SO_4 solution, no matter how long the time. But above a critical potential, called the *critical pitting potential* (CPP), there is a rapid increase in the current, accompanied by random formation of pits (Fig. 6.8). The CPP for 18–8 moves toward less noble values as the Cl^- concentration increases, and it moves toward more noble values with increasing pH and with lower temperatures [25]. The CPP is also moved in the noble direction by adding extraneous salts to the NaCl solution—for example, Na_2SO_4, $NaNO_3$, and $NaClO_4$ (Fig. 6.8). If sufficient extraneous salt is added to shift the CPP to a value more noble than the prevailing corrosion potential, pitting does not occur at all on exposure of the stainless steel to the aqueous salt mixture. The added salts under these conditions become effective inhibitors. Addition of 3% $NaNO_3$ to 10% $FeCl_3$, for example, has prevented pitting of 18–8 stainless steel or any appreciable weight loss for a period of more than 25 years, whereas in the absence of $NaNO_3$ the alloy is observed to corrode by pitting within hours [26]. The 3% $NaNO_3$ addition moves the CPP to a value more noble than the open-circuit cathode potential or the reversible potential for $Fe^{3+} + e^- \rightarrow Fe^{2+}$.

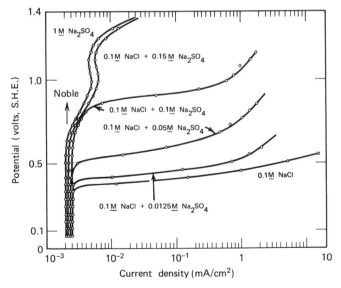

Figure 6.8. Potentiostatic polarization curves for 18–8 stainless steel in $0.1M$ NaCl showing increasingly noble values of the critical pitting potential with additions of Na_2SO_4, 25°C.

TABLE 6.1. Critical Pitting Potentials in 0.1 N NaCl at 25°C

	Potential (V, s.h.e.)	Reference
Al[a]	−0.37	[29]
Ni	0.28	[30]
Zr	0.46	[31]
18–8 stainless steel[b]	0.26	[30]
30% Cr–Fe	0.62	[30]
12% Cr–Fe	0.20	[30]
Cr	>1.0	[32]
Ti	>1.0 (1N NaCl)	[33]
	≈1.0 (1N NaCl, 200°C)	

[a]$\phi_{critical}$ (V, s.h.e.) = −0.124 log (Cl⁻) − 0.504, where (Cl⁻) is activity of Cl⁻.
[b]$\phi_{critical}$ (V, s.h.e.) = −0.088 log (Cl⁻) + 0.168.

Along the same lines, Leckie [27] polarized 18–8 stainless steel at a constant 0.1 V below the CPP in 0.1N NaCl for 14 weeks without observable pitting.

CPPs of several metals in 0.1N NaCl are listed in Table 6.1, values of which were derived from anodic polarization curves. Most of the data were obtained by allowing 5 minutes or more at a given potential and observing whether the resultant current increases or decreases with time. The CPP is the most noble potential for which the current decreases or remains constant; it is usually confirmed by holding the potential at the critical value for 12 hours or more and observing absence of pits under a low-power microscope.

CPPs can be used to compare the relative susceptibility of different alloys to pitting corrosion; however, because CPPs are the result of short-term experiments, they should be used with caution in predicting immunity to pitting corrosion in long-term service. Pitting corrosion has been reported in long-term exposure at potentials below the CPP [28].

It is observed that an increase in the chromium content of stainless steels—and, to a lesser extent, an increase of nickel content—shifts the critical potential to more noble values, corresponding to increased resistance to pitting [30, 31]. The noble critical potentials for chromium and titanium [more noble than the oxygen electrode potential in air (0.8 V) in accord with $O_2(0.2\,atm) + 4H^+ (10^{-7})$ $+ 4e^- \rightarrow 2H_2O$] indicate that these metals are not expected to undergo pitting corrosion in aerated saline media at normal temperatures. At elevated temperatures and high Cl⁻ concentrations, however, the critical potentials become more active, so that pitting of titanium, for example, is observed in concentrated hot $CaCl_2$ solution despite its immunity to pitting in seawater.

The critical potential has been explained, from one point of view, as that value needed to build up an electrostatic field within the passive or oxide film sufficient to induce Cl⁻ penetration to the metal surface [34]. Other anions may also penetrate the oxide, depending on size and charge, contaminating the oxide and making it a better ionic conductor favoring oxide growth. Eventually, either the oxide is undermined by condensation of migrating vacancies, or cations of the oxide undergo dissolution at the electrolyte interface; in both cases, pitting results. The induction period preceding pitting is related to the time required for the supposed penetration of Cl^- through the oxide film.

Alternatively, the critical potential is explained in terms of competitive adsorption of Cl^- with oxygen of the passive film (adsorption theory). The metal has typically greater affinity for oxygen than for Cl^-, but, as the potential is made more noble, the concentration of Cl^- ions at the metal surface increases, eventually reaching a value that allows Cl^- to displace adsorbed oxygen. The observed induction period is the time required for successful competitive adsorption at favored sites of the metal surface, as well as time for penetration of the passive film. Adsorbed Cl^-, compared to adsorbed oxygen, results in lower anodic overvoltage for metal dissolution, which accounts for a higher rate of corrosion at any site where the exchange has taken place. Extraneous anions, such as NO_3^- or SO_4^{2-}, which do not break down the passive film or cause pitting, compete with Cl^- for sites on the passive surface, making it necessary to shift the potential to a still more noble value in order to increase Cl^- concentration sufficient for successful exchange with adsorbed oxygen.*

Below the CPP, Cl^- cannot displace adsorbed oxygen so long as the passive film remains intact; hence, pitting is predicted not to occur. Should passivity break down because of factors other than those described [e.g., reduced oxygen or depolarizer concentration at a crevice (crevice corrosion), or cathodic polarization of local shielded areas], pitting could then initiate independent of whether the overall prevailing potential is above or below the critical value. But under conditions of uniform passivity for the entire metal surface, application of cathodic protection to avoid pitting corrosion need only shift the potential of the metal below the critical value. This is in contrast to the usual procedure of cathodic protection, which requires polarization of a metal to the much more active open-circuit anode potential.

The relation between minimum anion activity necessary to inhibit pitting of 18–8 stainless steel in a solution of given Cl^- activity follows the relation $\log (Cl^-) = k \log (anion) + const$. The same relation applies to inhibition of pitting of aluminum. The equation can be derived assuming that ions adsorb competitively in accord with the Freundlich adsorption isotherm [26].

6.7 CRITICAL PITTING TEMPERATURE

Pitting tendency increases with increasing temperature, with the result that the critical pitting potential decreases as temperature increases. The *critical pitting temperature* (CPT) is defined as the temperature below which an alloy does not

---

*Rosenfeld and Maximtschuk [*Z. Physik. Chem.* (D.D.R.) **215**, 25 (1960)] reported a quantitative study showing that the more noble the applied potential, the more Cl^- in aqueous solution adsorbs on metallic Cr, and they also showed that added SO_4^{2-} or OH^- displaces adsorbed Cl^-.

pit regardless of potential and exposure time. CPTs are usually measured in an aggressive Cl⁻ solution, such as a $FeCl_3$ solution. The CPT can be used to evaluate and compare pitting susceptibilities of alloys [35].

6.8 PASSIVITY OF ALLOYS

Several metals, such as chromium, are naturally passive when exposed to the atmosphere, and they remain bright and tarnish-free for years, in contrast to iron and copper, which corrode or tarnish in short time. It is found that the passive property of chromium is conferred on alloys of Cr–Fe, provided that ≥12 wt.% Cr is present. Iron-base alloys containing a minimum of 12 wt.% Cr are known as the *stainless steels*.

Typical corrosion, potential, and critical-current density behavior of Cr–Fe alloys is shown in Figs. 6.9–6.11. Note in Fig. 6.11 that $i_{critical}$ for passivation of Cr–Fe alloys at pH 7 reaches a minimum at about 12% Cr in the order of 2 μA/ cm². This value is so low that corrosion currents in aerated aqueous media easily achieve or exceed this value, illustrating why >12% Cr–Fe alloys are self-passivating. In addition, the passive film becomes more stable with increasing chromium content of the alloy.

Several other alloy systems exhibit critical compositions for passivity, as was first described by Tammann [36]. Examples of approximate critical compositions

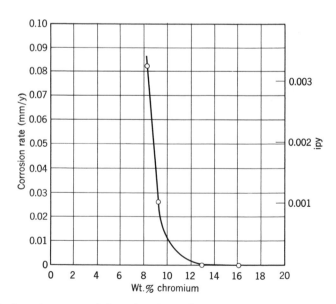

Figure 6.9. Corrosion rates of chromium–iron alloys in intermittent water spray at room temperature. [Reprinted with permission from W. Whitman and E. Chappell, *Ind. Eng. Chem.* **18**, 533 (1926). Copyright 1926, American Chemical Society.]

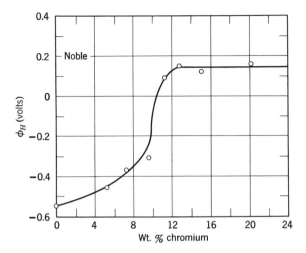

Figure 6.10. Potentials of chromium–iron alloys in 4% NaCl. [Reprinted with permission from H. Uhlig, N. Carr, and P. Schneider, *Trans. Electrochem. Soc.* **79**, 111 (1941). Copyright 1941, The Electrochemical Society.]

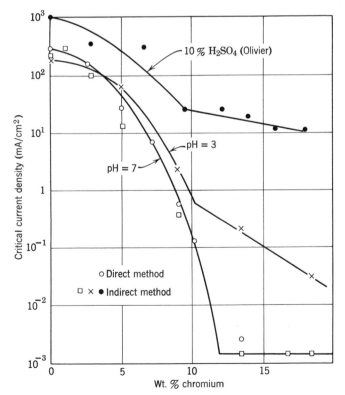

Figure 6.11. Critical current densities for passivation of chromium–iron alloys in deaerated 3% Na_2SO_4 at pH 3 and 7, 25°C [8]; data in 10% H_2SO_4, room temperature, are from R. Olivier, *Proceedings, 6th Meeting, International Committee on Electrochemistry, Thermodynamics, and Kinetics*, Poitiers, Butterworths, London, 1955, p. 314.

determined from plots of $i_{critical}$ versus alloy composition are 35% Ni–Cu, 15% Mo–Ni, 8% Cr–Co, and 14% Cr–Ni. Critical alloy compositions for passivity have also been observed in three- and four-component alloys—for example, Fe–Cr–Ni–Mo [37], Fe–Ni–Mo, and Cr–Ni–Fe [38].

Critical compositions based on $i_{critical}$ values are not sensitive to the pH of the electrolyte in which the polarization data are measured (Fig. 6.11). On the other hand, critical compositions based on corrosion data usually vary with the environment to which the alloys are exposed. In 33% HNO_3, for example, the critical chromium concentration for passivity of Cr–Fe alloys decreases to 7% Cr, whereas in $FeSO_4$ solution it increases to 20% Cr; only in aqueous solutions of about pH 7 is the critical composition equal to 12%. This effect of environment is to be expected because whether an alloy becomes passive depends on whether the corrosion rate equals or exceeds the specific $i_{critical}$ for passivity in the given environment. As seen in Fig. 6.11, increasing chromium content reduces $i_{critical}$; on the other hand, decreasing pH and increasing temperature increase $i_{critical}$. The corrosion rate, in turn, depends on factors of metal and environment affecting the rate of both cathodic and anodic reactions.

It can be shown from first principles [39] that $i_{critical} = K(H^+)^\lambda$ where K and λ are constants, the values of which depend on the anion. The cathodic reaction rate in an aerated solution when controlled by oxygen reduction, on the other hand, is given by the diffusion current i_{diff} corresponding to the reaction $\frac{1}{2}O_2 + H_2O \rightarrow 2OH^- - 2e^-$. For an unstirred air-saturated solution, it can be shown that $i_{diff} = 0.039\,mA/cm^2$ $(0.39\,A/m^2)$. Hence, a critical pH exists for each alloy composition at which $i_{critical} = i_{diff}$ and above which, but not below, passivity is stable. The observed critical pH values, for example, of 18% Cr, 8% Ni, and of 12% Cr stainless steels in $0.1M$ Na_2SO_4 [40] are 1.4 and 5.0, respectively, in close agreement with the calculated values. With iron, for which $i_{critical}$ is much higher than for the stainless steels, the critical pH is approximately 10.

The structure of the passive film on alloys, as with passive films in general, has been described both by the oxide-film theory and by the adsorption theory. It has been suggested that protective oxide films form above the critical alloy composition for passivity, but nonprotective oxide films form below the critical composition. The preferential oxidation of passive constituents (e.g., chromium) may form protective oxides (e.g., Cr_2O_3) above a specific alloy content, but not below. No quantitative predictions have been offered based on this point of view, and the fact that the passive film on stainless steels can be reduced cathodically, but not stoichiometric Cr_2O_3 itself, remains unexplained.

By the adsorption theory, it is considered that, in the presence of water, oxygen chemisorbs on chromium–iron alloys above the critical composition corresponding to passivity, but immediately reacts below the critical composition to form a less protective or nonprotective oxide film. Whether the alloy favors a chemisorbed or reaction-product film depends on the electron configuration of the alloy surface, in particular on d-electron interaction. The *electron configuration theory* describes the specific alloying proportions corresponding to a favorable d-electron configuration accompanying chemisorption and passivity. It suggests the nature of electron interaction that determines which alloyed component dominates in conferring its chemical properties on those of the alloy—for example, why the properties of nickel dominate above those of copper in the nickel–copper alloys at >30–40% Ni.

6.8.1 Nickel–Copper Alloys

Nickel, containing 0.6 d-electron vacancy per atom (as measured magnetically), when alloyed with copper, a nontransition metal containing no d-electron vacancies, confers passivity on the alloy above approximately 30–40 at.% Ni. Initiation of passivity beginning at this composition is indicated by corrosion rates in sodium chloride solution (Figs. 6.12 and 6.13), by corrosion pitting behavior in seawater (Fig. 6.13), and, more quantitatively, by measured values of $i_{critical}$ and $i_{passive}$ (Fig. 6.14), [41–43] or by decay (Flade) potentials (Fig. 6.15) [44] in $1N$ H_2SO_4.

Corrosion pitting in seawater is observed largely above 40% Ni because pit growth is favored by passive–active cells (see Section 6.5), and such cells can operate only when the alloy is passive—that is, in the range of high nickel compositions. Practically, this distinction is observed in the specification of materials for seawater condenser tubes in which pitting attack must be rigorously avoided. The cupro nickel alloys are used (10–30% Ni), but not Monel (70% Ni–Cu).

Similarly, marine fouling organisms are much less successful in establishing themselves on the surface of nonpassive nickel–copper compositions because the

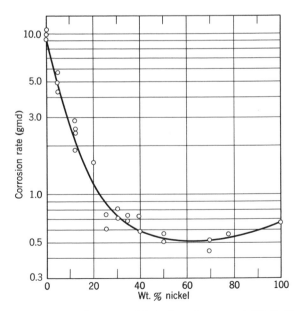

Figure 6.12. Corrosion rates of copper–nickel alloys in aerated 3% NaCl, 80°C, 48-h tests (M.I.T. Corrosion Lab.).

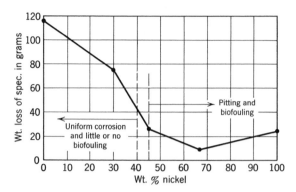

Figure 6.13. Behavior of copper–nickel alloys in seawater [F. LaQue, *J. Am. Soc. Nav. Eng.* **53**, 29 (1941)].

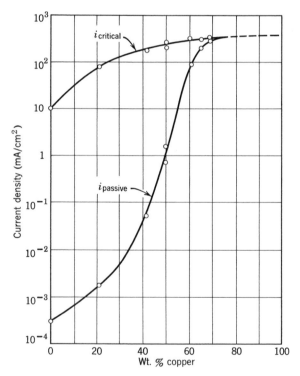

Figure 6.14. Values of critical and passive current densities obtained from potentiostatic anodic polarization curves for copper–nickel alloys in 1*N* H_2SO_4, 25°C [42]. (Reproduced with permission. Copyright 1961, The Electrochemical Society.)

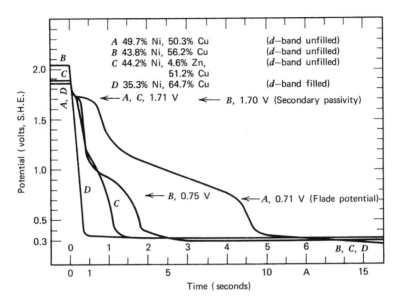

Figure 6.15. Potential decay curves for nickel–copper and nickel–copper–zinc alloys in 1N H$_2$SO$_4$, 25°C (two time scales). Pure copper behaves like Alloy D [36].

latter corrode uniformly at rates that release enough Cu^{2+} to poison fouling organisms.* But on passive nickel–copper compositions, for which the overall corrosion rate is much less, fouling organisms in general can gain a foothold and flourish (Fig. 6.13).

Potentiostatic anodic polarization behavior of nickel–copper alloys in 1N H$_2$SO$_4$ (Fig. 6.14) establishes that the passive current density largely disappears above 60% Cu and vanishes completely at about 70% Cu. Polarization curves of alloys containing >70% Cu or <30% Ni resemble those of pure copper; hence, such alloys are not passive. Potential decay curves (Fig. 6.15) confirm that a passive film is formed on anodically passivated alloys containing >40% Ni, but not otherwise. In other words, alloys containing copper above the critical composition lose their transition-metal characteristics; that is, they no longer contain d-electron vacancies. In this connection, the magnetic saturation moment, which is also a function of d-electron vacancies in the alloy, becomes zero at >60% Cu. This observation is interpreted as a filling of vacancies in the d band of electron energy levels of nickel by electrons donated by copper. Physicists [45] have made the assumption that, if copper and nickel atoms are considered to be alike, except that copper contains one more electron per atom than nickel, then the 0.6 d-electron vacancy per nickel atom is expected, as observed, to be just filled by electrons from copper at 60 at.% Cu.

---

*The minimum concentration of Cu^{2+} required to poison marine organisms corresponds to a corrosion rate for copper of about 0.001 ipy or 0.5 gmd [F. LaQue and W. Clapp, *Trans. Electrochem. Soc.* **87**, 103 (1945)].

One can also start with the alternative assumption that the two atoms maintain their individuality and that the vacancies per atom of nickel are a function of alloy concentration [46]. In the gaseous state, nickel has the configuration of $3d^84s^2$, corresponding to two d-electron vacancies or to the equivalent of two uncoupled d-electrons in the third shell of the atom. The maximum number of d-electrons that can be accommodated is 10, corresponding to copper: $3d^{10}4s$. In the process of condensing to a solid and forming the metallic bond, the uncoupled electrons of any single nickel atom tend to couple with uncoupled electrons of neighboring atoms. This results in a smaller number of electron vacancies in the solid compared to the gas, accounting for the measured 0.6 vacancy or 0.6 uncoupled electron per nickel atom. If we assume that the intercoupling of d-electrons increases with proximity of nickel atoms in the alloy and is a linear function of nickel concentration, then the vacancies per nickel atom can be set equal to $2 - (2 - 0.6)$ at.% Ni/100, corresponding to two vacancies for 0% Ni and 0.6 vacancy for 100% Ni. The alloy loses its transition-metal characteristics beginning at the composition for which the total number of d-vacancies equals the total number of donor electrons (one per copper atom), or at.% Ni $(2 - 0.014$ at.% Ni$) = 1\times$ at.% Cu. Setting at.% Cu $= (100 -$ at.% Ni$)$ and solving, the critical composition is found to be 41 at.% Ni. This value corresponds closely to the observed value derived from magnetic saturation data.

This model was checked by alloying small amounts of other nontransition elements Y, or transition elements Z, with nickel–copper alloys and noting the specific compositions at which $i_{critical}$ and $i_{passive}$ merged or at which Flade potentials disappeared. Non-transition-metal additions of valence >1 should shift the critical composition for passivity to higher percentages of nickel, whereas transition-metal additions should have the opposite effect. For example, one zinc atom of valence 2 or one aluminum atom of valence 3 should be equivalent in the solid solution alloy to two or three copper atoms, respectively. This has been confirmed experimentally [47]. The relevant equations become

$$\text{at.\% Ni } (2 - 0.014 \text{ at.\% Ni}) = 1 \times \text{ at.\% Cu} + n \text{ at.\% Y}$$

and

$$\text{at.\% Ni } (2 - 0.014 \text{ at.\% Ni}) + v \text{ at.\% Z} = 1 \times \text{at.\% Cu}$$

where n is the number of electrons donated per atom of Y and v is the number of vacancies introduced per atom of Z.

By plotting $1\times$ at.% Cu$-$ at.% Ni $(2 - 0.014$ at.% Ni$)$ with n at.% Y, a straight line is predicted of unit negative slope for nontransition-element additions. If plotted instead with v at.% Z for transition-metal additions, a unit positive slope should result. A plot for the alloying additions so far studied is shown by data summarized in Fig. 6.16 [46]. In order for the line to pass through the origin with unit slope, it was necessary to assume approximately 80% instead of 100% donation of valence electrons. This means that valence electrons from copper and

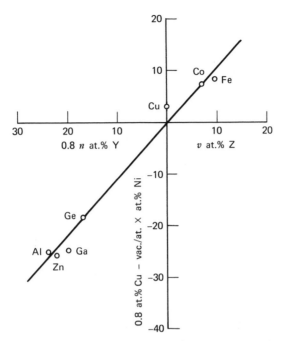

Figure 6.16. Plot of excess electrons or d-electron vacancies in nickel–copper alloys at their critical compositions versus electron vacancies, or electrons donated by alloying additions, unit slope.

from other nontransition elements are presumably donated in major part to nickel, but not entirely. Assuming 0.8 electron donor per copper atom in binary nickel–copper alloys, the critical nickel composition below which the d-band is filled becomes 35 at.% instead of 41 at.% as calculated earlier.* This value is consistent with the composition at which $i_{passive}$ and $i_{critical}$ intersect in Fig. 6.14.

Values of n for germanium, aluminum, and zinc shown in Fig. 6.16 are 4, 3, and 2, respectively, in accord with their normal valence. For gallium, a valence of 2 is in better accord with the other data, reflecting perhaps the known tendency of gallium to form chemical compounds having a valence lower than 3, but other explanations are also possible. For iron and cobalt, the number of vacancies per atom is equated to $v_g - (v_g - v_s)$ at.% $Z/100$, where v_g and v_s are d-electron vacancies per atom in the gas and solid, respectively. Values of v_g for iron and cobalt are 4 and 3, and for v_s they are 2.2 and 1.7, respectively.

There are no observed phase changes or major discontinuities in the thermodynamic properties of nickel–copper alloys at 60–70% Cu, whereas chemisorption on any metal is known to be favored by an unfilled d-band configuration [48]. The good agreement, therefore, between observed and predicted critical

*The calculated number of d vacancies per Ni atom at this composition equals $2 - 0.014 \times 35 = 1.51$, which is close to the value 1.6 assumed earlier [33, 39].

alloy compositions supports not only an effect of electron configuration on passivity, but also an adsorbed structure of the passive film.

6.8.2 Other Alloys

Because present-day theory of the metallic state does not treat the situation, the electron configuration of alloys made up of two or more transition metals with relation to their passive behavior is not as well understood as for the copper–nickel system. Nevertheless, useful simplifying assumptions can be made. For example, the most passive component of an alloy is assumed to be the acceptor element, which tends to share electrons donated by the less passive components.

Accordingly, for stainless steels, the d-electron vacancies of chromium are assumed to fill with electrons from alloyed iron [41]. At the critical composition at which vacancies of Cr are apparently filled, which occurs for alloys containing less than 12% Cr, the corrosion behavior of the alloy is like that of iron. Above 12% Cr, the d-electron vacancies of chromium are unfilled, and the alloy behaves more like chromium.

The critical compositions for passivity in the Cr–Ni and Cr–Co alloys, equal to 14% Cr and 8% Cr, respectively, can also be related to the contribution of electrons from nickel or cobalt to the unfilled d-band of chromium [49]. In the ternary Cr–Ni–Fe solid solution system, electrons are donated to chromium mostly by nickel above 50% Ni, but by iron at lower nickel compositions [50]. Similarly, molybdenum alloys retain in large part the useful corrosion resistance of molybdenum (e.g., to chlorides) so long as the d-band of energy levels for molybdenum remains unfilled. In Type 316 stainless steel (18% Cr, 10% Ni, 2–3% Mo), for example, the weight ratio of Mo/Ni is best maintained at or above 15/85, corresponding to the observed critical ratio for passivity in the binary molybdenum–nickel alloys equal to 15 wt.% Mo [51]. At this ratio or above, passive properties imparted by molybdenum appear to be optimum.

6.9 EFFECT OF CATHODIC POLARIZATION AND CATALYSIS

When chromium, stainless steels, and passive iron are cathodically polarized, passivity is destroyed. This occurs by reduction of the passive oxide or adsorbed oxygen film, according to whichever viewpoint of passivity is adopted. In addition, according to the adsorption theory, hydrogen ions discharged on transition metals tend to enter the metal as hydrogen atoms. Hydrogen so dissolved is partly dissociated into protons and electrons, the latter being available to fill vacancies in the d-band of the metal. A transition metal containing sufficient hydrogen, therefore, is no longer able to chemisorb oxygen or become passive, because the d-band is filled.

Often, catalytic properties also depend on ability of a metal or alloy to chemisorb certain components of its environment. It is not surprising, therefore, that

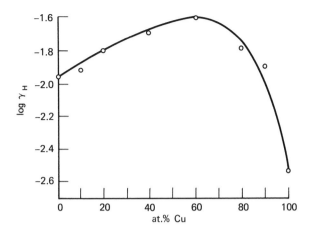

Figure 6.17. Catalytic efficiency of nickel–copper alloys for $2H \rightarrow H_2$ as a function of alloy composition (efficiency increases with values of γ_H) [53]. (Reproduced with permission of The Royal Society of Chemistry.)

transition metals are typically good catalysts and that the principles of electron configuration in alloys favoring catalytic activity are similar to those favoring passivity. For example, when palladium, which contains 0.6 d-electron vacancy per atom in the metallic state, is charged with hydrogen cathodically, it loses its efficiency as a catalyst for the ortho–para hydrogen conversion [52]. This behavior is explained by filling of the d-band by electrons of dissolved hydrogen, as a result of which the metal no longer chemisorbs hydrogen. Correspondingly, the catalytic efficiencies of palladium–gold alloys resemble palladium until the critical concentration of 60 at.% Au is reached, at which composition and above the alloys become poor catalysts. Gold, a non-transition metal, supplies electrons to the unfilled d-band of palladium, and magnetic measurements confirm that the d-band is just filled at the critical gold concentration.

The catalytic effect of copper–nickel alloys as a function of composition for the reaction $2H \rightarrow H_2$ is shown in Fig. 6.17 [53]. Above 60 at.% Cu, the filled d-band is less favorable to hydrogen adsorption; hence, favorable collisions of gaseous H with adsorbed H are less probable, and the reaction rate decreases. The similarity to passive behavior of copper–nickel alloys, which also decreases above 60 at.% Cu, can be noted. The parallel conditions affecting passivity and catalytic activity support the viewpoint that the passive films on transition metals and their alloys are chemisorbed.

REFERENCES

1. H. Uhlig, History of passivity, experiments and theories, *Passivity of Metals, Proceedings, 4th Symposium on Passivity*, R. Frankenthal and J. Kruger, editors, Electrochemical Society, Princeton, NJ, 1978, p. 1.

2. C. Schönbein, *Pogg. Ann.* **37**, 390 (1836).
3. M. Faraday, *Experimental Researches in Electricity*, Vol. II, Dover, New York, reprinted 1965.
4. C. Wagner, *Corros. Sci.* **5**, 751 (1965).
5. K. Bonhoeffer, *Z. Metallk.* **44**, 77 (1953).
6. F. Flade, *Z. Physik. Chem.* **76**, 513 (1911).
7. U. F. Franck, *Z. Naturforschung.* **4A**, 378 (1949).
8. P. King and H. Uhlig, *J. Phys. Chem.* **63**, 2026 (1959)
9. H. Rocha and G. Lennartz, *Arch. Eisenhüttenw.* **26**, 117 (1955).
10. G. Cartledge and R. Simpson, *J. Phys. Chem.* **61**, 973 (1957).
11. H. Uhlig and P. King, *J. Electrochem. Soc.* **106**, 1 (1959).
12. R. M. Burns, *J. Appl. Phys.* **8**, 398 (1937).
13. H. Gatos and H. Uhlig, *J. Electrochem. Soc.* **99**, 250 (1952).
14. H. Uhlig and T. O'Connor, *J. Electrochem. Soc.* **102**, 562 (1955).
15. N. Tomashov, *Corrosion* **14**, 229t (1958).
16. M. Stern and H. Wissenberg, *J. Electrochem. Soc.* **106**, 759 (1959).
17. R. Frankenthal, *J. Electrochem. Soc.* **114**, 542 (1967).
18. R. W. Revie, B. G. Baker, and J. O'M. Bockris, *J. Electrochem. Soc.* **122**, 1460 (1975); R. W. Revie, J. O'M. Bockris, and B. G. Baker, *Surface Science* **52**, 664 (1975).
19. M. J. Graham, *Corrosion* **59**(6), 475 (2003).
20. H. Uhlig, *Corros. Sci.* **7**, 325 (1967).
21. H. H. Uhlig, *Z. Elektrochem.* **62**, 626 (1958).
22. H. Uhlig and S. Lord, *J. Electrochem. Soc.* **100**, 216 (1953).
23. A. MacRae, *Surface Sci.* **1**, 319 (1964).
24. G. Okamoto and T. Shibata, *Nature* **206**, 1350 (1965).
25. H. Leckie and H. Uhlig, *J. Electrochem. Soc.* **113**, 1262 (1966).
26. H. Uhlig and J. Gilman, *Corrosion* **20**, 289t (1964).
27. H. Leckie, *J. Electrochem. Soc.* **117**, 1152 (1970).
28. M. A. Streicher, Austenitic and Ferritic Stainless Steels, in *Uhlig's Corrosion Handbook*, 2nd edition, R. W. Revie, editor, Wiley, New York, 2000, p. 620; *Mater. Perform.* **36**, 65 (1997).
29. H. Kaesche, *Z. Physik. Chem. N. F.* (F.R.D.) **34**, 87 (1962); H. Böhni and H. Uhlig, *J. Electrochem. Soc.* **116**, 906 (1969).
30. J. Horvath and H. Uhlig, *J. Electrochem. Soc.* **115**, 791 (1968).
31. Y. Kolotyrkin, *Corrosion* **19**, 261t (1963).
32. N. Greene, C. Bishop, and M. Stern, *J. Electrochem. Soc.* **108**, 836 (1961).
33. C. Hall, Jr., and N. Hackerman, *J. Phys. Chem.* **57**, 262 (1953); F. Poser and E. Bohlmann, *Second Symposium on Fresh Water from the Sea*, Athens, Greece, May, 1967.
34. T. Hoar, D. Mears, and G. Rothwell, *Corros. Sci.* **5**, 279 (1965).
35. H. Böhni, Localized Corrosion of Passive Metals, in *Uhlig's Corrosion Handbook*, 2nd edition, R. W. Revie, editor, Wiley, New York, 2000, pp. 175–176.
36. G. Tammann, *Z. Anorg. Allgem. Chem.* **169**, 151 (1928); (with E. Sotter), *ibid.* **127**, 257 (1923).

37. H. Uhlig, *Trans. Electrochem. Soc.* **85**, 307 (1944).
38. H. Feller and H. Uhlig, *J. Electrochem. Soc.* **107**, 864 (1960).
39. H. H. Uhlig, *J. Electrochem. Soc.* **108**, 327 (1961).
40. H. Leckie, *Corrosion* **24**, 70 (1968).
41. H. Uhlig, *Z. Elektrochem.* **62**, 700 (1958).
42. J. Osterwald and H. Uhlig, *J. Electrochem. Soc.* **108**, 515 (1961).
43. F. Mansfeld and H. Uhlig, *J. Electrochem. Soc.* **117**, 427 (1970).
44. F. Mansfeld and H. Uhlig, *Corros. Sci.* **9**, 377 (1969).
45. C. Kittel, *Introduction to Solid State Physics* 3rd edition, Wiley, New York, 1968, p. 581.
46. H. H. Uhlig, *Electrochim. Acta* **16**, 1939 (1971).
47. F. Mansfeld and H. Uhlig, *J. Electrochem. Soc.* **115**, 900 (1968).
48. D. Hayward and B. Trapnell, *Chemisorption*, Butterworths, Washington, D.C., 1964, p. 8.
49. A. Bond and H. Uhlig, *J. Electrochem. Soc.* **107**, 488 (1960).
50. H. Feller and H. Uhlig, *J. Electrochem. Soc.* **107**, 864 (1960).
51. H. Uhlig, P. Bond, and H. Feller, *J. Electrochem. Soc.* **110**, 650 (1963).
52. A. Couper and D. Eley, *Discussions Faraday Soc.* **8**, 172 (1950).
53. W. Hardy and J. Linnett, *Trans. Faraday Soc.* **66**, 447 (1970).

GENERAL REFERENCES

H. Böhni, Localized corrosion of passive metals, in *Uhlig's Corrosion Handbook*, 2nd edition, R. W. Revie, editor, Wiley, New York, 2000, pp. 173–190.

G. S. Frankel, Pitting corrosion, in *ASM Handbook*, Vol. 13A, *Corrosion: Fundamentals, Testing, and Protection*, ASM International, Materials Park, OH, 2003, pp. 236–241.

J. Kruger, Passivity, in *Uhlig's Corrosion Handbook*, 2nd edition, R. W. Revie, editor, Wiley, New York, 2000, pp. 165–171.

J. Kruger, Passivity, in *ASM Handbook*, Vol. 13A, *Corrosion: Fundamentals, Testing, and Protection*, ASM International, Materials Park, OH, 2003, pp. 61–67.

Passivation of Metals and Semiconductors, and Properties Of Thin Oxide Layers, A Selection of Papers from the 9th International Symposium, Paris, France, 27 June–1 July 2005, P. Marcus and V. Maurice, editors, Elsevier, Amsterdam, 2006.

Passivity and Localized Corrosion, An International Symposium in Honor of Professor Norio Sato, M. Seo, B. MacDougall, H. Takahashi, and R. G. Kelly, editors, Electrochemical Society, Pennington, NJ, 1999.

Passivity of Metals and Semiconductors, Proceedings of the Eighth International Symposium, May 9–15, 1999, Jasper, Alberta, Canada, M. B. Ives, J. L. Luo, and J. R. Rodda, editors, Electrochemical Society, Pennington, NJ, 2001.

Proceedings of the Symposium on Passivity and Its Breakdown, P. M. Natishan, H. S. Isaacs, M. Janik-Czachor, V. A. Macagno, P. Marcus, and M. Seo, editors, Electrochemical Society, Pennington, NJ, 1997.

PROBLEMS

1. Calculate the minimum concentration of oxygen in milliliters per liter necessary to passivate iron in 3% Na_2SO_4; also for 12% Cr–Fe alloy. (D for O_2 at $25°C = 2 \times 10^{-5} cm^2/s$) (*Hint*: Equate limiting diffusion current for reduction of oxygen to critical current density for passivity.)

2. Calculate the standard Flade potential at 25°C for iron, assuming that the passive-film substance represented schematically as O·M in Eq. (6.2) in Section 6.2 is (*a*) Fe_2O_3, (*b*) Fe_3O_4, and (*c*) chemisorbed oxygen. (*Data*: $\Delta G°$ of formation for $Fe_2O_3 = -741.0 kJ/mole$ Fe_2O_3; for Fe_3O_4, $\Delta G° = -1014 kJ/$ mole Fe_3O_4; for H_2O (l), $\Delta G° = -237.2 kJ/mole$ H_2O; and for H^+, $\Delta G° = 0$. For chemisorption of oxygen on Fe, $\Delta H° = -314 kJ/mole$ O_2; $\Delta S° = -193 J/$ mole O_2-°C (estimated from $\Delta S°$ value for chemisorbed N_2.).

3. Calculate the absolute emf of the cell:

$$\text{stainless steel; } O_2(0.2\,atm), 0.001M\ ZnSO_4(pH = 3.0); Zn$$

assuming that passive stainless steel acts as a reversible oxygen electrode. What is the change of emf per 1 atm O_2 pressure?

4. From the standard Flade potential for iron, calculate the apparent free energy of formation of the passive film per gram-atom of oxygen. Do the same for nickel and chromium.

5. A pit in 18–8 stainless steel exposed to seawater grows to a depth of 6.5 cm in one year. To what average current density at the base of the pit does this rate correspond?

6. The steady-state passive current density for iron in $1N$ H_2SO_4 is $7\,\mu A/cm^2$. How many atom layers of iron are removed from a smooth electrode surface every minute?

7. Derive the relation $\log(Cl^-) = k\ \log(\text{anion}) + \text{const}$, where (anion) is the minimum activity of anion necessary to inhibit pitting of a passive metal in a chloride solution of activity (Cl^-). Assume that the amount of ion a adsorbed per unit area follows the Freundlich adsorption isotherm, $a = k_1(\text{anion})^{1/n_1}$, where k_1 and n_1 are constants, and that at a critical ratio of adsorbed Cl^- to adsorbed anion the passive film is displaced by Cl^-, allowing a pit to initiate.

Answers to Problems

1. 31,800; 0.32 mL/liter.
2. (*a*) −0.051 V; (*b*) −0.085 V; (*c*) 0.57 V.
3. 1.90 V; 0.032 V/atm at 0.2 atm O_2 pressure.
4. For iron, −115 kJ/g-atom O.
5. 6×10^{-3} A/cm^2.
6. 0.5.

7

IRON AND STEEL

7.1 INTRODUCTION

The electrochemical theory of corrosion, as described earlier, relates corrosion to a network of short-circuited galvanic cells on the metal surface. Metal ions go into solution at anodic areas in an amount chemically equivalent to the reaction at cathodic areas. At anodic areas, the following reaction takes place:

$$Fe \rightarrow Fe^{2+} + 2e^- \tag{7.1}$$

This reaction is rapid in most media, as shown by lack of pronounced polarization when iron is made anode employing an external current. When iron corrodes, the rate is usually controlled by the cathodic reaction, which, in general, is much slower (cathodic control). In deaerated solutions, the cathodic reaction is

$$H^+ \rightarrow \frac{1}{2}H_2 - e^- \tag{7.2}$$

This reaction proceeds rapidly in acids; as pH is reduced, the equilibrium potential of the hydrogen electrode becomes more noble; hence, $\phi_C - \phi_A$ becomes larger, and the resulting corrosion current increases. On the other hand, this reaction proceeds slowly in alkaline or neutral aqueous media. The corrosion

Corrosion and Corrosion Control, by R. Winston Revie and Herbert H. Uhlig
Copyright © 2008 John Wiley & Sons, Inc.

rate of iron in deaerated water at room temperature, for example, is less than 0.005 mm/y (0.1 gmd). The rate of hydrogen evolution at a specific pH depends on the presence or absence of low hydrogen overpotential impurities in the metal. For pure iron, the metal surface itself provides sites for H_2 evolution; hence, high-purity iron continues to corrode in acids, but at a measurably lower rate than does commercial iron.

The cathodic reaction can be accelerated by dissolved oxygen in accord with the following reaction, a process called *depolarization*:

$$2H^+ + \frac{1}{2}O_2 \rightarrow H_2O - 2e^- \tag{7.3}$$

Dissolved oxygen reacts with hydrogen atoms adsorbed at random on the iron surface, independent of the presence or absence of impurities in the metal. The oxidation reaction proceeds as rapidly as oxygen reaches the metal surface.

Adding (7.1) and (7.3), using the reaction $H_2O \rightarrow H^+ + OH^-$, we obtain

$$Fe + H_2O + \frac{1}{2}O_2 \rightarrow Fe(OH)_2 \tag{7.4}$$

Hydrous ferrous oxide ($FeO \cdot nH_2O$) or ferrous hydroxide [$Fe(OH)_2$] composes the diffusion-barrier layer next to the iron surface through which O_2 must diffuse. The pH of saturated $Fe(OH)_2$ is about 9.5, so that the surface of iron corroding in aerated pure water is always alkaline. The color of $Fe(OH)_2$, although white when the substance is pure, is normally green to greenish black because of incipient oxidation by air. At the outer surface of the oxide film, access to dissolved oxygen converts ferrous oxide to hydrous ferric oxide or ferric hydroxide, in accord with

$$Fe(OH)_2 + \frac{1}{2}H_2O + \frac{1}{4}O_2 \rightarrow Fe(OH)_3 \tag{7.5}$$

Hydrous ferric oxide is orange to reddish brown in color and comprises most of ordinary rust. It exists as nonmagnetic αFe_2O_3 (hematite) or as magnetic γFe_2O_3, with the α form having the greater negative free energy of formation, that is, greater stability. Saturated $Fe(OH)_3$ is nearly neutral in pH. A magnetic hydrous ferrous ferrite, $Fe_3O_4 \cdot nH_2O$, often forms a black intermediate layer between hydrous Fe_2O_3 and FeO. Hence rust films normally consist of three layers of iron oxides in different states of oxidation.

7.2 AQUEOUS ENVIRONMENTS

7.2.1 Effect of Dissolved Oxygen

7.2.1.1 Air-Saturated Water. In neutral or near-neutral water at ambient temperatures, dissolved oxygen is necessary for appreciable corrosion of iron. In air-saturated water, the initial corrosion rate may reach a value of about 10 gmd. This rate diminishes over a period of days as the iron oxide (rust) film is formed and acts as a barrier to oxygen diffusion. The steady-state corrosion rate may be 1.0–2.5 gmd, tending to be higher the greater the relative motion of water with respect to iron. Since the diffusion rate at steady state is proportional to oxygen concentration, it follows from (7.3) that the corrosion rate of iron is also proportional to oxygen concentration. The concentration of dissolved oxygen in air-saturated, natural fresh waters at ordinary temperatures is 8–10 ppm. In the absence of dissolved oxygen, the corrosion rate at room temperature is negligible both for pure iron and for steel.

In the absence of diffusion-barrier films of corrosion products on the surface, the theoretical corrosion current density, i (A/cm^2), of steel in stagnant, air-saturated fresh water is [cf. Eq. (5.5), Section 5.4]

$$i = \left(\frac{DnF}{\delta}\right)c \times 10^{-3} \tag{7.6}$$

where D is the diffusion coefficient for dissolved oxygen in water (cm^2/s), δ is the thickness of the stagnant layer of electrolyte next to the electrode surface, c is the concentration of oxygen, and the other terms have their usual significance. Using $D = 2 \times 10^{-5}$ cm^2/s, $n = 4$ eq/mole, $F = 96{,}500$ C/eq, $\delta = 0.05$ cm, and $c = 8/32 \times 10^{-3}$ mole/liter at 25 °C, we obtain $i = 38.6 \times 10^{-6}$ A/cm^2, corresponding to a corrosion rate of 0.45 mm/y [1].

7.2.1.2 Higher Partial Pressures of Oxygen.

Although increase in oxygen concentration at first accelerates corrosion of iron, it is found that, beyond a critical concentration, the corrosion drops again to a low value [2]. In distilled water, the critical concentration of oxygen above which corrosion decreases again, is about 12 mL O$_2$/liter (Fig. 7.1). This value increases with dissolved salts and with temperature, and it decreases with an increase in velocity and pH. At pH of about 10, the critical oxygen concentration reaches the value for air-saturated water (6 mL O$_2$/liter) and is still less for more alkaline solutions.

The decrease of corrosion rate is caused by passivation of iron by oxygen, as shown by potentials of iron in air-saturated H$_2$O of –0.4 to –0.5 V versus S.H.E. and 0.1 to 0.4 V in oxygen-saturated H$_2$O (28 mL O$_2$/liter). Apparently, at higher partial pressures, more oxygen reaches the metal surface than can be reduced by the corrosion reaction—the excess, therefore, is available to form the passive film; increasing the cathodic reaction rate by higher oxygen concentration increases polarization of anodic areas until the critical current density for passivity is reached (see Fig. 6.1, Section 6.2) [3].

Because passivity accompanies higher oxygen pressures, passive–active cells are established in the event that passivity breaks down locally (e.g., at crevices). Such breakdown is accompanied by severe pitting, particularly at higher temperatures, in the presence of halide ions, or at a critical pressure of oxygen where passivity is on the verge of either forming or breaking down. This behavior limits the practical use of high partial oxygen pressures as a means of reducing

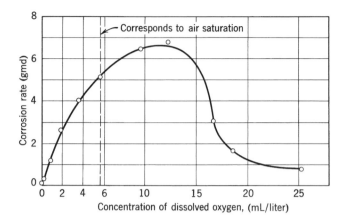

Figure 7.1. Effect of oxygen concentration on corrosion of mild steel in slowly moving distilled water, 48-h test, 25 °C [2]. (Reproduced with permission. Copyright 1955, The Electrochemical Society.)

corrosion of steel. In appreciable concentration of chlorides, as in seawater, passivity of iron is not established at all, and in such media increased oxygen pressure results in an increased corrosion rate.

7.2.1.3 Microbiologically Influenced Corrosion.
Microbiologically influenced corrosion (MIC) is corrosion that is caused by the presence and activities of microorganisms—that is, organisms that cannot be seen individually with the unaided human eye, including microalgae, bacteria, and fungi [4]. Microbiologically influenced corrosion can cause various forms of localized corrosion, including pitting, dealloying, enhanced erosion corrosion, enhanced galvanic corrosion, stress corrosion cracking, and hydrogen embrittlement [4]. As a result of MIC, corrosion can occur at locations where it would not be predicted, and it can occur at very high rates. All engineering alloys, with the exception of titanium and high chromium–nickel alloys, have been reported to undergo MIC. Furthermore, MIC has been documented to take place in seawater, fresh water, distilled/demineralized water, hydrocarbon fuels, process chemicals, foods, soils, human plasma, saliva, and sewage [4].

Although sulfate-reducing bacteria (SRBs), active only in anaerobic (oxygen-free) environments, are a very common cause of corrosion and have been extensively studied, MIC can also be caused by other types of microorganisms—for example, *Thiobacilli* [sulfur-oxidizing bacteria (SOB), which oxidize sulfur compounds to sulfuric acid] and other acid-producing microorganisms, including both bacteria and fungi.

Microbes can adhere to metal surfaces forming a biofilm, consisting of a community of microorganisms, leading to corrosion [4, 5]. When the acidic products of bacterial action are trapped at the biofilm–metal interface, their impact on corrosion is intensified [6].

Although iron does not corrode appreciably in deaerated water, the corrosion rate in some natural deaerated environments is found to be abnormally high. These high rates have been traced to the presence of sulfate-reducing bacteria (e.g., *Desulfovibrio desulfuricans*). Their relation to an observed accelerated corrosion rate in soils low in dissolved oxygen was first observed by von Wolzogen Kühr in Holland [7]. The bacteria are curved, measuring about $1 \times 4\,\mu m$, and are found in many waters and soils. They thrive only under anaerobic conditions in the pH range of about 5.5–8.5. Certain varieties multiply in fresh waters and in soils containing sulfates, others flourish in brackish waters and seawater, and still others are stated to exist in deep soils at temperatures as high as 60–80 °C (140–175 °F).

Sulfate-reducing bacteria easily reduce inorganic sulfates to sulfides in the presence of hydrogen or organic matter, and they are aided in this process by the presence of an iron surface. The aid that iron provides in this reduction is probably to supply hydrogen, which is normally adsorbed on the metal surface and which the bacteria use in reducing SO_4. For each equivalent of hydrogen atoms they consume, one equivalent of Fe^{2+} enters solution to form rust and FeS. The bacteria, therefore, probably act essentially as depolarizers.

A possible reaction sequence can be outlined as follows:

Anode: $4Fe \rightarrow 4Fe^{2+} + 8e^-$ $\qquad\qquad$ (7.7)

Cathode: $8H_2O + 8e^- \rightarrow 8H_{ads\ on\ Fe} + 8OH^-$ $\qquad\qquad$ (7.8)

$8H_{ads} + Na_2SO_4 \xrightarrow{\text{bacteria}} 4H_2O + Na_2S$ $\qquad\qquad$ (7.9)

$Na_2S + 2H_2CO_3 \rightarrow 2NaHCO_3 + H_2S$ $\qquad\qquad$ (7.10)

Summary: $4Fe + 2H_2O + Na_2SO_4 + 2H_2CO_3 \rightarrow 3Fe(OH)_2 + FeS + 2NaHCO_3$ $\qquad$ (7.11)

Ferrous hydroxide and ferrous sulfide are formed in the proportion of 3 moles to 1 mole. Analysis of a rust in which sulfate-reducing bacteria were active shows this approximate ratio of oxide to sulfide. Qualitatively, the action of sulfate-reducing bacteria as the cause of corrosion in a water initially free of sulfides can be detected by adding a few drops of hydrochloric acid to the rust and noting the smell of hydrogen sulfide.

In addition to the cathodic reaction listed in (7.8), other cathodic reactions could be considered [8, 9], such as the reduction of H_2S:

$$2H_2S + 2e^- \rightarrow 2HS^- + H_2 \tag{7.12}$$

Severe damage by sulfate-reducing bacteria has occurred particularly in oil-well casing, buried pipelines, water-cooled rolling mills, and pipe from deep water wells. Within 2 years, well water in the U.S. Midwest caused failure of a galvanized water pipe 50 mm (2 in.) in diameter by action of sulfate-reducing bacteria, whereas municipal water using similar wells, but which was chlorinated before-hand, was much less corrosive.

A combination of low temperature and low humidity is one approach to controlling the growth of bacteria, but fungi may be capable of growing under such conditions [10]. Regular cleaning is a good practice to prevent biofilm for-mation and subsequent corrosion. Chlorination is used to eliminate bacteria that cause corrosion, but this treatment can produce byproducts that are environmen-tally unacceptable [10]. Aeration of water reduces activity of anaerobic bacteria since they are unable to thrive in the presence of dissolved O_2. Addition of certain biocides can be beneficial, but microorganisms are capable of becoming resistant to specific chemicals after long-term use. Eradication of microbial popu-lations may be achieved by combining several chemicals or by increasing the concentration of a biocide [10].

7.2.2 Effect of Temperature

When corrosion is controlled by diffusion of oxygen, the corrosion rate at a given oxygen concentration approximately doubles for every 30 °C (55 °F) rise in tem-perature [11]. In an open vessel, allowing dissolved oxygen to escape, the rate increases with temperature to about 80 °C (175 °F) and then falls to a very low value at the boiling point (Fig. 7.2). The lower corrosion rate above 80 °C is related to a marked decrease of oxygen solubility in water as the temperature is raised, and this effect eventually overshadows the accelerating effect of tempera-ture alone. In a closed system, on the other hand, oxygen cannot escape, and the corrosion rate continues to increase with temperature until all the oxygen is consumed.

When corrosion is attended by hydrogen evolution, the rate increase is more than double for every 30 °C rise in temperature. The rate for iron corroding in hydrochloric acid, for example, approximately doubles for every 10 °C rise in temperature.

7.2.3 Effect of pH

The effect of pH of an aerated pure, or soft, water on corrosion of iron at room temperature is shown in Fig. 7.3 [12]; however, the effect of pH may be different in a hard water, in which a protective film of $CaCO_3$ forms on the metal surface,

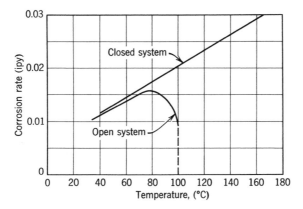

Figure 7.2. Effect of temperature on corrosion of iron in water containing dissolved oxygen. (Reprinted with permission from F. N. Speller, *Corrosion, Causes and Prevention*, 3rd edition, McGraw-Hill, New York, 1951, p. 168.)

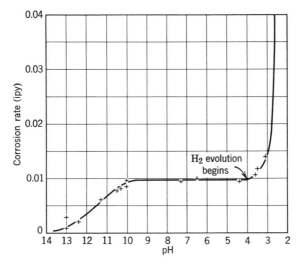

Figure 7.3. Effect of pH on corrosion of iron in aerated soft water, room temperature [12]. [Reprinted with permission from W. Whitman, R. Russell, and V. Altieri, *Ind. Eng. Chem.* **16**, 665 (1924). Copyright 1924, American Chemical Society].

as will be discussed in Section 7.2.6.1. In obtaining the data shown in Fig. 7.3, sodium hydroxide and hydrochloric acid were used to adjust pH in the alkaline and acid ranges.

Within the range of about pH 4–10, the corrosion rate is independent of pH and depends only on how rapidly oxygen diffuses to the metal surface. The major diffusion barrier of hydrous ferrous oxide is continuously renewed by the corrosion process. Regardless of the observed pH of water within this range, the surface of iron is always in contact with an alkaline solution of saturated hydrous ferrous oxide, the observed pH of which is about 9.5.*

Within the acid region, pH < 4, the ferrous oxide film is dissolved, the surface pH falls, and iron is more or less in direct contact with the aqueous environment. The increased rate of reaction is then the sum of both an appreciable rate of hydrogen evolution and oxygen depolarization.

Above pH 10, an increase in alkalinity of the environment raises the pH of the iron surface. The corrosion rate correspondingly decreases because iron becomes increasingly passive in the presence of alkalies and dissolved oxygen, as explained in Section 7.2.1.2. Confirming the occurrence of passivity by Definition 1 in Section 6.1, the potential of iron changes from an active value of −0.4 to −0.5 in water of pH < 10, to a noble value of 0.1 V in 1 N NaOH, with an accompanying decrease in the corrosion rate. If the alkalinity is markedly increased, for example, to 16 N NaOH (43%), passivity is disrupted, and the potential achieves the very active value of −0.9 V. The corrosion rate correspondingly increases slightly to 0.003–0.1 mm/y (0.05–2.0 gmd), but that is still a relatively low rate. In this region, iron corrodes with formation of soluble sodium ferrite ($NaFeO_2$). In the absence of dissolved oxygen, the reaction proceeds with hydrogen evolution forming sodium hypoferrite, Na_2FeO_2 [14]. The fact that Fe^{2+} is complexed by OH^- in strong alkalies to form FeO_2^- with accompanying reduction in activity of Fe^{2+} accounts for the observed active potential of iron. Although the rate of formation of FeO_2^- in concentrated alkalies at room temperature is low, caused by marked polarization of probably both anodic and cathodic areas, the rate becomes excessively high at boiler temperatures.

In the region of pH 4–10, the corrosion rate depends only on the rate of diffusion of oxygen to the available cathodic surface. The extent of the cathodic surface is apparently not important. This was established in an experiment of Whitman and Russell [15], who exposed steel specimens that were copper-plated over three-quarters of the surface to tap water of Cambridge, Massachusetts. The total weight loss of these specimens compared to control specimens not plated was found to be the same. All oxygen reaching the copper surface, acting as cathode, was reduced in accord with the reaction

$$H_2O + \frac{1}{2}O_2 \rightarrow 2OH^- - 2e^-$$

and an equivalent amount of Fe^{2+} was formed at the iron surface acting as anode.†
In addition, all oxygen reaching cathodic areas of the iron surface itself produced

---

*On the other hand, some investigators, using different natural waters and different chemicals to control pH, have observed that the corrosion rate does change with pH in the pH range 6–9. This behavior has been attributed to reduced buffer capacity of HCO_3^- (an inhibitor) as pH increases, causing a pH at local sites that is lower than the pH that would otherwise exist in a solution of saturated hydrous ferrous oxide. The reader may wish to consult the review by Matsushima on this matter [13].

†This equivalence is valid in waters of low conductivity only if anode and cathode areas are in close proximity. In seawater, anodes and cathodes may be several feet apart.

an equivalent amount of Fe^{2+}. Thus, the total amount of iron corroding was the same regardless of whether copper was plated over part of the specimens. However, penetration of iron in the case of the plated specimens was four times that of the bare specimens.

Therefore, it follows that so long as oxygen diffusion through the oxide layer is controlling, which is the case within pH 4–10, any small variation in composition of a steel and its heat treatment, or whether it is cold worked or annealed, has no bearing on corrosion behavior, provided that the diffusion-barrier layer remains essentially unchanged. Oxygen concentration, temperature, and velocity of the water alone determine the reaction rate. These facts are important because almost all natural waters fall within the pH range 4–10. This means that whether a high- or low-carbon steel, low-alloy steel (e.g., 1–2% Ni, Mn, Mo, etc.), wrought iron, cast iron, or cold-rolled mild steel is exposed to fresh water or seawater, all the observed corrosion rates in a given environment are essentially the same. Many laboratory and service data obtained with a variety of irons and steels support the validity of this conclusion [16]. A few typical data are summarized in Table 7.1. These observations answer the once vociferous argument that wrought iron, for example, is supposedly more corrosion resistant than steel.

In the acid range, pH < 4, oxygen is not controlling, and the corrosion reaction is established, in part, by the rate of hydrogen evolution. The latter, in turn, depends on the hydrogen overpotential of various impurities or phases present in specific steels or irons. The rate becomes sufficiently high in this pH range to make anodic polarization a possible contributing factor (i.e., mixed control). Because cementite, Fe_3C, is a phase of low hydrogen overpotential, a low-carbon steel corrodes in acids at a lower rate than does a high-carbon steel. Similarly, heat treatment affecting the presence and size of cementite particles can appreciably alter the corrosion rate. Furthermore, cold-rolled steel corrodes more rapidly in acids than does an annealed or stress-relieved steel because cold working produces finely dispersed low-overpotential areas originating largely from interstitial nitrogen and carbon.

Since iron is not commonly used in strongly acid environments, the factors governing its corrosion in media of low pH are less important than those in natural waters and soils. Nevertheless, there are certain applications where such factors must be considered—for example, in steam-return lines containing carbonic acid, as well as in food cans where fruit and vegetable acids corrode the container with accompanying hydrogen evolution.

Less is known about the effect of impurities and metallurgical factors on the corrosion rate in the very alkaline region (pH ~14) where corrosion is again accompanied by hydrogen evolution. In the passive region (pH ~10–13), the effect on passivity by impurities in their usual concentrations, as well as the effect of metallurgical factors, is not expected to be pronounced. In general, any condition producing a large cathode-to-anode area ratio facilitates achievement of passivity and increases the stability of the passive state once it is achieved.

TABLE 7.1. Corrosion Rates of Various Steels when Oxygen Diffusion Is Controlling

% Carbon	Treatment	Environment	Temperature of Test	Corrosion Rate	
				ipy	mm/y
Effect of Heat Treatment					
0.39	Cold drawn, annealed 500 °C	Distilled water	65 °C	0.0036	0.091
0.39	Normalized 20 min at 900 °C	Distilled water	65 °C	0.0034	0.086
0.39	Quenched 850 °C, various specimens tempered at 300–800 °C	Distilled water	65 °C	0.0033	0.084
Effect of Carbon					
0.05	Not stated	3% NaCl	Room	0.0014	0.036
0.11	Not stated	3% NaCl	Room	0.0015	0.038
0.32	Not stated	3% NaCl	Room	0.0016	0.041
Effect of Alloying					
0.13	Not stated	Seawater		0.004	0.10
0.10, 0.34% Cu	Not stated	Seawater		0.005	0.13
0.06, 2.2% Ni	Not stated	Seawater		0.005	0.13
Wrought iron	Not stated	Seawater		0.005	0.13

Source: F. N. Speller, *Corrosion, Causes and Prevention*, McGraw-Hill, New York, 1951.

7.2.3.1 Corrosion of Iron in Acids.

We have seen that, in strong acids, such as hydrochoric and sulfuric acids, the diffusion-barrier oxide film on the surface of iron is dissolved below pH 4. In weaker acids, such as acetic or carbonic acids, dissolution of the oxide occurs at a higher pH; hence, the corrosion rate of iron increases accompanied by hydrogen evolution at pH 5 or 6. This difference is explained [12] by the higher total acidity or neutralizing capacity of a partially dissociated acid compared with a totally dissociated acid at a given pH. In other words, at a given pH, there is more available H^+ to react with and dissolve the barrier oxide film using a weak acid compared to a strong acid.

The increased corrosion rate of iron as pH decreases is not caused by increased hydrogen evolution alone; in fact, greater accessibility of oxygen to the metal surface on dissolution of the surface oxide favors oxygen depolarization, which is often the more important reason. The sensitivity of the corrosion rate of iron or steel in nonoxidizng acids to dissolved oxygen concentration is shown

TABLE 7.2. Effect of Dissolved Oxygen on Corrosion of Mild Steel in Acids [17]

Acid	Corrosion Rate (mm/y)		Ratio
	Under O_2	Under H_2	
6% acetic	13.9	0.16	87
6% H_2SO_4	9.1	0.79	12
4% HCl	12.2	0.79	15
0.04% HCl	9.9	0.14	71
1.2% HNO_3	46	40	1.2

by data of Table 7.2. In 6% acetic acid at room temperature, the ratio of corrosion rate with oxygen present to corrosion rate with oxygen absent is 87. In oxidizing acids, such as nitric acid, which act as depolarizers and for which the corrosion rate is, therefore, independent of dissolved oxygen, the ratio is almost unity. In general, the ratios are larger the more dilute the acid. In more concentrated acids, the rate of hydrogen evolution is so pronounced that oxygen has difficulty in reaching the metal surface. Hence, depolarization in more concentrated acids contributes less to the overall corrosion rate than in dilute acids, in which diffusion of oxygen is impeded to a lesser extent.

Traces of oxygen in dilute H_2SO_4—or substantial amounts in more concentrated acid, in which the corrosion rate is higher—inhibit the corrosion reaction. For zone-refined iron, the average corrosion rate in aerated $1.0 N$ H_2SO_4 at 25 °C was found to be 41.5 gmd, whereas in hydrogen-saturated acid the rate was 68.0 gmd [18]. Similar effects are shown by pure 9.2% Co–Fe alloy in $1.0 N H_2SO_4$, both aerated and deaerated, for which the corrosion rates are high and the diffusion of oxygen is impeded by rapid hydrogen evolution (Fig. 7.4) [18]. Potential and polarization measurements indicate that oxygen in small concentrations at the metal surface increases cathodic polarization, thereby decreasing corrosion; in higher concentrations, oxygen acts mainly as a depolarizer, increasing the rate.

The important depolarizing action of dissolved oxygen suggests that the velocity of an acid should markedly affect the corrosion rate. This effect is observed, particularly with dilute acids, for reasons previously stated (Fig. 7.5). In addition, the inhibiting effect of dissolved oxygen is observed within a critical velocity range, with the critical velocity becoming higher the more rapid the inherent reaction rate of steel with the acid. Relative motion of acid with respect to metal sweeps away hydrogen bubbles and reduces the thickness of the stagnant liquid layer at the metal surface, allowing more oxygen to reach the metal surface. Accordingly, the corrosion rate of steel in the presence of air in $0.0043 N$ H_2SO_4 at a velocity of 3.7 m/s (12 ft/s) is the same as that in $5 N$ acid at the same velocity. At rest, the ratio of corrosion rates is about 12 [19]. In the absence of dissolved oxygen, only hydrogen evolution occurs at cathodic areas, and an effect of velocity is no longer observed (Fig. 7.5) [20]. This result is expected because hydrogen overpotential (activation polarization) is insensitive to velocity of the electrolyte.

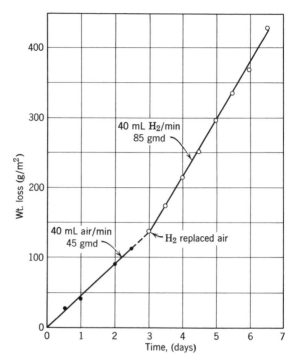

Figure 7.4. Corrosion of 9.2% Co–Fe alloy in 1 N H₂SO₄, showing an inhibiting effect of dissolved oxygen, 25 °C [18].

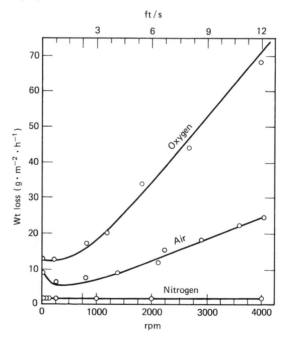

Figure 7.5. Effect of velocity on corrosion of mild steel (0.12% C) in 0.33 N sulfuric acid under air and oxygen [19], 23 ± 2 °C, 45-min test, rotating spec., 18 mm diameter, 56 mm long; and 0.009% C steel under nitrogen [20].

For aerated acid, the minimum rate occurs at higher velocities the more concentrated the acid because the rate of hydrogen evolution is more pronounced, thereby impeding oxygen diffusion to the metal surface. Similarly, the minimum shifts to higher velocities the higher the carbon content of the steel because the corrosion rate and the accompanying hydrogen evolution both increase with carbon content.

7.2.4 Effect of Galvanic Coupling

The classic experiment of Whitman and Russell, which showed that weight loss of iron coupled to copper is the same as if the entire surface had been iron, also showed that the actual penetration of iron increases when iron is coupled to a more noble metal. This experiment, therefore, provides information about the effect of coupling on the corrosion rate of the less noble component of a couple. For the situation where diffusion of a depolarizer is controlling, the general relation between penetration, p (proportional to corrosion rate), of a metal having area A_a coupled to a more noble metal of area A_c, is given by

$$p = p_0\left(1 + \frac{A_c}{A_a}\right) \qquad (7.13)$$

where p_0 is the normal penetration of the uncoupled metal.

If the ratio of areas A_c/A_a is large, the increased corrosion caused by coupling can be considerable. Conductivity of the electrolyte and geometry of the system enter the problem because only that part of the cathode area is effective for which resistance between anode and cathode is not a controlling factor. In soft tap water, the critical distance between copper and iron may be 5 mm; in seawater, it may be several decimeters. The critical distance is greater the larger the potential difference between anode and cathode. All more noble metals accelerate corrosion similarly, except when a surface film (e.g., on lead) acts as a barrier to diffusion of oxygen or when the metal is a poor catalyst for reduction of oxygen.

In the case of coupled metals exposed to deaerated solution for which corrosion is accompanied by hydrogen evolution, increased area of the more noble metal also increases corrosion of the less noble metal. Figure 7.6 shows polarization curves for an anode that polarizes little in comparison to a cathode at which hydrogen is evolved (cathodic control). Slope 1 represents polarization of a noble metal area having high hydrogen overpotential. Slopes 2 and 3 represent metals with lower hydrogen overpotential. The corresponding galvanic currents are given by projecting the intersection of anode–cathode polarization curves to the log I axis. In general, any metal on which hydrogen discharges acts as a hydrogen electrode with an equilibrium potential at 1 atm hydrogen pressure of -0.059pH volt. When a corroding metal is coupled to a more noble metal of variable area, the situation is shown in Fig. 7.7, where log current density is plotted instead of log total current. If the anode of area A_a is coupled to the more noble

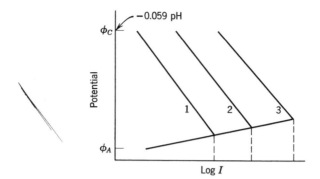

Figure 7.6. Effect of hydrogen overpotential of cathode on galvanic corrosion in deaerated nonoxidizing acids.

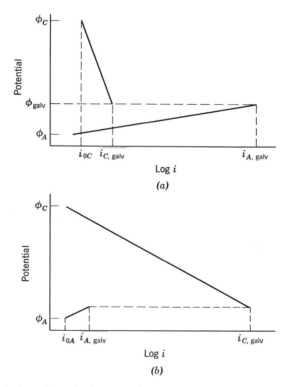

Figure 7.7. Effect of anode–cathode area ratio on corrosion of galvanic couples in deaerated nonoxidizing acids. (a) Large cathode coupled to small anode. (b) Large anode coupled to small cathode.

metal of area A_c, then the galvanic current density at the anode produced by coupling can be shown to be

$$\log i_{A,galv} = \frac{-\phi_{galv} - 0.059\text{pH}}{\beta} + \log \frac{A_c}{A_a} i_0 \tag{7.14}$$

where ϕ_{galv} is the corrosion potential (S.H.E.) of the galvanic couple (measured at a large distance compared to dimensions of the couple), β and i_0 are the Tafel constant and exchange current density, respectively, for hydrogen ion discharge on the noble metal, and $\dfrac{I_{galv}}{A_a} = i_{A,galv}$.

7.2.5 Effect of Velocity on Corrosion in Natural Waters

In natural fresh water, the pH is usually too high for hydrogen evolution to play an important role, and relative motion of the water at first increases the corrosion rate by bringing more oxygen to the surface. At sufficiently high velocities, enough oxygen may reach the surface to cause partial passivity. If this happens, the rate decreases again after the initial increase (Fig. 7.8). Should the velocity increase still further, the mechanical erosion of passive or corrosion-product films again increases the rate. The maximum corrosion rate preceding passivity occurs at a velocity that varies with smoothness of the metal surface and with impurities in the water. In the presence of high concentration of Cl⁻, as in seawater, passivity is not established at any velocity, and the corrosion rate increases without a

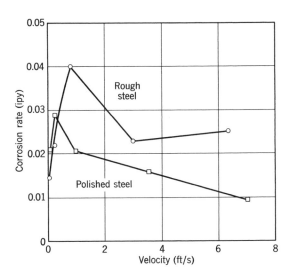

Figure 7.8. Effect of velocity on corrosion of mild steel tubes containing Cambridge water, 21°C, 48-h tests [21]. [Reprinted with permission from R. Russell et al., *Ind. Eng. Chem.* **19**, 65 (1927). Copyright 1927, American Chemical Society.]

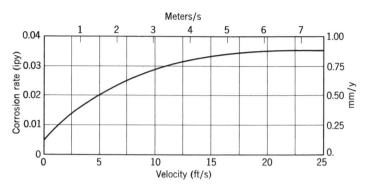

Figure 7.9. Effect of velocity on corrosion of steel in seawater [22].

decrease at any intermediate velocity (Fig. 7.9). The same behavior is expected at elevated temperatures that preclude the possibility of passivity by dissolved oxygen.

7.2.5.1 Cavitation–Erosion. If conditions of velocity are such that repetitive low-pressure (below atmospheric) and high-pressure areas are developed, bubbles form and collapse at the metal–liquid interface. This phenomenon is called cavitation. The damage to a metal caused by cavitation is called cavitation–erosion, or cavitation damage. The damage can be reproduced in the laboratory using metal probes undergoing forced high-frequency mechanical oscillations normal to the metal surface. The surface becomes deeply pitted and appears to be spongy (Fig. 7.10), or like a honeycomb-like structure [24]. Damage from this cause may be purely mechanical, as is experienced with glass or plastics or when damage to a metal occurs in organic liquids. The damage, however, may involve both chemical and mechanical factors, particularly if protective films are destroyed, allowing corrosion to proceed at a higher rate. The part played by chemical factors is apparent, for example, from increased metal loss in laboratory tests using seawater compared to fresh water.

Cavitation–erosion occurs typically on rotors of pumps, on the trailing faces of propellers and of water-turbine blades, and on the water-cooled side of diesel-engine cylinders [25]. Cavitation attack is usually controlled by design and materials selection [26]. Damage can be reduced by operating rotary pumps at the highest possible head of pressure in order to avoid bubble formation. For turbine blades, aeration of water cushions the damage caused by collapse of bubbles. There is a wide range of polymers with good resistance to cavitation–erosion and excellent resistance to corrosion—for example, high-density polyethylene has a cavitation–erosion resistance similar to that of nickel-based and titanium alloys [26, 27]. Since 18–8 stainless steel is one of the relatively resistant alloys, it is used as a facing for water-turbine blades. To reduce damage of diesel-engine cylinder liners, addition of inhibitor to the cooling water has proved effective [28].

Figure 7.10. Cavitation–erosion damage to cylinder liner of a diesel engine [23]. (Copyright NACE International, 1950.)

7.2.6 Effect of Dissolved Salts

The effect of sodium chloride concentration on corrosion of iron in air-saturated water at room temperature is shown in Fig. 7.11. The corrosion rate first increases with salt concentration and then decreases, with the value falling below that for distilled water when saturation is reached (26% NaCl).

Since oxygen depolarization controls the rate throughout the sodium chloride concentration range, it is important to understand why the rate first increases, reaching a maximum at about 3% NaCl (seawater concentration), and then decreases. Oxygen solubility in water decreases continuously with sodium chloride concentration, explaining the lower corrosion rates at the higher sodium chloride concentrations. The initial rise appears to be related to a change in the protective nature of the diffusion-barrier rust film that forms on corroding iron. In distilled water having low conductivity, anodes and cathodes must be located

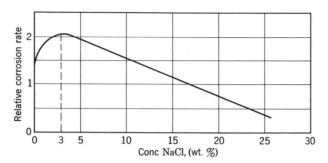

Figure 7.11. Effect of sodium chloride concentration on corrosion of iron in aerated solutions, room temperature (composite data of several investigations).

relatively near each other. Consequently, OH⁻ ions forming at cathodes in accordance with [cf. Eqs. (7.3) and (7.4), Sec. 7.1]

$$\frac{1}{2}O_2 + H_2O \rightarrow 2OH^- - 2e^- \qquad (7.15)$$

are always in the proximity of Fe^{2+} ions forming at nearby anodes, resulting in a film of $Fe(OH)_2$ adjacent to and adherent to the metal surface. This film provides an effective diffusion-barrier film.

In sodium chloride solutions, on the other hand, the conductivity is greater; hence, additional anodes and cathodes can operate much further removed one from the other. At such cathodes, NaOH does not react immediately with $FeCl_2$ formed at anodes; instead, these substances diffuse into the solution and react to form $Fe(OH)_2$ away from the metal surface. Any $Fe(OH)_2$ so formed does not provide a protective barrier layer on the metal surface. Hence, iron corrodes more rapidly in dilute sodium chloride solution because more dissolved oxygen can reach cathodic areas. Above 3% NaCl, the continuing decreased solubility of oxygen becomes more important than any change in the diffusion-barrier layer; hence, the corrosion rate decreases.

Alkali-metal salts (e.g., KCl, LiCl, Na_2SO_4, KI, NaBr, etc.) affect the corrosion rate of iron and steel in approximately the same manner as sodium chloride. Chlorides appear to be slightly more corrosive in the order Li, Na, and K [29]. Alkaline-earth salts (e.g., $CaCl_2$, $SrCl_2$) are slightly less corrosive than alkali-metal salts. Nitrates appear to be slightly less corrosive than chlorides or sulfates at low concentrations (0.2–0.25 N), but not necessarily at higher concentrations [30]. The small differences for all these solutions may arise, for example, from their specific effect on the $Fe(OH)_2$ diffusion-barrier layer, or perhaps from the different adsorptive properties of the ions at a metal surface resulting in differing anode–cathode area ratios or differing overvoltage characteristics for oxygen reduction.

Acid salts, which are salts that hydrolyze to form acid solutions, cause corrosion with combined hydrogen evolution and oxygen depolarization at a rate paralleling that of the corresponding acids at the same pH value. Examples of such salts are $AlCl_3$, $NiSO_4$, $MnCl_2$, and $FeCl_2$. *Ammonium salts* (e.g., NH_4Cl) are also acid, but produce a higher corrosion rate than corresponds to their pH (at NH_4Cl concentrations >0.05 N) [30]. Increased corrosivity is accounted for by the ability of NH^{4+} to complex iron ions, thereby reducing activity of Fe^{2+} and increasing the tendency of iron to corrode. Ammonium nitrate in high concentration is more corrosive (as much as eight times more) than the chloride or sulfate, in part because of the depolarizing ability of NO_3^-.

In the presence of excess NH_3, common to some synthetic fertilizer solutions, the corrosion rate in ammonium nitrate at room temperature may reach the very high value of 50 mm/y (2 ipy). The complex formed in this case has the structure $[Fe(NH_3)_6](NO_3)_2$ [31]. Since coupling mild steel to an equal area of platinum has no effect on the rate, the reaction is apparently anodically controlled. Metallurgical structure affects the rate, a cold-worked mild steel reacting much more rapidly than one quenched from elevated temperature. This indicates that the reaction is not diffusion-controlled, but depends instead on the rate of metal-ion formation at the anode and perhaps also to some extent on the rate of depolarization at the cathode.

Alkaline salts, which hydrolyze to form solutions of pH > 10, act as corrosion inhibitors. They passivate iron in the presence of dissolved oxygen in the same manner as NaOH (Fig. 7.3, Section 7.2.3). Examples of such salts are trisodium phosphate (Na_3PO_4), sodium tetraborate ($Na_2B_2O_7$), sodium silicate (Na_2SiO_3), and sodium carbonate (Na_2CO_3). In addition to favoring passivation of iron by dissolved oxygen, they may form corrosion-product layers of ferrous or ferric phosphates in the case of Na_3PO_4, or analogous compounds in the case of Na_2SiO_3, with such layers acting as more efficient diffusion barriers than hydrous FeO. They may, on this account, also inhibit corrosion below pH 10 and may provide better inhibition above pH 10 than NaOH or Na_2CO_3.

Oxidizing salts are either (a) good depolarizers, and therefore corrosive, or (b) passivators and efficient inhibitors. Examples of the first class are $FeCl_3$, $CuCl_2$, $HgCl_2$, and sodium hypochlorite. They represent the most difficult class of chemicals to handle in metal equipment. Examples of the second class are Na_2CrO_4, $NaNO_2$, $KMnO_4$, and K_2FeO_4. The differences in properties accounting for an oxidizing salt being either a depolarizer or a passivator are discussed in Chapter 17.

7.2.6.1 Natural-Water Salts.

Natural fresh waters contain dissolved calcium and magnesium salts in varying concentrations, depending on the source and location of the water. If the concentration of such salts is high, the water is called hard; otherwise, it is called soft. For many years before the causes were clearly understood, it was recognized that a soft water was more corrosive than a hard water. For example, a galvanized-iron hot-water tank was observed to last 10–20 years before failing by pitting in Chicago Great Lakes water (34 ppm Ca^{2+}, 157 ppm dissolved solids), whereas in Boston water (5 ppm Ca^{2+}, 43 ppm dissolved solids), a similar tank lasted only one to two years.

The two parameters that control corrosivity of soft waters are the pH and the dissolved oxygen concentration. In hard waters, however, the natural deposition on the metal surface of a thin diffusion-barrier film composed largely of calcium carbonate ($CaCO_3$) protects the underlying metal. This film retards diffusion of dissolved oxygen to cathodic areas, supplementing the natural corrosion barrier of $Fe(OH)_2$ mentioned earlier (Section 7.2.3). In soft water, no such protective film of $CaCO_3$ can form. But hardness alone is not the only factor that determines whether a protective film is possible. Ability of $CaCO_3$ to precipitate on the metal surface also depends on total acidity or alkalinity, pH, and concentration of dissolved solids in the water. For given values of hardness, alkalinity, and total dissolved salt concentration, a value of pH, given the symbol pH_s, exists at which the water is in equilibrium with solid $CaCO_3$. When pH > pH_s, the deposition of $CaCO_3$ is thermodynamically possible.

Langelier divided natural fresh waters into two groups: those oversaturated with $CaCO_3$ and those undersaturated [32]. Since only near- or oversaturated waters tended to form a protective film of $CaCO_3$ on iron, an estimate of the corrosivity of a water was established through analytical criteria for under- or oversaturation. Using certain simplifications, Langelier showed that the value of pH_s—at which a water is in equilibrium with solid $CaCO_3$ (neither tends to dissolve nor precipitate)—can be calculated from the relation*

$$pH_s = (pK_2' - pK_s') + p_{Ca}^{2+} + p_{alk} \qquad (7.16)$$

where K_2' is the ionization constant $(H^+)(CO_3^{2-})/(HCO_3^-)$, K_s' is the solubility product of calcium carbonate $[(Ca^{2+})(CO_3^{2-})]$, the concentration of calcium ions (Ca^{2+}) is in moles/1000 g H_2O, and alk (alkalinity) represents the equivalents per liter of titratable base to the methyl orange end point (often reported as ppm $CaCO_3$) according to the relation

$$(alk) + (H^+) \rightarrow 2CO_3^{2-} + HCO_3^- + OH^-$$

The letter p refers to the negative logarithm of all these quantities. The saturation index, also known as the Langelier index, is defined as the difference between the measured pH of a water and the equilibrium pH_s for $CaCO_3$, or

$$\text{Saturation index} = pH_{measured} - pH_s$$

A positive value of the saturation index indicates a tendency for the protective $CaCO_3$ film to form, whereas a negative value indicates that it will not form. Charts (see Appendix, Section 29.3) and nomographs have been prepared to obtain values of pH_s for waters varying widely in composition and at various temperatures [33].

---

*For a derivation of this and a more exact equation, see Section 29.3 in the Appendix.

Tabulated data, originally organized by Nordell, for calculating the Langelier saturation index are presented in Table 7.3 [34, 35].

After studying hundreds of concentrated cooling waters, Puckorius and Brooke [36] proposed a new index, the practical saturation index (PSI),

$$PSI = 2\,pH_s - pH_e$$

$$pH_e = 1.485 \times \log(\text{total alkalinity}) + 4.54$$

$CaCO_3$ precipitates at PSI < 6.0, according to Puckorius and Brooke.

An estimate of over- or undersaturation can also be obtained in the laboratory by measuring the pH of a water before and after exposure to pure $CaCO_3$ powder for a time adequate to achieve equilibrium. An increase in pH corre-

TABLE 7.3. Data for Calculating the Langelier Saturation Index [34, 35]

Total Dissolved Solids (ppm)	A	Ca Hardness (ppm as CaCO₃)	C	M. O. Alkalinity (ppm as CaCO₃)	D
50–300	0.1	10–11	0.6	10–11	1.0
400–1000	0.2	12–13	0.7	12–13	1.1
		14–17	0.8	14–17	1.2
		18–22	0.9	18–22	1.3
Temperature (°C)	B	23–27	1.0	23–27	1.4
		28–34	1.1	28–35	1.5
0–1.1	2.6	35–43	1.2	36–44	1.6
2.2–5.6	2.5	44–55	1.3	45–55	1.7
6.7–8.9	2.4	56–69	1.4	56–69	1.8
10.0–13.3	2.3	70–87	1.5	70–88	1.9
14.4–16.7	2.2	88–110	1.6	89–110	2.0
17.6–21.1	2.1	111–138	1.7	111–139	2.1
21.2–26.7	2.0	139–174	1.8	140–176	2.2
27.8–31.1	1.9	175–220	1.9	177–220	2.3
32.2–36.7	1.8	230–270	2.0	230–270	2.4
37.8–43.3	1.7	280–340	2.1	280–350	2.5
44.4–50.0	1.6	350–430	2.2	360–440	2.6
51.1–55.6	1.5	440–550	2.3	450–550	2.7
56.7–63.3	1.4	560–690	2.4	560–690	2.8
64.4–71.1	1.3	700–870	2.5	700–880	2.9
72.2–81.1	1.2	880–1000	2.6	890–1000	3.0

1. Obtain values of A, B, C, and D from this table.
2. $pH_s = (9.3 + A + B) - (C + D)$.
3. Langelier saturation index = $pH - pH_s$.

sponds to undersaturation. Because the composition of the water is altered by any $CaCO_3$ taken up or precipitated during the test, the final pH is not necessarily the same as the calculated pH_s. Alternatively, the water after saturation with $CaCO_3$ can be titrated with acid, with the increase in required milliliters of acid compared to that required by the original water being a measure of undersaturation (marble test) [37].

Any fresh water can be placed in one of the following categories:

Saturation Index	Saturation Level	Characteristic of Water
Positive	Oversaturated with respect to $CaCO_3$	Protective film of $CaCO_3$ forms
Zero	In equilibrium	
Negative	Undersaturated with respect to $CaCO_3$	Corrosive

Chicago water, for example, has an index of 0.2, whereas the value for Boston water is −3.0.

A soft water with negative saturation index can usually be treated with lime, sodium carbonate, or sodium hydroxide, in order to raise the saturation index to a positive value, thereby making the water less corrosive. For very soft waters, the required pH may be too high for uses such as tap water unless Ca^{2+} ions are added simultaneously. A saturation index of +0.1 to about +0.5 is considered to be satisfactory to provide corrosion protection. At higher values of the saturation index, excessive deposition of $CaCO_3$ (scaling) can occur, particularly at elevated temperatures. Waters with a slightly negative Langelier index may deposit $CaCO_3$ because of pH fluctuations. In general, for waters of negative saturation index, the more negative the index, the more corrosive the water. The relationship between the Langelier index and the corrosion rates of water pipes from a field test at a city water works is presented in Fig. 7.12 [38, 39].

At above-room temperatures, the saturation index may become more positive, and, in any case, the rate of $CaCO_3$ deposition is higher, should there be any tendency toward deposition. Hence, corrosion protection at all temperatures without excessive scaling requires adjustment of water composition to a saturation index that is at least constant over the entire operating temperature range. Powell et al. [33, 40] showed that, for any specific alkalinity, there is a corresponding pH value at which the decrease of measured pH with temperature is almost exactly compensated by a decrease in the factor $(pK_2' - pK_s')$. Under these conditions, the saturation index is nearly constant with temperature, and scaling tends to be the same in hot or cold water (Table 7.4). The amount of possible scaling, of course, depends on the value of the saturation index. The alkalinity and the pH of waters can be adjusted to bring them into the proper composition range using $Ca(OH)_2$, Na_2CO_3, $NaOH$, H_2SO_4, or CO_2.

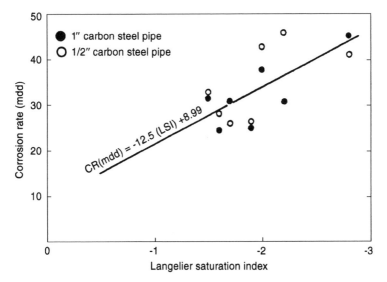

Figure 7.12. Relationship between the Langelier index and the corrosion rates of water pipes. (Reproduced from Fujii [39] with permission of Japan Association of Corrosion Control.)

TABLE 7.4. Alkalinity–pH Limits for Uniform Scale Deposition at Various Temperatures [40]

Alkalinity (ppm as $CaCO_3$)	pH Measured at Room Temperature
50	8.10–8.65
100	8.60–9.20
150	8.90–9.50
200	8.90–9.70

LIMITATIONS OF SATURATION INDEX. If a natural water contains colloidal silica or organic matter (e.g., algae), $CaCO_3$ may precipitate on the colloidal or organic particles instead of on the metal surface. If such is the case, the corrosion rate will remain high even though the saturation index is positive. For waters high in dissolved salts, such as NaCl, or for waters at elevated temperatures, the $CaCO_3$ film may lose its protective character at local areas, resulting in corrosion pitting. Furthermore, if complexing ions are added to a chemically treated water, such as polyphosphates, which retard precipitation of $CaCO_3$, the saturation index may no longer apply as an index of corrosivity.

Other than these exceptions, the saturation index is a useful qualitative guide to the relative corrosivity of a fresh water in contact with metals for which the corrosion rate depends on diffusion of dissolved oxygen to the surface, such as iron, copper, brass, and lead. The index does not apply to corrosivity of a water in contact with passive metals that corrode less the *higher* the surface concentration of oxygen, such as aluminum and the stainless steels.

7.3 METALLURGICAL FACTORS

7.3.1 Varieties of Iron and Steel

As discussed in Section 7.2.3, the corrosion rate of iron and steel in natural waters is controlled by diffusion of oxygen to the metal surface. Hence, whether a steel is manufactured by the Bessemer, oxygen furnace, or open-hearth process, as well as whether it is a wrought iron or a cast iron, makes little or no difference to the corrosion rate in natural waters, including seawater [16]. The same statement applies to corrosion in a variety of soils, because factors determining the corrosion rate underground are similar to those of total submersion. In general, therefore, the least expensive steel or iron of a given cross-sectional thickness having the required mechanical properties is the one that should be specified for those environments.

On the other hand, in the acid range of pH (approximately <4) and probably also in the extreme alkaline range (>13.5) where impurities play a role in the hydrogen evolution reaction, differences in manufacture affect the corrosion rate. A relatively pure iron corrodes in acids at a much lower rate than an iron or steel high in residual elements, such as carbon, nitrogen, sulfur, and phosphorus.

The high nitrogen content of Bessemer steel makes it more sensitive than open-hearth steels to stress-corrosion cracking in hot caustic or nitrate solution. For this reason, open-hearth steel is usually specified for boilers.

Cast iron in natural waters or in soils corrodes initially at the expected normal rate, but may provide much longer service life than steel. Other than the factor of considerable metal thickness common to cast structures, this advantage occurs because cast iron, composed of a mixture of ferrite phase (almost pure iron) and graphite flakes, forms corrosion products that, in some soils or waters, cement together the residual graphite flakes. The resulting structure (e.g., water pipe), although corroded completely, may have sufficient remaining strength, despite low ductility, to continue functioning under the required operating pressures and stresses. This type of corrosion is called *graphitic corrosion*. It occurs only with gray cast iron (or "ductile" cast iron containing spheroidal graphite) but not with white cast iron (cementite + ferrite). Graphitic corrosion can be reproduced in the laboratory over a period of weeks or months by exposing gray cast iron to a very dilute sulfuric acid that is renewed periodically.

7.3.2 Effects of Composition

Composition of an iron or steel within the usual commercial limits of carbon and low-alloy steels has no practical effect on the corrosion rate in natural waters or soils (see Table 7.1, Section 7.2.3; Table 10.1, Section 10.3). Only when a steel is alloyed in the proportions of a stainless steel (>12% Cr) or a high-silicon iron or high-nickel iron alloy for which oxygen diffusion no longer controls the rate, is corrosion appreciably reduced. For atmospheric exposures, the situation is changed because the addition of certain elements in small amounts (e.g., 0.1–1% Cr, Cu, or Ni) has a marked effect on the protective quality of naturally formed rust films (see Chapter 9).

Although *carbon* content of a steel has no effect on the corrosion rate in fresh waters, a slight increase in rate (maximum 20%) has been observed in sea-water as the carbon content is raised from 0.1 to 0.8% [41]. The cause of this increase is probably related to greater importance of the hydrogen evolution reaction in chloride solution (with complexing of Fe^{2+} by Cl^-) supplementary to oxygen depolarization as the cathodic surface of cementite (Fe_3C) increases.

In acids, the corrosion rate depends on the composition as well as the structure of the steel and increases with both carbon and nitrogen content. The extent of the increase depends largely on prior heat treatment (see Section 7.3.3) and is greater for cold-worked steels (Fig. 8.2, Section 8.1). Statistical techniques have been used to investigate the effects of minor alloying elements on the corrosion characteristics of commercial carbon and low-alloy steels in $0.1 N$ H_2SO_4 at 30 °C [42]. For the particular steels studied, corrosion rates were increased with increasing carbon content, especially in the range 0.5–0.7% C, and by phosphorus.

Both alloyed *sulfur* and *phosphorus* markedly increase the rate of attack in acids. These elements form compounds that apparently have low hydrogen over-potential; in addition, they tend to decrease anodic polarization so that the corrosion rate of iron is stimulated by these elements at both anodic and cathodic sites. Rates in deaerated citric acid are given in Table 7.5 [43]. In strong acids, the effects of these elements are still more pronounced [44] (see Fig. 7.13 and Table 7.5).

Sulfide inclusions have been found to act as initiation sites for pitting corrosion of mild steels in neutral-pH solutions [45, 46]. On the other hand, sulfur content has been found to have no significant effect on corrosion rates in acids of steels containing more than 0.01% Cu [42].

TABLE 7.5. Corrosion Rates of Iron Alloys in Deaerated Citric Acid and in 4% NaCl + HCl at 25 °C [43]

	$0.1 M$ Citric Acid pH = 2.06	4% NaCl + HCl pH = 1
Pure iron (0.005% C)	2.9 gmd	3.0 gmd
+0.02% P	16.5	100
+0.015% S	70.6	283
+0.11% Cu	4.1	39
+0.10% Cu + 0.03% P	37.6	60.6
+0.08% Cu + 0.02% S	3.2	18.6

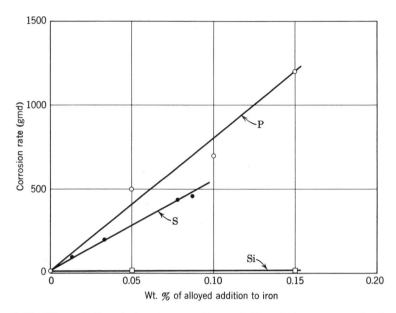

Figure 7.13. Effect of alloyed phosphorus, sulfur, and silicon in iron on corrosion in deaerated 0.1 N HCl, annealed specimens, 25 °C [44]. (Reproduced with permission. Copyright 1965, The Electrochemical Society.)

Measurements of hydrogen permeation rates through cathodically polarized mild steels containing elongated (FeMn)S inclusions show that H_2S produced at the metal surface by dissolution of inclusions promotes hydrogen entry into the steels. Permeation rates increase with sulfur content of the steel in the range 0.002–0.24% S, but only where the H_2S supply can be maintained by dissolution of inclusions [47]. Elongated inclusions, most commonly manganese sulfide inclusions, provide sites for initiation of *hydrogen-induced cracking* (HIC), also called *stepwise cracking* (SWC), in steels that are used for high-pressure pipelines transporting oil and natural gas [48, 49]. HIC is discussed further in Section 8.4.

Arsenic is present in some steels in small amounts. In quantities up to 0.1%, it increases the corrosion rate in acids (less than sulfur and phosphorus); in large amounts (0.2%), it decreases the rate [44].

Manganese, in amounts normally present, effectively decreases acid corrosion of steel containing small amounts of sulfur. Inclusions of MnS have low electrical conductivity compared to FeS; in addition, manganese reduces the solid solubility of sulfur in iron, thereby probably restoring the anodic polarization of iron which is lowered by the presence of sulfur [50]. *Silicon* only slightly increases the corrosion rate in dilute hydrochloric acid (Fig. 7.13).

Copper alloyed with pure iron to the extent of a few tenths of 1% moderately increases the corrosion rate in acids. However, in the presence of phosphorus or sulfur, which are normal components of commercial steels, copper counteracts the accelerating effect of these elements. Copper-bearing steels, therefore, usually corrode in nonoxidizing acids at lower rates than do copper-free steels [51]. The data of Table 7.5 for several relatively pure iron alloys show that 0.1% Cu reduces corrosion in 4% NaCl + HCl when 0.03% P or 0.02% S is present in the iron, but not for the phosphorus alloy in citric acid. Addition of 0.25% Cu to a low-alloy steel caused reduction of the corrosion rate from 1.1 to 0.8 mm/y in 0.5% acetic acid–5% sodium chloride solution saturated with hydrogen sulfide at 25 °C [52]. These particular relations apply only to the specific compositions and experimental conditions reported; they are not necessarily general. Steels containing a few tenths of 1% of copper are more resistant to the atmosphere, but show no advantage over copper-free steels in natural waters or buried in the soil where oxygen diffusion controls the rate.

An amount up to 5% *chromium* (0.08% C) was reported to decrease weight losses in seawater at the Panama Canal [53] at the end of one year. A sharp increase in rates was observed between 2 and 4 years; after 16 years, the chromium steels lost 22–45% more weight than did 0.24% C steel. Depth of pitting was less for the chromium steels after one year, but comparable to pit depth in carbon steel after 16 years. Hence, for long exposures to seawater, low-chromium steels apparently offer no advantage over carbon steel. By comparison, however, low-alloy chromium steels (<5% Cr) have improved resistance to corrosion fatigue in oil-well brines free of hydrogen sulfide.

An amount up to 5% *nickel* (0.1% C) did not appreciably change weight losses of steels exposed to seawater in tests at the Panama Canal [53] extending up to 16 years. Depth of pits, although less for one year of exposure, was greater for long-exposure times for the nickel steels compared to 0.24% C steel (77% deeper for 5% Ni steel after 8 years). Low-nickel steels (<5% Ni) have improved resistance to corrosion fatigue in oil-well brines containing hydrogen sulfide [54]. Nickel also decreases the corrosion rate of steels exposed to alkalies, and the effect increases with nickel content [55].

7.3.2.1 Galvanic Effects through Coupling of Different Steels.

Unimportant though low-alloy components may be in determining overall corrosion rates in waters or in soils, composition is nevertheless of considerable importance to the galvanic relations and consequent corrosion of steels coupled to each other. For example, because of increased anodic polarization, a low-nickel, low-chromium steel is cathodic to mild steel in most natural environments. The reason is obvious from relations shown in Fig. 7.14. Both mild steel and a low-alloy steel, not coupled, corrode at about the same rate, i_{corr}, established by the limiting rate of oxygen reduction. When coupled, however, the potentials of both steels, which are originally different, become equal to ϕ_{galv}. The corrosion rate of mild steel, estimated from the extension of its anodic polarization curve, is now increased to i_2, and that of the low-alloy steel is decreased to i_1. The precise value of ϕ_{galv} depends on the relative areas of the two steels. Accordingly, steel bolts and nuts used to couple underground mild-steel pipes, or a weld rod used for steel plates on the hull of a ship, should always be of a low-nickel, low-chromium

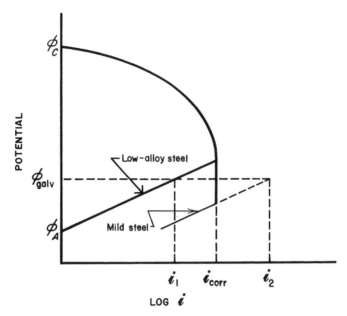

Figure 7.14. Polarization diagram showing effect of coupling of a low-alloy steel to mild steel on subsequent corrosion rates.

steel or similar composition that is cathodic to the major area of the structure (small cathode, large anode). Should the reverse polarity occur, serious corrosion damage would be caused quickly either to the bolts or to the critical area of weld metal [56, 57].

Cast iron is initially anodic to low-alloy steels and not far different in potential from mild steel. As cast iron corrodes, however, especially if graphitic corrosion takes place, exposed graphite on the surface shifts the potential in the noble direction. After some time, therefore, depending on the environment, cast iron may achieve a potential cathodic to both low-alloy steels and mild steel. This behavior is important in designing valves, for example. The trim of valve seats must maintain dimensional accuracy and be free of pits; consequently, the trim must always be chosen cathodic to the valve body making up the major internal area of the valve. For this reason, valve bodies of steel are often preferred to cast iron for aqueous media of high electrical conductivity.

7.3.3 Effect of Heat Treatment

A carbon steel quenched from high temperatures has a structure called *martensite*—a supersaturated solution of carbon in iron—a single metastable phase with carbon in solid solution in interstitial positions of the body-centered tetragonal lattice of iron atoms. Random distribution of carbon atoms accompanied by electronic interaction of carbon atoms with neighboring iron atoms limits their effectiveness as cathodes of local-action cells; consequently, in dilute acid the corrosion rate of martensite is relatively low. Interstitial carbon reacts in large part with acid to form a hydrocarbon gas mixture (accounting for the odor of pickled steel) and some residual amorphous carbon, which is observed as a black smut on the steel surface.

Heating martensite to low temperatures (below 727 °C) and then air-cooling (called tempering) results in the formation of tempered martensite, which consists of α-iron and many dispersed particles of cementite (i.e., iron carbide, Fe_3C). This two-phase structure of α-iron and cementite sets up galvanic cells that accelerate the corrosion reaction. The amount of finely divided cementite that forms depends on the tempering time and temperature; for a 2-h heat treatment of 0.95% C steel, the maximum cementite forms at about 400 °C (300 °C for 0.07% C steel). After tempering at this temperature, cementite, acting as cathode, offers maximum peripheral surface adjoining ferrite, and galvanic action during subsequent exposure to an aqueous environment is at a maximum. Above this temperature, cementite coalesces to larger particle size, and the subsequent corrosion rate is lower. The particles of cementite are now large enough to resist complete dissolution in acid and can be detected in the residue of corrosion products. At the same time, there is a corresponding decrease in hydrocarbon gas formation.

On slowly cooling a carbon steel from the austenite (face-centered cubic lattice) region, above 727 °C, cementite, in part, assumes a lamellar shape, forming a structure called *pearlite*. This structure again corrodes at a comparatively low rate because of the relatively massive form of cementite formed by decomposition of austenite compared with smaller-size cementite particles resulting from decomposition of martensite. Corrosion rate increases as the size of iron carbide particles decreases. Pearlitic structures corrode faster than spheroidized ones, and steels containing fine pearlite corrode more rapidly than those with coarse pearlite [42]. The importance of both the amount of cementite acting as cathode and its state of subdivision supports the electrochemical mechanism of corrosion. The rate of corrosion is cathodically controlled and, hence, depends on hydrogen overpotential and interfacial area of cathodic sites.

In practice, an effect of heat treatment on corrosion is seldom observed, because oxygen diffusion controls the rate in the usual environments. However, in handling acid oil-well brines, marked localized corrosion is sometimes found near welds or "up-set" ends of steel oil-well tubing. This observed increase in corrosion encircling a limited inner region of the tube is called *ring-worm corrosion*. It is caused by heat treatment incidental to joining and fabrication techniques, and it can be minimized by a final heat treatment of the tubing or by addition of inhibitors to the brine [58].

7.4 STEEL REINFORCEMENTS IN CONCRETE

Steel-reinforced concrete is an industrially very important system in which steel is exposed to an alkaline (pH > 12) environment and, hence, is normally passive.

Nevertheless, corrosion of steel reinforcements does occur in structures such as concrete bridge decks and parking garages. These problems have been studied for many years and result from the breakdown of passivity caused, for example, by salt in the environment or in the concrete [59]. Passivity may also be lost after several years if air diffuses through the concrete to the reinforcements, converting alkaline $Ca(OH)_2$ to less alkaline $CaCO_3$. Corrosion processes for steel in concrete are illustrated in Fig. 7.15 [60].

There are several methods that can be used to control corrosion of steel reinforcements in concrete. First, the design of the structure should provide for drainage of salt-containing waters away from the reinforced concrete. Second, concrete of adequate thickness, high quality, and low permeability should be specified to protect the reinforcements from the environment. Third, chloride content of the concrete mix should be kept to a minimum. For further protection, the steel reinforcements can be epoxy-coated. In many parts of North America, steel reinforcements used in bridge decks are now epoxy-coated as a standard construction procedure. Cathodic protection is also being used, both with impressed current anodes and with sacrificial anodes [61]. (See Chapter 13.)

Steel reinforcements in concrete, being passive, are noble in potential with respect to steel outside the concrete that is galvanically coupled to the reinforcements. The measured potential difference is in the order of 0.5 V [62]. The effect of large cathode area and small anode area has caused premature failures of buried steel pipe entering a concrete building [63]. In this situation, use of epoxy-coated reinforcements (to coat the cathode) or insulated couplings should prove beneficial.

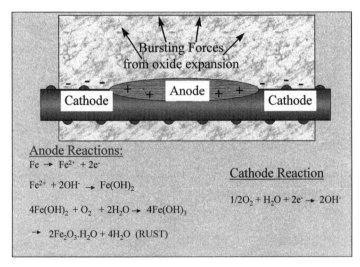

Anode Reactions:

$Fe \rightarrow Fe^{2+} + 2e^-$

$Fe^{2+} + 2OH^- \rightarrow Fe(OH)_2$

$4Fe(OH)_2 + O_2 + 2H_2O \rightarrow 4Fe(OH)_3$

$\rightarrow 2Fe_2O_3.H_2O + 4H_2O$ (RUST)

Cathode Reaction

$1/2O_2 + H_2O + 2e^- \rightarrow 2OH^-$

Figure 7.15. Corrosion process for steel in concrete [59]. (Reprinted with permission of John Wiley & Sons, Inc.)

There is evidence that high-strength steel reinforcements, such as rod or bar, heat-treated to high hardness (Rockwell C > 40) and subjected to a tensile stress, have failed by SCC or hydrogen cracking (see Chapter 8) through exposure to moisture penetrating the concrete coating. Heat-treating processes sometimes produce a decarburized (hence softer) surface layer. In this situation, SCC may be delayed until the outer metal layer undergoes general corrosion, exposing the SCC-susceptible harder core underneath. Either moisture must be excluded from contact with such steels, or a steel of lower strength should be specified.

REFERENCES

1. I. Matsushima, in *Uhlig's Corrosion Handbook*, 2nd edition, R. W. Revie, editor, Wiley, New York, 2000, p. 529.
2. H. Uhlig, D. Triadis, and M. Stern, *J. Electrochem. Soc.* **102**, 59 (1955).
3. H. Uhlig, *J. Electrochem. Soc.* **108**, 327 (1961).
4. B. J. Little and J. S. Lee, *Microbiologically influenced Corrosion*, Wiley Series in Corrosion, R. W. Revie, series editor, Wiley, New York, 2007, p. 1.
5. J.-D. Gu, T. E. Ford, and R. Mitchell, in *Uhlig's Corrosion Handbook*, 2nd edition, edited by R. W. Revie, Wiley, New York, 2000, pp. 349–365, 915–927.
6. Ref. 4, p. 41.
7. C. von Wolzogen Kühr, *Water and Gas* **7**, 26, 277 (1923); *Corrosion* **17**, 293t (1961).
8. Ref. 4, pp. 28–30.
9. R. Schweisfurth and E. Heitz, Mikrobiologische Materialzerstoerung und Materialschutz, DECHEMA, Frankfurt am Main, Germany, 1989.
10. Ref. 5, pp. 359–360.
11. G. Skaperdas and H. Uhlig, *Ind. Eng. Chem.* **34**, 748 (1942).
12. W. Whitman, R. Russell, and V. Altieri, *Ind. Eng. Chem.* **16**, 665 (1924).
13. I. Matsushima, Ref. 1, pp. 536–538.
14. R. Scholder, *Angew. Chem.* **49**, 255 (1936).
15. W. Whitman and R. Russell, *Ind. Eng. Chem.* **16**, 276 (1924).
16. F. LaQue, *Am. Soc. Testing Mater. Proc.* **51**, 495 (1951).
17. W. Whitman and R. Russell, *Ind. Eng. Chem.* **17**, 348 (1925).
18. A. P. Bond, Sc.D. thesis, Department of Metallurgy, M.I.T., Cambridge, MA, 1958.
19. W. Whitman, R. Russell, C. Welling, and J. Cochrane, *Ind. Eng. Chem.* **15**, 672 (1923).
20. Z. Foroulis and H. Uhlig, *J. Electrochem. Soc.* **111**, 13 (1964).
21. R. Russell et al., *Ind. Eng. Chem.* **19**, 65 (1927).
22. F. LaQue, in *Corrosion Handbook*, H. H. Uhlig, editor, Wiley, New York, 1948, p. 391.
23. F. Speller and F. LaQue, *Corrosion* **6**, 209 (1950).
24. J. Postlethwaite and S. Nesic, in *Uhlig's Corrosion Handbook*, 2nd edition, R. W. Revie, editor, Wiley, New York, 2000, p. 261.
25. Ibid., p. 262.

26. Ibid., pp. 267–269.

27. B. Angell, in *Corrosion*, 3rd edition, L. L. Shreir, R. A. Jarman, and G. T. Burstein, editors, Butterworth-Heinemann, Oxford, U.K., 1994, pp. 8:197–8:207.

28. W. Leith and A. Thompson, *Trans. Am Soc. Mech. Eng., J. Basic Eng.* December, 795 (1960).

29. C. Borgmann, *Ind. Eng. Chem.* **29**, 814 (1937).

30. O. Bauer, O. Kröhnke, and G. Masing, *Die Korrosion metallischer Werkstoffe*, Vol. 1, S. Hirzel, Leipzig, 1936, p. 240.

31. G. Schick and H. Uhlig, *J. Electrochem. Soc.* **111**, 1211 (1964).

32. W. F. Langelier, *J. Am. Water Works Assoc.* **28**, 1500 (1936).

33. S. Powell, H. Bacon, and J. Lill, *Ind. Eng. Chem.* **37**, 842 (1945).

34. Ref. 1, p. 534.

35. E. Nordell, *Water Treatment for Industrial and Other Uses*, 2nd edition, Reinhold, New York, 1961, p. 287.

36. P. R. Puckorius and J. M. Brooke, *Corrosion* **47**, 280 (1991).

37. C. P. Hoover, *J. Am. Water Works Assoc.* **30**, 1802 (1938).

38. Ref. 1, p. 535.

39. T. Fujii, *Bosei Kanri (Rust Prevention and Control)* **27**, 85 (1983).

40. S. Powell, H. Bacon, and J. Lill, *J. Am. Water Works Assoc.* **38**, 808 (1946).

41. C. Chappell, *J. Iron Steel Inst.* **85**, 270 (1912).

42. H. Cleary and N. D. Greene, *Corros. Sci.* **7**, 821 (1967).

43. M. Stern, *J. Electrochem. Soc.* **102**, 663 (1955).

44. Z. Foroulis and H. Uhlig, *J. Electrochem. Soc.* **112**, 1177 (1965).

45. L. J. Gainer and G. R. Wallwork, *Corrosion* **36**, 348 (1980).

46. Y.-Z. Wang, R. W. Revie, and R. N. Parkins, *CORROSION/99*, Paper No. 143, NACE International, Houston, TX, 1999.

47. P. H. Pumphrey, *Corrosion* **36**, 537 (1980).

48. M. Elboujdaini, in *Uhlig's Corrosion Handbook*, 2nd edition, R. W. Revie, editor, Wiley, New York, 2000, pp. 205–220.

49. B. E. Wilde, C. D. Kim, and E. H. Phelps, *Corrosion* **36**, 625 (1980).

50. G. Wranglen, *Corros. Sci.* **9**, 585 (1969).

51. E. Williams and M. Komp, *Corrosion* **21**, 9 (1965).

52. Y. Yoshino, *Corrosion* **38**, 156 (1982).

53. C. Southwell and A. Alexander, *Mater. Prot.* **9**, 14 (1970).

54. B. B. Wescott, *Mech. Eng.* **60**, 813 (1938).

55. F. L. LaQue and H. R. Copson, editors, *Corrosion Resistance of Metals and Alloys*, Reinhold, New York, 1963, p. 334.

56. E. Uusitalo in *Proc. 2nd International Congress on Metallic Corrosion*, National Association of Corrosion Engineers, Houston, Texas, 1966, p. 812.

57. R. J. Brigham, M. McLean, V. S. Donepudi, S. Santyr, L. Malik, and A. Garner, *Can. Metall. Q.* **27**(4), 311 (1988).

58. M. Holmberg, *Corrosion* **2**, 278 (1946); R. Manuel, *Corrosion* **3**, 415 (1947).

59. J. P. Broomfield, in *Uhlig's Corrosion Handbook*, 2nd edition, R. W. Revie, editor, Wiley, New York, 2000, pp. 581–600.

60. Ibid., p. 582.

61. J. Broomfield, *Mater. Perf.* **46**(1), 26 (2007).

62. J. Vrable, *Mater. Perf.* **21**(3), 51 (1982).

63. I. Matsushima, private communication.

GENERAL REFERENCES

H. Ahluwalia, *Highway Bridges Gap Analysis*, Report for NACE International, Houston, Texas, August 14, 2007, http://www.nace.org/nace/images_ndx_story/highwaybridgegapanalysis.pdf.

Neal S. Berke, Edward Escalante, Charles K. Nmai, and David Whiting, Editors, *Techniques to Assess the Corrosion Activity of Steel Reinforced Concrete Structures*, STP 1276, ASTM, West Conshohocken, PA, 1996.

S. W. Borenstein, *Microbiologically Influenced Corrosion Handbook*, Woodhead Publishing, Cambridge, England, 1994.

J. P. Broomfield, Corrosion of steel in concrete, in *Uhlig's Corrosion Handbook*, 2nd edition, R. W. Revie, editor, Wiley, New York, 2000, pp. 581–600.

J. P. Broomfield, *Corrosion of Steel in Concrete: Understanding, Investigation and Repair*, 2nd edition, Taylor & Francis, London, 2007.

Corrosion of ferrous metals, in *ASM Handbook*, Vol. 13B: *Corrosion: Materials*, ASM International, Materials Park, OH, 2005, pp. 1–53.

J.-D. Gu, T. E. Ford, and R. Mitchell, Microbial degradation of materials: General processes, in *Uhlig's Corrosion Handbook*, 2nd edition, R. W. Revie, editor, Wiley, New York, 2000, pp. 349–365.

J. R. Kearns and B. J. Little, editors, *Microbiologically Influenced Corrosion Testing*, STP 1232, ASTM, Philadelphia, PA, 1994.

B. J. Little and J. S. Lee, *Microbiologically influenced Corrosion*, Wiley Series In Corrosion, R. W. Revie, series editor, Wiley, Hoboken, NJ, 2007.

P. R. Roberge, *Corrosion inspection and MonitorIng*, Wiley Series In Corrosion, R. W. Revie, series editor, Wiley, New York, 2007, pp. 373–376.

Uhlig's Corrosion Handbook, 2nd edition, R. W. Revie, editor, Wiley, New York, 2000, pp. 515–600.

PROBLEMS

1. Five iron rivets, 0.5 in.^{2} (3.2 cm^2) of total exposed area each, are inserted in a copper sheet of 8 ft^2 (7430 cm^2) exposed area. The sheet is immersed in an aerated, stirred, conducting solution in which uncoupled iron corrodes at 0.0065 ipy (0.165 mm/y).

(a) What is the corrosion rate of the rivets in ipy (mm/y)?

(b) What is the corrosion rate of an iron sheet in which five copper rivets are placed, with the same dimensions?

2. Water entering a steel pipeline at the rate of 40 liters/min contains 5.50 mL O_2 per liter (25 °C, 1 atm). Water leaving the pipe contains 0.15 mL O_2 per liter. Assuming that all corrosion is concentrated at a heated section 30 m² in area forming Fe_2O_3, what is the corrosion rate in gmd?

3. Two steels immersed in dilute deaerated sulfuric acid corrode at different rates. When they are arranged as a cell, a small potential difference is observed between them. Which is anode? Illustrate by a polarization diagram.

4. Calculate the saturation index of each of the following waters at:

(a) 25 °C (77 °F) using Eq. (7.16) in Section 7.2.6.1;

(b) 25 °C (77 °F) using chart Fig. 29.4, Section 29.3;

(c) 75 °C (167 °F) using charts. (Note that the calcium concentration is listed in the chart as ppm $CaCO_3$.)

Water	Alkalinity (ppm $CaCO_3$)	Ca (ppm)	$pK_2' - pK_s'$	pH	Total Solids (ppm)
Boston, MA	13	5	2.2	6.90	43
Chicago, IL	120	34	2.3	8.0	157
Detroit, MI	73	26	2.2	7.7	153

Answers to Problems

1. (a) 3.0 ipy (77 mm/y).
2. 31.3 gmd.
3. Slower corroding steel is anode.
4.

Water	Saturation Index (Using Charts)	
	25 °C	75 °C
Boston, MA	−2.7	−2.2
Chicago, IL	0.2	0.6
Detroit, MI	−0.5	−0.1

8

EFFECT OF STRESS

8.1 COLD WORKING

As discussed in Section 7.2, a cold-worked commercial steel corrodes in natural waters at the same rate as an annealed steel. In acids, however, cold working increases the corrosion rate severalfold, as shown in Fig. 8.1 [1]. Many authors have traditionally ascribed this effect to residual stress within the metal, which serves to increase the corrosion tendency. But the residual energy produced by cold working, as measured in a calorimeter (usually <30 J/g), is less than sufficient to account for an appreciable change in free energy; hence, this intuitive concept is probably not correct [2]. The observed increase of rate is apparently caused instead by segregation of carbon or nitrogen atoms at imperfection sites produced by plastic deformation rather than by the presence of the imperfections themselves (Fig. 8.2). Such sites exhibit lower hydrogen overpotential than does either cementite or iron [1] and probably constitute the most important factor. Entering to a lesser extent are (1) increased surface area of metal exposed at slip steps, (2) the increased surface area of cementite lamellae fractured by the cold-work process, and (3) the preferred orientation of ferrite grains, with the latter either increasing or decreasing corrosion depending on the particular crystal faces that lie parallel to the metal surface.

Corrosion and Corrosion Control, by R. Winston Revie and Herbert H. Uhlig
Copyright © 2008 John Wiley & Sons, Inc.

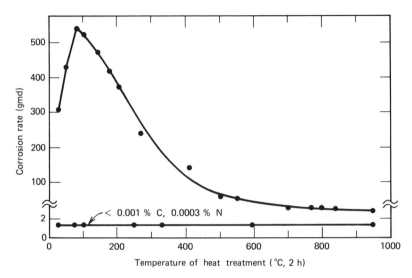

Figure 8.1. Effect of heat treatment of cold-worked 0.076% C steel (85% reduction of thickness) and zone-refined iron (50% reduction of thickness) on corrosion in deaerated 0.1N HCl, 25°C [1]. (Reproduced with permission. Copyright 1964, The Electrochemical Society.)

Subsequent heat treatment of cold-worked steel at 100°C induces additional diffusion of interstitial carbon atoms to imperfections in the metal lattice, thereby increasing the cathodic area of low hydrogen overvoltage and also accelerating the corrosion rate. Heat treatment at higher temperatures reduces the density of the imperfection sites, accompanied by precipitation of carbides or nitrides of particle size that increases with heat treatment temperature. Hence, heat treatment above 100°C results in cathodic sites of higher hydrogen overpotential and of decreasing peripheral area, leading to a corresponding reduction of corrosion rate. A severely cold-worked, zone-refined (pure) iron, on the other hand, lacking sufficient carbon or nitrogen, corrodes in dilute, deaerated hydrochloric acid at no higher rate than when annealed (Figs. 8.1 and 8.2) [1].

When commercial nickel is severely cold-worked, unlike mild steel, it does not corrode appreciably more rapidly in acids [2], which suggests that similar formation of low-overpotential cathodic sites derived from segregated, interstitial impurities does not occur.

8.2 STRESS-CORROSION CRACKING

Stress-corrosion cracking (S.C.C.) is a type of cracking that occurs when a material that is susceptible to S.C.C. is simultaneously stressed in tension and exposed to an environment that causes S.C.C. On a macroscopic level, S.C.C. failures appear to be brittle; that is, the usual ductility of the material (e.g., when stressed in air) is considerably reduced. The tensile stress can be applied or residual, or

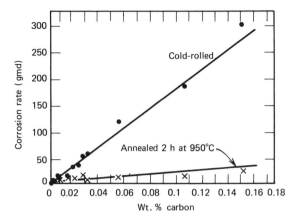

Figure 8.2. Effect of carbon in steels cold-rolled 50%, or subsequently annealed, on corrosion in deaerated 0.1N HCl, 25°C [1]. (Reproduced with permission. Copyright 1964, The Electrochemical Society.)

both. Residual stresses result from fabrication processes, such as deformation (e.g., forming of a pipe) and welding. Stress-corrosion cracks can be intergranular or transgranular, or a combination of the two. In general, there are three stages in the stress-corrosion cracking process:

1. Generation of the environment that causes S.C.C. [3]
2. Initiation of S.C.C.
3. Propagation of S.C.C. until failure occurs

Depending on the metal–environment combination and the stressing condition, the time to failure can vary from minutes to many years. For this reason, inspection of stressed metals that are exposed to a corrosive environment during service is essential to establish whether cracks have initiated and to develop mitigation procedures before failure occurs.

Table 8.1 lists some alloy–environment combinations in which S.C.C. occurs.

Stress-corrosion cracking is caused by the interaction of metallurgical, mechanical, and environmental factors, with the result that there is a multiplicity of possible S.C.C. control measures that can be implemented. In addition, the complexity of S.C.C. has led to a very large number of hypotheses, models, theories, and controversies on the mechanisms by which S.C.C. occurs. These matters will be summarized in this section.

8.2.1 Iron and Steel

When mild steel is stressed in tension to levels near or beyond the elastic limit and is exposed simultaneously to hot concentrated alkaline or hot concentrated

TABLE 8.1. Some Alloy–Environment S.C.C. Systems

Metal	Environment
Mild steel	NO_3^-, OH^-
High-strength steels[a]	H_2O
Stainless steels, austenitic	Cl^-, OH^-, Br^-
α-Brass	NH_3, amines
Titanium alloys	
8% Al, 1% Mo, 1% V	Cl^-, Br^-, I^-
Aluminum alloys	H_2O, NaCl solutions
Zirconium alloys	Cl^-
Magnesium alloys	Cl^-
<40 at.% Au–Cu	$FeCl_3$, aqua regia

[a]>180,000 psi (>1240 MPa) yield strength or approximately >Rockwell C 40 hardness value. Includes high-strength martensitic and precipitation-hardening stainless steels.

nitrate solutions [such as boiling 60% $Ca(NO_3)_2$ + 3% NH_4NO_3], it cracks along intergranular paths by S.C.C. The mechanism is quite different from that described in Section 8.1.

Stress-corrosion cracking of steel was first encountered in a practical way in riveted steam boilers. Stresses at rivets always exceed the elastic limit, and boiler waters are normally treated with alkalies to minimize corrosion. Crevices between rivets and boiler plate allow boiler water to concentrate, until the concentration of alkali suffices to induce S.C.C., sometimes accompanied by explosion of the boiler. Because alkalies were recognized as one of the causes, failures of this kind were first called *caustic embrittlement*. With the advent of welded boilers and with improved boiler-water treatment, S.C.C. of boilers has become less common. Its occurrence has not been eliminated entirely, however, because significant stresses, for example, may be established at welded sections of boilers or in tanks used for storing concentrated alkalies.

The presence of silicates in alkaline boiler water above 225°C accelerates S.C.C., whereas at lower temperatures silicates may act as inhibitors [4]. Laboratory tests in boiling 50% NaOH have shown that at atmospheric pressure, 0.2–1.0% PbO, $KMnO_4$, Na_2CrO_4, or $NaNO_3$ are also accelerators [4]. At the higher temperatures and pressures common to boiler conditions, nitrates added in amounts equivalent to 20–40% of NaOH alkalinity act as inhibitors of S.C.C. [4] and are used for this purpose in practice. On the other hand, addition of 2% NaOH to boiling nitrate solutions may inhibit cracking [5]. This is one of several illustrations that can be cited regarding the necessity to understand thoroughly the fundamentals of a laboratory test before extrapolating results to large-scale applications.

In addition to failures in the chemical and nuclear-chemical [6, 7] industries in handling hot nitrate solutions, stressed steel may also fail in contact with nitrates at room temperature. Such failure occurred, for example, in 0.7% C steel cables of the Portsmouth, Ohio, bridge after 12 years in service [8]. The cables cracked at their base where rain water, presumably containing trace amounts of ammonium nitrate from the atmosphere, had accumulated and concentrated. Subsequent tests conducted at the U.S. National Bureau of Standards (now the National Institute for Standards and Technology) showed that specimens of the cable stressed in tension failed within $3\frac{1}{2}$–$9\frac{1}{2}$ months when immersed in $0.01N$ NH_4NO_3 or $NaNO_3$ at room temperature. No cracking occurred after similar tests in distilled water, or in $0.01N$ solutions of NaCl, $(NH_4)_2SO_4$, $NaNO_2$, or NaOH. There was evidence in this case that the steel cable was unusually susceptible to S.C.C.

Stress-corrosion cracking of steel also occurs when the metal is in contact with anhydrous liquid ammonia at room temperature—for example, at cold-formed heads and at welds of steel tanks used to contain the liquefied gas. Cracks are intergranular, but they can also be transgranular. Cracking can be avoided by stress-relief heat treatment of the steel, by avoiding air contamination of the NH_3, or by addition of about 0.2% H_2O, which acts as an inhibitor [9]. Intergranular cracking of stressed steel has also been reported in contact with $SbCl_3$ + HCl + $AlCl_3$ in a hydrocarbon solvent [10].

Transgranular S.C.C. was reported at room temperature on exposure of 0.1–0.2% C steel to an aqueous mixture of carbon dioxide and carbon monoxide at 100 psi [11]. Cathodic polarization in the latter solution prevented damage. Explosions have been reported involving cracks in steel tanks containing compressed illuminating gas. The failures at stresses below the elastic limit were transgranular and were traced to small amounts of HCN contained in the gas [12]. The problem was solved by removing traces of HCN and moisture from the gas.

Underground steel pipelines that transmit oil and gas and that operate at high internal pressures have been reported to fail by S.C.C. Both transgranular and intergranular S.C.C. of pipeline steels have been reported [13, 14].

Based on tests in boiling nitrate solution [15], it was found that severely cold-worked mild steel (0.06% C, 0.001% N) is resistant to S.C.C. (Fig. 8.3). Along these lines, it is recognized in practice that cold-drawn steel wire is more resistant to S.C.C. than is oil-tempered wire having equal mechanical properties. Heat treatment of cold-rolled mild steel at 600°C (1110°F) for $^1/_2$ h, at 445°C (830°F) for 48h, or at lower temperatures for correspondingly longer times induces susceptibility again. Therefore, plastically deformed steel, stress-relieved in the range of 400–650°C (750–1200°F) and subsequently stressed, is *more* rather than less susceptible as a result of heat treatment. Mild steel quenched at 900–950°C (1650–1740°F) is susceptible, but can be made resistant by tempering at 250°C (480°F) for $\frac{1}{2}$ h (Fig. 8.3), or at 200°C (390°F) for 48h, even if the steel exposed to nitrates is highly stressed after heat treatment. This resistant state, however, is temporary; on further heating (in the unstressed state) at 445°C (830°F) for 70h or at 550°C (1020°F) for 3h, as well as for correspondingly shorter times at higher temperatures, the steel becomes susceptible again and remains susceptible.

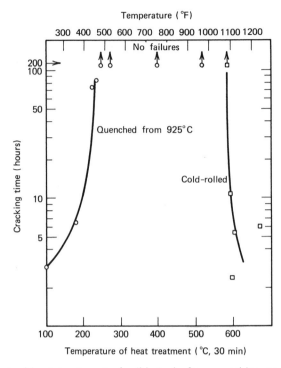

<u>Figure 8.3.</u> Effect of heat treatment of mild steel after quenching or cold rolling (70% reduction of thickness) on resistance to stress-corrosion cracking in boiling nitrate solution [15]. (Reprinted with permission of ASM International®. All rights reserved. www.asminterna tional.org.)

The overall evidence suggests that grain boundaries become suitable crack paths only when carbon (or nitrogen) atoms (not Fe_3C) segregate at grain-boundary regions. Pure iron is immune to S.C.C. Iron (>0.002% C) [16] or mild steel (0.06% C) quenched from about 925°C contains a sufficient concentration of carbon atoms along the grain boundaries to induce susceptibility. Low-temperature heating (e.g., 250°C for $\frac{1}{2}$ h) randomly nucleates iron carbides, which rob the grain boundaries of carbon, accounting for increased resistance or immunity. Longer heating at higher temperatures (e.g., 70 h at 445°C) allows slow-moving lattice imperfections within the grain to migrate to the grain boundaries, transporting carbon atoms with them, and the steel again becomes susceptible. Cold working, on the other hand, produces immunity by destroying continuous grain-boundary paths, and, more important, by generating imperfections that have high affinity for carbon and that deplete any continuous paths along which carbon atoms tend to segregate.

Remedial measures to avoid S.C.C. of steel in nitrate solutions include the following:

1. *Severe Cold Working.* Cold rolling to >50% reduction of thickness imparts relative immunity to a stressed mild steel in boiling nitrate solution. The resistant state is predicted to persist at low temperatures [e.g., 100–200°C (200–400°F)], for thousands of hours.

2. *Heat Treatment.* Mild steel slowly cooled at 900–950°C (1650–1740°F), or quenched from this temperature range and then tempered at about 250°C (480°F) for $\frac{1}{2}$ h, or at higher temperatures for shorter times, is resistant. This resistance is temporary and short-lived (200h) at temperatures on the order of 400°C (750°F), but it is predicted to reach the order of thousands of hours at 300°C (570°F) or below [15].

3. *Surface Peening or Shot Blasting.* Compressive stresses are produced at the surface of the metal that are effective in avoiding damage so long as the compressive layers are continuous, remain intact, and are not dissolved by corrosion.

4. *Cathodic Protection.* Schroeder and Berk [4] found that cathodic polarization of steel stressed in hot sodium hydroxide–sodium silicate solution greatly delayed or prevented cracking. Parkins [17] found similar protection in hot nitrate solution. Bohnenkamp [18] reported maximum susceptibility of various carbon steels (0.003–0.11% C) in 33% NaOH boiling at 120°C at –0.66 to –0.75 V (S.H.E.) with orders-of-magnitude longer life at potentials at 0.1 V either noble or active to this range. Both anodic and cathodic protection were found to be effective (see Fig. 8.7, Section 8.3.3).

5. *Special Alloys.* Steels containing small alloying additions of aluminum, titanium, or niobium plus tantalum [19], which react preferentially with carbon and nitrogen, exhibit improved resistance to S.C.C., but are not immune. Alloyed additions of <2% Ni increase susceptibility of low-carbon steels in nitrates; >1% Cr or Mo decrease susceptibility. Furnace-cooled (pearlitic) steels of >0.2% C are resistant [20].

6. *Inhibitors.* Sodium nitrate has already been mentioned as a practical inhibitor for steel exposed to boiler waters. Crude quebracho extract and waste sulfite liquor are also used. Buffer ions, such as PO_4^{3-}, are useful because they avoid high concentrations of OH^- in concentrated boiler water. In accelerated tests, the addition of 3% NaCl or 2% sodium acetate to boiling 60% $Ca(NO_3)_2$ + 3% NH_4NO_3, at 108°C was found to inhibit cracking (>200h) [21] of mild steel.

The various approaches to S.C.C. control are presented in Table 8.2 [22]. The dependence of S.C.C. on metallurgical, mechanical, and environmental variables results in three main avenues for S.C.C. control:

Metallurgical: Use an alloy that is resistant to S.C.C.

Mechanical: Reduce the applied stress and/or stress-relief anneal to reduce the residual stress.

TABLE 8.2. Stress-Corrosion Cracking Control Measures [22]

Mechanical	Metallurgical	Environmental
Avoid stress concentrators	Change alloy composition	Modify the environment
Relieve residual stresses	Change alloy structure	Apply anodic or cathodic
Introduce surface	Use metallic or conversion	protection
compressive stresses	coating	Add inhibitor
Reduce operating stresses		Use organic coating
Design for nondestructive		Modify temperature
inspection		

Environmental: Change the environment or eliminate the environment from contact with the metal surface by using a coating; apply electrochemical protection (cathodic or anodic, depending on the system).

8.3 MECHANISMS OF STRESS-CORROSION CRACKING OF STEEL AND OTHER METALS

The distinguishing characteristic of S.C.C. is the requirement of tensile stress acting conjointly with a specific environment. Typical specific environments for several metals are listed in Table 8.1 in Section 8.2. The damaging anions have no clear relation to general corrosion rates of the metals that they affect. Chloride ions, but not nitrate ions, cause S.C.C. of 18–8 austenitic stainless steels (containing 74% Fe), but the opposite situation applies to mild steel. The necessity for a tensile stress is illustrated by the sensitivity of 70% Cu–30% Zn brass to intergranular corrosion (not requiring stress) in a variety of electrolytes, such as dilute H_2SO_4, $Fe_2(SO_4)_3$, or $BiCl_3$ solutions [23]. The much more rapid S.C.C., which is usually also intergranular, occurs in the presence of NH_3 or amines. Furthermore, an improperly heat-treated 18–8 stainless steel (e.g., sensitized at 650°C for 1 h) fails intergranularly in a wide variety of electrolytes regardless of whether it is stressed; but the same heat-treated steel stressed in tension and placed in a boiling solution of magnesium chloride fails *transgranularly* by S.C.C. despite well-defined corrosion paths along grain boundaries [24].

Following are five characteristics of S.C.C.:

1. Specificity of damaging chemical environments—*not* all environments that cause corrosion also cause S.C.C.
2. General resistance or immunity of all pure metals. For many years, it was understood that pure metals did not undergo S.C.C.; however, S.C.C. of pure copper under aggressive laboratory conditions has been reported [25, 26]. Pure alloys, such as Cu–Zn, Cu–Au, and Mg–Al, on the other hand, are susceptible.

3. Successful use of cathodic polarization to avoid initiating S.C.C.

4. Inhibiting effect of various extraneous anions added to damaging environments. For example, 2% $NaNO_3$, or 1% sodium acetate, or 3.5% NaI added to magnesium chloride solution boiling at 130°C (33 g/100 mL) inhibits S.C.C. of 18–8 stainless steel (>200 h) [27]. Similarly, Cl⁻ or acetates inhibit S.C.C. of mild steel in boiling nitrates [21], and SO_4^{2-} or NO_3^- can be used to inhibit a titanium alloy (8% Al, 1% Mo, 1% V) that otherwise cracks in 3.5% NaCl solution at room temperature [28].

5. An appreciable effect of metallurgical structure; for example, ferritic stainless steels (body-centered cubic) are much more resistant to Cl⁻ than are austenitic stainless steels (face-centered cubic). Also, β and γ brass (>40% Zn) may crack in water alone, but α brass (70% Cu, 30% Zn) fails in NH_3 or an amine. Any metal of large grain size, whether failure is inter- or transgranular, is more susceptible to S.C.C. than is the same metal of small grain size.

Many models have been proposed to explain the mechanisms of S.C.C. at atomic levels, but each of these models has its own limitations in that it can be used to explain S.C.C. in a limited number of metal–environment systems. Notwithstanding the absence of an accepted generalized model of S.C.C., a few models are summarized in the following paragraphs.

8.3.1 Electrochemical Dissolution

Electrochemical dissolution is the basis for the *film rupture model*, also called the *slip–dissolution* model. In 1940, Dix [29] proposed that galvanic cells are set up between metal and anodic paths established by heterogeneous phases (e.g., $CuAl_2$ precipitated from a 4% Cu–Al alloy) along grain boundaries or along slip planes. When the alloy, stressed in tension, is exposed to corrosive environments, the ensuing localized electrochemical dissolution of metal opens up a crack; in addition, the applied stress effectively ruptures brittle oxide films at the tip of the crack, thereby exposing fresh anodic material to the corrosive medium. Supporting this mechanism was a measured potential of metal at grain boundaries that was negative, or active, to the potential of grains. Furthermore, cathodic polarization prevented S.C.C. The film rupture model was used to explain transgranular S.C.C. of copper in $1M$ $NaNO_2$ solution [30].

If the crack growth rate is controlled by the rate of dissolution at the crack tip, then the following equation can be used to estimate the crack growth rate, da/dt:

$$\frac{da}{dt} = i_a \times \frac{M}{zFd}$$

where i_a is the anodic current density, M is the atomic weight of the metal, z is the valency of the solvated species, F is Faraday's constant, and d is the density of the metal. Indeed, a correlation has been established for some systems between crack growth rates and dissolution current densities on bare surfaces, providing some basis for the credibility of this equation. The crack growth rate represented by this equation is an upper limit of the crack growth rate by continuous dissolution since neither dissolution nor cracking is likely to take place continuously, but rather will be interrupted by film growth at the crack tip until film rupture takes place [31].

8.3.2 Film-Induced Cleavage

The film-induced cleavage model of S.C.C. is based on the following five steps:

1. A thin surface layer forms on the surface.
2. A brittle crack initiates in this layer.
3. The brittle crack crosses the film–metal-substrate interface with little loss in velocity.
4. The brittle crack continues to propagate in the metal substrate.
5. After this crack becomes blunt and stops growing, this process repeats itself.

This model can also be used to explain the crack arrest markings and cleavage-like facets on the fracture surface, as well as the discontinuous nature of crack propagation [32, 33], although more research into surface films and brittle fracture is required before this model can be thoroughly evaluated [34].

8.3.3 Adsorption-Induced Localized Slip

The proponents of this model reasoned that similar fracture processes occur in liquid-metal embrittlement, hydrogen embrittlement, and S.C.C., with chemisorption facilitating the nucleation of dislocations at the crack tip and promoting the shear processes that result in brittle, cleavage-like fracture. It was found that cleavage fracture occurs by alternate slip at the crack tip and formation of voids ahead of the crack tip [35–37].

8.3.4 Stress Sorption

Figures 8.4 and 8.5 indicate the critical potentials noble to which S.C.C. of 18–8 stainless steel initiates when exposed to magnesium chloride solution boiling at 130°C with and without inhibiting anion additions [27]. Anodic polarization induces shorter cracking times the more noble the controlled potential; cathodic polarization, on the other hand, extends the observed cracking times. Below the critical value of –0.145 V (S.H.E.), the alloy becomes essentially immune (Fig. 8.4). Addition of various salts, such as sodium acetate, to the magnesium chloride solution shifts the critical potential to more noble values. When the amount of

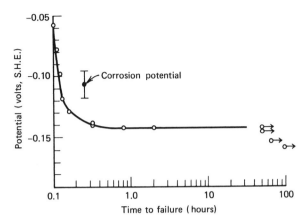

Figure 8.4. Effect of applied potential on time to failure of stressed moderately cold-rolled 18–8 stainless steel in magnesium chloride solution boiling at 130°C [27]. (Reproduced with permission. Copyright 1969, The Electrochemical Society.)

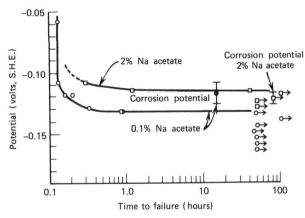

Figure 8.5. Effect of applied potential on time to failure of stressed moderately cold-rolled 18–8 stainless steel in magnesium chloride solution with sodium acetate additions, boiling at 130°C (2% sodium acetate addition is inhibiting) [27]. (Reproduced with permission. Copyright 1969, The Electrochemical Society.)

added salt shifts the critical potential to a value that is noble to the corrosion potential, S.C.C. no longer occurs (Fig. 8.5).

It was proposed [38, 39] that S.C.C., in general, proceeds not by electrochemical dissolution of metal, but instead by weakening of the cohesive bonds between surface metal atoms through adsorption of damaging components of the environment. The name suggested for this mechanism is *stress-sorption cracking*. Since chemisorption is specific, damaging components are also specific. The surface energy of the metal is said to be reduced, increasing the tendency of the metal to form a crack under tensile stress. Hence, the mechanism is related to the Griffith criterion [40] of crack formation in glass and similar brittle solids, for which the strain energy of the stressed solid must be greater than the total increased surface energy generated by an incipient crack. Adsorption of any kind that reduces surface energy should favor crack formation; hence, water adsorbed on glass understandably reduces the stress needed to cause fracture.

The critical potential for S.C.C. is interpreted as that value above which damaging ions adsorb on defect sites and below which they desorb. Such a potential, in principle, can be either active or noble to the corrosion potential. Inhibiting anions, which of themselves do not initiate cracks, compete with damaging species for adsorption sites, thereby making it necessary to apply a more positive potential in order to reach a surface concentration of the damaging species adequate for adsorption and cracking. Whenever the inhibiting anion drives the critical potential above the corrosion potential, cracking no longer occurs; this is because the damaging ion can no longer adsorb. The mechanism of competitive adsorption parallels that describe earlier applying to the critical pitting potential, which is also shifted in the noble direction by the presence of extraneous anions (Section 6.6).

Stress-sorption cracking is the basic mechanism applying to stress cracking of plastics by specific organic solvents [41, 42] and to *liquid-metal embrittlement*—the cracking of solid metals by specific liquid metals. It is also the mechanism proposed earlier by Petch and Stables [43] to account for stress cracking of steel induced by interstitial hydrogen (see Section 8.4).

The mechanism of liquid-metal embrittlement is analogous to that of S.C.C. in that only certain combinations of liquid metals and stressed solid metals result in intergranular failure (Table 8.3) [44]. This specificity has important consequences, one being that large-scale mercury boilers can be constructed of carbon steel, but not of titanium or its alloys or of brass, because catastrophic intergranular failure would result. It is, apparently, possible for adsorbed mercury atoms to reduce the metal bond strength within grain-boundary regions of stressed titanium or brass adequate to cause failure, but this reduction of bond strength does not occur for iron in this application.

TABLE 8.3. Susceptibility[a] of Solid Metals to Embrittlement by Liquid Metals [44]

Liquid →	Li	Hg	Bi	Ga	Zn
Steel	C	NC	NC	NC	C
Copper alloys	C	C	C	—	—
Aluminum alloys	NC	C	NC	C	C
Titanium alloys	NC	C	NC	NC	NC

[a]C = cracking; NC = no cracking.

8.3.5 Initiation of Stress-Corrosion Cracking and Critical Potentials

For various metals and solutions, values of critical potentials, immediately above which (or noble to) S.C.C. initiates, are listed [45] in Table 8.4. Based on data obtained for 18–8 stainless steel in $MgCl_2$ at 130°C, cracks initiated to a depth not greater than 0.013–0.025 cm (0.005–0.01 in.) stop propagating at a potential 5 mV or less below the critical value [46]. For deeper cracks, the imposed potential necessary to stop propagation is increasingly negative (active), accounted for by metallic shielding effects within the crack and by solution composition changes resulting from factors like accumulation of anodic dissolution products within the crack. In other words, the environmental conditions required to initiate a crack are also those required for crack propagation.

For some metal–solution combinations, S.C.C. is avoided by polarizing not only below a certain critical potential or potential range, but also at potentials not far above (noble to) such a range. Failure times are most rapid at potentials between those at which S.C.C. occurs. In the relatively narrow potential range over which S.C.C. occurs, adsorption of damaging ions on mobile defect sites is optimum, according to the stress-sorption viewpoint. For example, mild steel undergoes pronounced intergranular S.C.C. in ammonium carbonate (170 g/liter) at 70°C within the potential range −0.26 to −0.35 V (S.H.E.), but much less so outside this range (Fig. 8.6) [47]. The susceptible range of potentials is still more restricted for slightly cold-rolled steel. Since the corrosion potential is more noble

TABLE 8.4. Some Critical Potentials for Initiation of S.C.C. [45]

Metal	Solution	Critical Potential (V,S.H.E.)
18% Cr, 8% Ni stainless Steel, WQ[a]	$MgCl_2$, 130°C	−0.128
	$MgBr_2$, 154°C	−0.04
18% Cr, 8% Ni stainless steel, cold-rolled 36% reduction area	$MgCl_2$, 130°C	−0.145
25% Cr, 20% Ni stainless steel, WQ	$MgCl_2$, 130°C	−0.113
Al, 5.5% Zn, 2.5% Mg	0.5M NaCl, room temperature	−1.11
Ti, 7% Al, 2% Cb, 1% Ta[b]	3% NaCl, room temperature	−1.1
Mild steel	60% $Ca(NO_3)_2$, 3% NH_4NO_3, 110°C	−0.055
63% Cu, 37% Zn brass[c]	1M $(NH_4)_2SO_4$, 0.05M $CuSO_4$, pH 6.5, room temperature	+0.095

[a]WQ, water-quenched from 1050°C.
[b]Precracked specimen.
[c]See Fig. 20.7, Section 20.2.3.

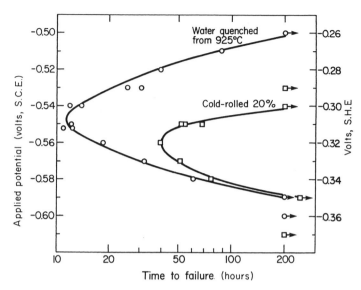

Figure 8.6. Effect of applied potential on stress-corrosion cracking of mild steel in 170 g ammonium carbonate per liter, 70°C [47]. [Reprinted with permission from *Corrosion* **32** (2), 57 (1976). Copyright NACE International 1976.]

[0.16 V(S.H.E.)] and well removed from the susceptible range, ammonium carbonate is not listed among the usually encountered damaging environments. However, the susceptible potential range may be overlapped under conditions of cathodic protection accompanied by risk of S.C.C. if carbonates are present. A similar restricted potential versus time-to-failure curve is exhibited by mild steel in 35% NaOH at 85–125°C (caustic embrittlement) (Fig. 8.7) [48]. Since the corrosion potential is at –0.90 V (S.H.E.), S.C.C. does not occur within 200 h or more unless dissolved O_2 or an oxidizing agent, such as PbO, is present that shifts the corrosion potential into the maximum susceptibility range at or near –0.71 V. In this event, either anodic or cathodic polarization extends time to failure.

8.3.6 Rate of Crack Growth (Fracture Mechanics)

Under conditions favorable to initiation of S.C.C., the subsequent rate of crack growth depends on prevailing stress; that is, the higher the stress, the higher the rate. An empirical linear relation is found between applied stress and log time to fracture by S.C.C. for smooth specimens of austenitic and martensitic stainless steels, carbon steels, brass, and aluminum alloys. The linear relation and usually greater resistance of small-grain-size alloys is shown for brass in Fig. 8.8 [49]. For some metals, the slope changes to one that is relatively shallow at low values of stress, indicating greater sensitivity of failure time to change of applied stress. It will be noted, however, that neither the empirical relation expressing fracture time versus applied stress nor the typical data of Fig. 8.8 support the concept of

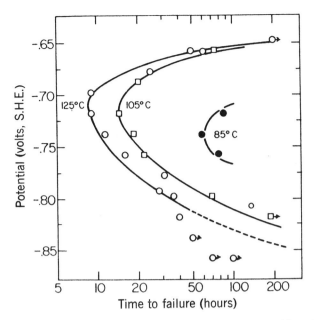

Figure 8.7. Effect of applied potential on failure times of 0.09% C mild steel at three temperatures in 35% sodium hydroxide solution [48]. [Reprinted with permission from *Corrosion* **28**, No. (11), 430 (1972). Copyright NACE International 1972.]

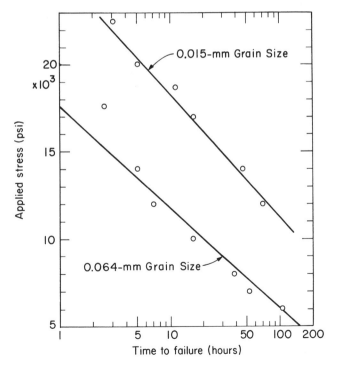

Figure 8.8. Relation of applied stress to time to fracture of 66% Cu, 34% Zn brass exposed to ammonia. (Data of Morris [49] replotted.)

a threshold stress below which S.C.C. does not occur. A low surface tensile stress means only that time to fracture is relatively long.

Since stress at a surface is highest at the root of a notch or imperfection, it is there that a crack grows most rapidly. Surface defects acting as effective stress raisers include corrosion pits and fatigue cracks. Additionally, at surface defects that are sufficiently deep, the electrochemical potential differs from that at the surface; also, the composition and pH of solution within the defect can be altered by electrochemical transport through operation of differential aeration cells. These changes plus increased local stress sometimes combine to initiate S.C.C. as well as accelerate crack growth [50].

The actual stress near the tip of a crack of length a in a stressed elastic solid can be calculated in terms of the elastic stress intensity factor, K_1. For the specimen geometry shown in Fig. 8.9, K_1 is given by the following expression [51]:

$$K_1 = \frac{Pa^{1/2}}{BW}\left[1.99 - 0.41\left(\frac{a}{W}\right) + 18.70\left(\frac{a}{W}\right)^2 + \cdots\right]$$

where P is applied uniform load, B is specimen thickness, and W is specimen width. K_1 takes into account the load and specimen geometry for a Mode 1 of crack opening. The dimensions of K_1 are ksi in.$^{1/2}$ or kg mm$^{-3/2}$. A stress intensity factor of 20 kg mm$^{-3/2}$ corresponds to a crack 1 mm deep in a solid subject to a uniform stress of 10.1 kg/mm^2.

If the stress intensity factor K_1 is plotted versus a measured S.C.C. crack velocity, three regions are observed (Fig. 8.10). At low levels of K_1, within region I, crack growth rate is highly sensitive to K_1 (unlike region II). A threshold value, $K_{1S.C.C.}$, is designated at which the measured S.C.C. crack growth rate is very low, or essentially zero, in short-time tests. From a practical standpoint, the rate cannot be assumed to be zero unless conditions are such that S.C.C. is not operative. For reference, a crack growth rate of 10^{-10} m/s corresponds to 0.1 nm/s, a very low rate requiring years to measure. In region II, the crack velocity is independent of K_1 but is strongly dependent on solution pH, viscosity, and temperature.

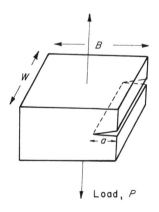

Load, P **Figure 8.9.** Mode 1 type of crack opening.

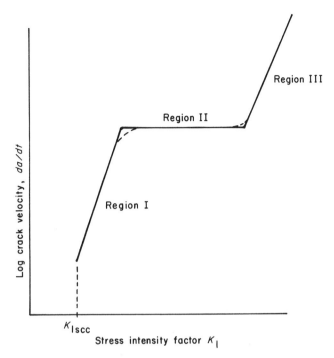

<u>Figure 8.10.</u> Effect of stress intensity at a crack tip on stress-corrosion crack velocity.

Figure 8.11 shows region I and II crack velocities for 7075 Al alloy (1.2–2.0% Cu, 2.1–2.9% Mg, 0.3% Cr, 5.5% Zn) in NaCl solution and also in liquid mercury (liquid-metal embrittlement) at room temperature [52]. Crack velocities in mercury, although orders of magnitude higher than in aqueous solution, show a similar dependence on stress intensity. Also, metallurgical factors affecting crack growth rates in one environment are similarly effective in the other. It is reasonable that some aspects of the crack mechanism overlap in the two different environments.

For high-strength aluminum alloys, in general, there is no crack growth in dry air whatever the value of K_1. With increase of relative humidity, region I is shifted to lower stress intensities the higher the humidity, and region II is shifted to higher velocities [53]. If one assumes that the applied stress is on the order of the yield strength (Y.S.) of the metal, then a critical crack depth, a_{cr}, exists above which the stress intensity factor exceeds $K_{1S.C.C.}$. Under this condition, the crack grows at an increasing rate until failure occurs. Based on the previous expression for K_1, the following approximate relation holds:

$$a_{cr} = 0.25 \left(\frac{K_{1S.C.C.}}{Y.S.} \right)^2$$

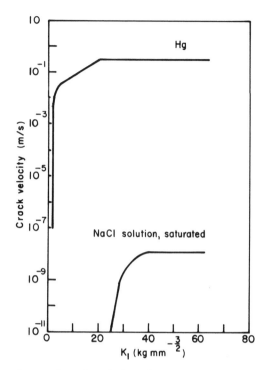

Figure 8.11. Dependence of crack velocity of 7075 Al alloy on stress intensity factor K_1 [52].

The lower the value of $K_{1S.C.C.}$ and the higher the value of Y.S., the smaller the allowable depth of surface flaws leading to failure.

8.4 HYDROGEN DAMAGE

Hydrogen damage is a generic term that includes:

- Hydrogen embrittlement (loss of ductility)
- Hydrogen cracking, also called hydrogen stress cracking
- Hydrogen blistering
- Hydrogen-induced cracking (H.I.C.) [also known as stepwise cracking (S.W.C.) because of its appearance]
- Stress-oriented hydrogen-induced cracking (S.O.H.I.C.)
- Sulfide stress cracking (S.S.C.)

Some metals, when stressed, crack on exposure to a variety of corrosive aqueous solutions with no evidence that the damaging solutions need be specific. For example, a stressed high-strength carbon steel or a martensitic stainless steel immersed in dilute sulfuric or hydrochloric acid may crack within a few minutes. The failures have the outward appearance of S.C.C.; but if the alloy is cathodically polarized, cracking still occurs or happens in shorter time. This is contrary to the behavior of austenitic stainless steels in boiling magnesium chloride; such steels are cathodically protected under these conditions. Failures have also occurred within hours when martensitic 12% Cr-steel self-tapping screws (cathode) were applied in contact with an aluminum roof (anode) in a moist atmosphere. Cracking of steel springs, sometimes observed during pickling in sulfuric acid or after electroplating, is another example. If catalyst poisons that favor entrance of hydrogen into the metal lattice, such as sulfur or arsenic compounds, are added to the acids, cracking is intensified. When H_2S is the poison, failures are described as *sulfide stress cracking* (S.S.C.). In practice, many stressed high-strength steels (e.g., carbon steels or 9% Ni steels) have failed from this source within days or weeks after exposure to oil-well brines or to natural gas containing H_2S [54].

In all these cases, cracking is caused by hydrogen atoms entering the metal interstitially either through a corrosion reaction or by cathodic polarization. Steels containing interstitial hydrogen are not always damaged. They almost always lose ductility (hydrogen embrittlement), but cracking usually takes place only under conditions of sufficiently high applied or residual tensile stress. Failures of this kind are called *hydrogen stress cracking* or *hydrogen cracking*. The cracks tend to be mostly transgranular. In martensite, they may follow former austenite grain boundaries [55].

Steels are less susceptible to hydrogen cracking above room temperature, with iron becoming a better catalyst for the reaction $H_{ads} + H^+ + e^- \rightarrow H_2$. Hence, more hydrogen escapes as molecular H_2 and less adsorbed H is available to enter the metal, contrary to the effects of catalyst poisons, which retard the above reaction.

Carbon steels are especially susceptible to hydrogen cracking when heat-treated to form martensite, but are less so if the structure is pearlitic. Carbon steel heat-treated to form a spheroidized carbide structure is less susceptible than pearlite, bainite, or martensite [56]. Austenitic steels—for example, 18–8 and 14% Mn steel (face-centered cubic), in which hydrogen is more soluble than in ferrite and the diffusion rate is lower—are immune under most conditions of exposure [57].

8.4.1 Mechanisms of Hydrogen Damage

The mechanism of hydrogen cracking has been explained by development of internal pressure [54] on the assumption that interstitial atomic hydrogen is released as molecular hydrogen at voids or other favored sites under extreme pressure. Such an effect certainly occurs, as is shown by the formation of visible blisters containing hydrogen when ductile metals are cathodically polarized or exposed to certain corrosive media. Under similar conditions, a less ductile metal would crack instead.

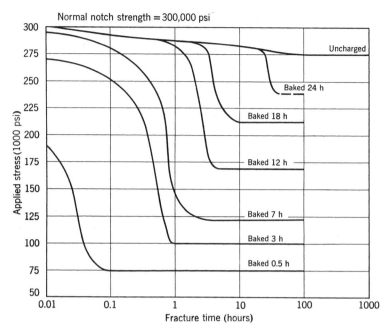

Figure 8.12. Delayed fracture times and minimum stress for cracking of 0.4% C steel as a function of hydrogen content. Specimen initially charged cathodically, baked at 150°C for varying times to reduce hydrogen content [59]. [Figure 5 from H. Johnson, J. Morlet, and A. Troiano, Hydrogen, crack initiation, and delayed failure in steel, *Trans. AIME* **212**, 531 (1958).]

An interesting characteristic of hydrogen cracking is a specific delay time for appearance of cracks after stress is applied. The delay time is only slightly dependent on stress; it decreases with increasing hydrogen concentration in the steel and with increase in hardness or tensile strength [58]. For small concentrations of hydrogen, fracture may occur some days after the stress is applied.

A critical minimum stress exists, below which delayed cracking does not take place in any time. The critical stress decreases with increase in hydrogen concentration. These effects are shown in Fig. 8.12 for SAE 4340 steel (0.4% C) charged with hydrogen by cathodic polarization in sulfuric acid, then cadmium-plated to help retain hydrogen, and finally subjected to a static stress [59]. The hydrogen concentration was reduced systematically by baking.

Delay in fracture apparently results because of the time required for hydrogen to diffuse to specific areas near a crack nucleus until the concentration reaches a damaging level. These specific areas are presumably arrays of imperfection sites produced by plastic deformation of metal just ahead of the crack. Hydrogen atoms preferably occupy such sites because they are then in a lower-energy state compared to their normal interstitial positions. The crack propagates discontinuously because plastic deformation occurs first, and then hydrogen dif-

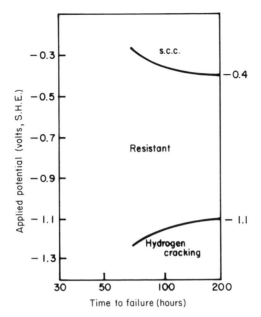

Figure 8.13 Effect of applied potential on time to failure of 4140 low-alloy steel, R_c46, in boiling 3% NaCl [60]. (Reproduced with permission. Copyright 1975, The Electrochemical Society.)

fuses to imperfection arrays produced by deformation, whereupon the crack propagates one step further. A short notch in a steel surface favors plastic deformation at its base, and hence it lowers the critical minimum stress and shortens the delay time. Below −110°C or at high deformation speeds, hydrogen embrittlement and cracking are minimized because diffusion of hydrogen is too slow.

It is sometimes assumed that S.C.C. of high-strength steels of hardness >R_c40 (Table 8.1) in water or moist air is caused by hydrogen resulting from the reaction of H_2O with iron. However, the effect of applied potential on failure times shows that, in boiling 3% NaCl, cracking occurs only above (noble to) a critical potential of −0.40 ± 0.02 V and below a potential of about −1.1 V (Fig. 8.13) with resistance to cracking between these potentials. Failure in the upper potential region is better interpreted as S.C.C., whereas failure at or below −1.1 V corresponds to hydrogen cracking [60]. In support, steels fail by S.C.C. in H_2O above room temperature in shorter times than at room temperature; failure times by hydrogen cracking (cathodic polarization), to the contrary, are longer the higher the temperature. Also, cold working of high-strength steels improves resistance to S.C.C. because the critical potential is shifted noble to the corrosion potential, whereas resistance to hydrogen cracking decreases. Accordingly, in practice, it is important that high-strength steel bridge cables be cold-drawn in order to avoid failure by S.C.C. in moist air. In the absence of cold work, they break prematurely despite more-than-adequate test strength, as has happened in the failure of bridges in the United States and other countries as well. Furthermore, a surface-decarburized (hence softer surface) high-strength steel does not fail in boiling water or 3% NaCl solution, but is readily hydrogen-cracked when cathodically polarized. The small amount of hydrogen produced by the H_2O–Fe reaction has no effect on the hard steel core. The significant effect of adsorbed H_2O rather than interstitial hydrogen as a cause of cracking of high-strength steels probably also applies to high-strength martensitic and precipitation-hardening stainless steels, Al alloys, Mg alloys,Ti alloys, and β and γ brass, all of which are sensitive to failure in the presence of moisture.

8.4.2 Effect of Metal Flaws

Nuclei for hydrogen cracking of steels form at the interface of precipitated phases (e.g., Fe_3C or intermetallic compounds, such as those occurring in maraging steels) by separation from the plastically deformed matrix. Nuclei of perhaps Fe_3C in low-carbon 10% Ni–Fe alloys can be oriented by cold rolling, thereby greatly improving resistance to hydrogen cracking of specimens stressed parallel to the rolling direction, but not of specimens stressed at right angles [61]. The role of internal lattice flaws on hydrogen cracking explains in part the resistance of stressed pure iron to cracking despite its becoming hydrogen-embrittled. On the other hand, pure carbon–iron alloys crack readily.

In steels, nonmetallic inclusions—for example, MnS—provide internal flaws where hydrogen blisters often nucleate. The blisters contribute to pipeline failures by H.I.C. [53], which occurs in three steps:

1. Formation of hydrogen atoms at the steel surface and adsorption on the surface.
2. Diffusion of hydrogen atoms into the steel substrate.
3. Accumulation of hydrogen atoms at hydrogen traps, such as voids around inclusions in the steel matrix, leading to increased internal pressure, crack initiation and propagation, and linkage of separate cracks.

Internal cracks are nucleated where stress intensity is highest, such as at elongated inclusions oriented parallel to the rolling direction; elliptically and spherically shaped inclusions are less damaging. Eventually, in final failure, cracks appear at right angles, linking the cracks that previously developed at the elongated inclusions. H.I.C. requires only a steel with a susceptible microstructure (typically elongated sulfide inclusions) and sufficient hydrogen to cause cracking. H.I.C. can occur in the absence of any applied or residual stress in the steel other than the hydrogen pressure in traps.

Under the influence of an applied stress, failure can occur by S.O.H.I.C., similar to H.I.C. except that the morphology of cracking is different. Whereas in H.I.C. the blister cracks form at widely distributed sites and then link in a stepwise pattern, in S.O.H.I.C. the blister cracks tend to form in an array that is per-

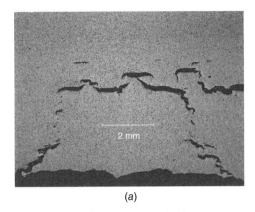

(a)

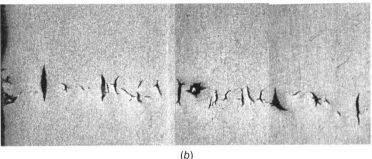

(b)

<u>Figure 8.14.</u> (a) Hydrogen-induced cracking (H.I.C.). (b) Stress-oriented hydrogen-induced cracking (S.O.H.I.C.). (Courtesy of Malcolm Hay.)

pendicular to the applied stress, and these individual blister cracks link leading to failure. Figure 8.14 presents illustrations of H.I.C. and S.O.H.I.C.

Any factors that reduce hydrogen absorption in the corrosion process are beneficial; for example, alloying with a small percentage of Pt or Pd, which catalyze formation of molecular hydrogen at the steel surface [62], or with copper, which forms an insoluble sulfide film that provides some protection at pH above about 4.5. Similarly, any processing of steel that minimizes sharp internal surfaces at the inclusion interface, for example by avoiding low rolling temperatures, reduces the tendency toward cracking. Increased resistance to H.I.C. can also be obtained by reducing the sulfur content in the steel and by controlling the shape of sulfides—for example, by adding calcium to the steel composition.

Surface flaws may influence the hydrogen cracking (or sulfide cracking) of moderate- or high-strength steels exposed to oil-well brines containing H_2S. Marked susceptibility to cracking in this environment makes it necessary to specify steels below the strength level at which failure occurs. Strength of a steel is proportional to its hardness. The empirically determined maximum hardness is specified as Rockwell C 22 (R_c22), corresponding to a yield strength of about 90 ksi [63]. It has been suggested [64], based on threshold values of stress intensity factors for steels exposed to aqueous H_2S solutions, that R_c22 corresponds to a critical surface-flaw depth of about 0.5 mm (0.02 in). At greater depths, flaws are likely to develop quickly into major cracks. Small-size flaws of this order, which are calculated to be still less tolerable for harder steels, are not readily avoided in practice, corresponding with the general experience of the oil industry that, in H_2S environments, steels of hardness $>R_c22$ should be avoided. In general, surface flaws become more important to cracking the higher the strength of the steel; internal flaws, on the other hand, affect both low- and high-strength steels.

8.5 RADIATION DAMAGE

Radiation damage is the change of properties of a material caused by exposure to ionizing radiation, such as X-rays, gamma rays, neutrons, heavy-particle radiation, or fission fragments in nuclear fuel material [65]. Metals exposed to intense radiation in the form of neutrons or other energetic particles undergo lattice changes resembling in many respects those produced by severe cold work. The porous structure produced by dealloying has been described as similar to that found in irradiated materials [66]. Lattice vacancies, interstitial atoms, and dislocations are produced, and these increase the diffusion rate of specific impurities or alloyed components. In copper and nickel at room temperature, radiation damage leads to increased hardness, caused by clusters of interstitials and vacancies [66]. During radiation, a local temperature rise, called "temperature spike," may occur. There are two kinds of spikes: *thermal spikes*, in which few or no atoms leave their lattice sites, and *displacement spikes*, in which many atoms move into interstitial positions.

Except for chemicals produced in the environment by radiation, such as HNO_3 and H_2O_2, which have a secondary effect on corrosion, or formation of localized displacement spikes during radiation, the effect of radiation may be expected to parallel that of cold work. That is, metals for which the corrosion rate is controlled by oxygen diffusion should suffer no marked change in rate after irradiation. In acids, on the other hand, irradiated steel (but not pure iron) would presumably have a greater increase in rate than would irradiated nickel, which is less sensitive to cold working.

Austenitic stainless steels often become more sensitive to S.C.C. after cold working; on this basis, they might be expected to become more sensitive after irradiation. Indeed, *irradiation-assisted stress-corrosion cracking* (I.A.S.C.C.) is a cause of failure of core components in both boiling water reactors (BWR) and pressurized water reactors (PWR), and it has been observed, as intergranular S.C.C., in austenitic stainless steels and nickel-base alloys [67]. The complexities of S.C.C. are, in I.A.S.C.C., compounded by the effects of radiation on microstructure, microchemistry, and deformation behavior of the material as well as on the chemistry and electrochemistry of the solution. I.A.S.C.C. of austenitic stainless steels is related to chromium depletion along grain boundaries. In addition to the grain-boundary chromium depletion that can result from sensitization during welding (see Figs. 19.2 and Fig. 19.3, Section 19.2.3.1), chromium depletion in the grain boundary can also occur because of neutron irradiation, further reducing resistance to S.C.C. [68, 69]. One materials solution to mitigate I.G.S.C.C. in nuclear reactors is to use nuclear-grade stainless steels, such as type 316NG and type 304NG, which have a maximum carbon content of 0.020% and a nitrogen content of 0.060 to 0.100% to maintain strength at the lower carbon content compared to the non-nuclear grades (see Table 19.2, Section 19.2.2) [69]. Niobium-stabilized, type 347 stainless steel, with low carbon called 347NG, has also been used successfully [70]. Of course, good corrosion design is also essential—for example, a design without crevices, which can provide sites for localized corrosion (see Sections 2.3 and 19.2.).

The effect of irradiation on corrosion of some uranium alloys is considerable. For example, a 3% Cb–U alloy having moderate resistance to water at 260°C disintegrated within 1 h after irradiation. Furthermore, the corrosion of a zirconium alloy (Zircaloy-2, see Section 26.2) at 250°C in dilute uranyl sulfate solution containing small amounts of H_2SO_4 and $CuSO_4$ was very much increased by reactor irradiation [71]. In a review of the subject, Cox [72] stated that both fast neutron irradiation and presence of dissolved oxygen or an oxidizing electrolyte must be present simultaneously for any observed acceleration of corrosion to occur in high-temperature water. Accelerated corrosion of Zircaloys induced by irradiation is not observed above 400°C (750°F). The effects have been explained in terms of changes in the physical properties of the protective oxide film.

8.6 CORROSION FATIGUE

A metal that progressively cracks on being stressed alternately (reverse bending) or repeatedly is said to fail by fatigue. The greater the applied stress at each cycle, the shorter is the time to failure. A plot of stress versus number of cycles to failure, called the *S–N* curve, is shown in Fig. 8.15. A number of cycles at the corresponding stress to the right of the upper solid line results in failure, but no failure occurs for an infinite number of cycles at or below the *endurance limit* or *fatigue limit*. For steels, but not necessarily for other metals, a true endurance limit exists which is approximately half the tensile strength. The *fatigue strength* of any metal, on the other hand, is the stress below which failure does not occur within a stated number of cycles. Frequency of stress application is sometimes also stated because this factor may influence the number of cycles to failure.

In general, a corrosive environment can decrease the fatigue properties of any engineering alloy, meaning that corrosion fatigue is not dependent on material and environment [73]. In a corrosive environment, failure at a given stress level usually occurs within fewer cycles, and a true fatigue limit is not observed (Fig. 8.15). In other words, failure occurs at any applied stress if the number of cycles is sufficiently large. Cracking of a metal resulting from the combined action of a corrosive environment and repeated or alternate stress is called *corrosion*

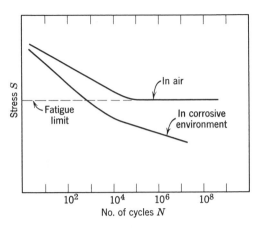

Figure 8.15. *S–N* curve for steels subjected to cyclic stress.

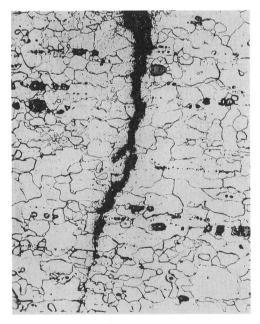

Figure 8.16. Corrosion-fatigue crack through mild steel sheet, resulting from fluttering of the sheet in a flue gas condensate (250×).

fatigue. The damage is almost always greater than the sum of the damage by corrosion and fatigue acting separately.

Corrosion-fatigue cracks are typically transgranular. They are often branched (Fig. 8.16), and several cracks are usually observed at the metal surface in the vicinity of the major crack accounting for failure. Fatigue cracks are similarly transgranular (exception: lead and tin), but rarely is there evidence of more than one major crack. In corrosion fatigue, corrosion pits may form at the metal surface at the base of which cracks initiate, but pitting is not a necessary precursor to failure.

Aqueous environments causing corrosion fatigue are numerous and are not specific, contrary to the situation for S.C.C., for which only certain ion–metal combinations result in damage. Corrosion fatigue of steel occurs in fresh waters, seawater, combustion-product condensates, general chemical environments, and so on, with the general rule that the higher the uniform corrosion rate, the shorter the resultant fatigue life.

Corrosion fatigue is a common cause of unexpected cracking of vibrating metal structures designed to operate safely in air at stresses below the fatigue limit. For example, the shaft of a ship propeller slightly out of line will operate satisfactorily until a leak develops, allowing water to impinge on the shaft in the area of maximum alternating stress. Cracks may then develop within a matter of days, resulting in eventual parting and failure of the shaft. Steel oil-well sucker rods, used to pump oil from underground, have limited life because of corrosion fatigue resulting from exposure to oil-well brines. Despite use of high-strength medium-alloy steels and oversized rods, failures from this source are a loss to the oil industry in the order of millions of dollars annually. Wire cables commonly fail by corrosion fatigue. Pipes carrying steam or hot liquids of variable temperature may fail similarly because of periodic expansion and contraction (thermal cycling).

The usual fatigue test conducted in air on a structural metal is influenced by both oxygen and moisture and, in part, therefore, always represents a measure of corrosion fatigue. In early tests, the fatigue strength for copper in a partial vacuum was found to be increased 14% over that in air. For mild steel the increase was only 5%, but for 70–30 brass it was 26% [74]. In later tests [75], fatigue life of oxygen-free, high-conductivity (OFHC) copper at 10^{-5} mmHg air pressure was found to be 20 times greater than at 1-atm air pressure. The main effect was attributed to oxygen; this had little effect on initiation of cracks, but had considerable effect on rate of crack propagation. Fatigue life of pure aluminum was also affected by air; but contrary to the situation for copper, water vapor in absence of air was equally effective. Gold, which neither chemisorbs oxygen nor oxidizes, had the same life whether fatigued in air or in a vacuum.

In some environments, fatigue cracking is supplemented by S.C.C., with the latter occurring under conditions of constant stress. This is seen in the behavior of a high-strengh low-alloy steel fatigued in both the absence and presence of moisture. Steels of greater than about 1140 MPa (165 ksi) yield strength (R_c37) are subject to S.C.C. in water at room temperature and have shorter fatigue life in moist air compared to dry air; whereas steels of lower strength, which do not undergo S.C.C. in water, have the same fatigue life (Fig. 8.17) [76].

Fresh waters and particularly brackish waters have a greater effect on the corrosion fatigue of steels than on that of copper, with the latter being a more corrosion-resistant metal. Stainless steels and nickel or nickel-base alloys are also better than carbon steels. In general, *resistance of a metal to corrosion fatigue is*

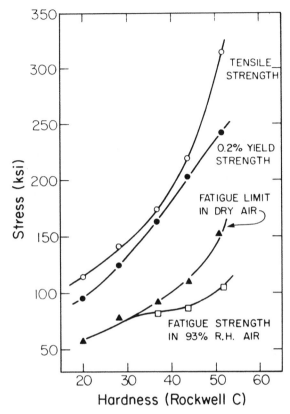

Figure 8.17. Tensile strength, yield strength, fatigue limit in dry air, and fatigue strength at 10^7 cycles in 93% relative humidity air, of 4140 steel heat-treated to various hardness values [76]. (With kind permission of Springer Science and Business Media.)

associated more nearly with its inherent corrosion resistance than with high mechanical strength.

A few values of corrosion fatigue strength determined by McAdam [77] in fresh and brackish waters are listed in Table 8.5. These values, besides varying with environment, are found to vary with rate of stressing, with temperature, and with aeration; hence, they are useful only for qualitative comparison of one metal with another. Unlike the fatigue limit in air, they are not usually reliable for engineering design. Conclusions from data of Table 8.5 and similar data are as follows:

1. There is no relation between corrosion fatigue strength and tensile strength.
2. Medium-alloy steels have only slightly higher corrosion fatigue strength than carbon steels.

TABLE 8.5. Fatigue Limit and Corrosion Fatigue Strength of Various Metals [77]

Metal	Fatigue Limit in Air (psi[a])	Corrosion Fatigue Strength (psi[a])		Damage Ratio (Corrosion Fatigue Strength)/ (Fatigue Limit)	
		Well Water[b]	Salt Water[c]	Well Water	Salt Water
		(10^7–10^8 cycles at 1450 cycles/min)			
0.11% C steel, annealed	25,000	16,000		0.64	
0.16% C steel, quenched and tempered	35,000	20,000	7,000	0.57	0.2
1.09% C steel, annealed	42,000	23,000		0.55	
3.5% Ni, 0.3% C steel, annealed	49,000	29,000		0.59	
0.9% Cr, 0.1% V, 0.5% C steel, annealed	42,000	22,000		0.52	
13.8% Cr, 0.1% C steel, quenched and tempered	50,000	35,000	18,000	0.70	0.36
17% Cr, 8% Ni, 0.2% C steel, hot-rolled	50,000	50,000	25,000	1.00	0.50
Nickel, 98.96%, annealed 760°C	33,000	23,500	21,500	0.71	0.65
Monel, 67.5% Ni, 29.5% Cu, annealed 760°C	36,500	26,000	28,000	0.71	0.77
Cupro-nickel, 21% Ni, 78% Cu, annealed 760°C	19,000	18,000	18,000	0.95	0.95
Copper, annealed 650°C	9,800	10,000	10,000	1.02	1.02
Aluminum, 99.4%, annealed	5,900		2,100		0.36
Aluminum, 98%, 1.2% Mn, hard	10,700	5,500	3,800	0.51	0.36
Duralumin, tempered	17,000	7,700	6,500	0.45	0.38
Brass, 60–40, annealed	21,000	18,000		0.86	

[a]To convert from psi to MPa, multiply by 6890.
[b]2 ppm $CaSO_4$, 200 ppm $CaCO_3$, 17 ppm $MgCl_2$, 140 ppm NaCl.
[c]Severn River water, with about one-sixth the salinity of seawater.

3. Heat treatment does not improve corrosion fatigue strength of either carbon or medium-alloy steels; residual stresses are deleterious.

4. Corrosion-resistant steels, particularly steels containing chromium, have higher corrosion fatigue strength than other steels.

5. Corrosion fatigue strength of all steels is lower in salt water than in fresh water.

8.6.1 Critical Minimum Corrosion Rates

In order for the corrosion process to affect fatigue life, the corrosion rate must exceed a minimum value. Such rates are conveniently determined by anodic polarization of test specimens in deaerated 3% NaCl, translating uniform

TABLE 8.6. Critical Minimum Corrosion Rates, 25°C, 30 Cycles/s

	i		Reference
	$\mu A/cm^{2\,a}$	gmd	
0.18% C mild steel	2.0	0.50	78
4140 low alloy steel			
R_c20	2.5	0.63	76
R_c37	2.0	0.50	76
R_c44	2.8	0.70	76
Nickel	1.2	<0.30	79
OFHC copper	100	28.5	80

^{a}To convert $\mu A/cm^2$ to A/m^2, divide by 100.

current densities by Faraday's Law into corrosion rates below which the fatigue life is unaffected. These measured current densities are found to be independent of the total area of specimen surface exposed to anodic currents. Various values at 30 cycles/s (1800 cycles/min) are listed in Table 8.6. It is expected that such values increase with cyclic frequency of the test. For the steels, critical corrosion rates are independent of (a) carbon content, (b) applied stress below the fatigue limit, and (c) heat treatment (hardness). The average value of 0.58 gmd (5.8 mdd) is less than the uniform corrosion rates of steels in aerated water and 3% NaCl (1–10 gmd, 10–100 mdd). But at pH 12, the uniform corrosion rate falls below the critical rate, and the fatigue life regains its value in air [78]. The existence of a critical rate in 3% NaCl explains why cathodic protection of steel against corrosion fatigue requires polarizing to only –0.49 V (S.H.E.), whereas the value required to attain zero corrosion rate lies 40 mV more active at –0.53 V.

For copper, the critical rate of 28.5 gmd (285 mdd) is much higher than the uniform corrosion rates in aerated water and 3% NaCl (0.4–1.5 gmd, 4–15 mdd); hence, the fatigue life of copper is observed to be about the same in air as in fresh and saline waters (Table 8.5).

8.6.2 Remedial Measures

There are several means available for reducing corrosion fatigue. In the case of mild steel, thorough deaeration of a saline solution restores the normal fatigue limit in air (Fig. 8.18) [81]. Cathodic protection to –0.49 V (S.H.E.) accomplishes the same result. Inhibitors are also effective [82, 83]. Sacrificial coatings (e.g., zinc or cadmium electrodeposited on steel) are effective because they cathodically protect the base metal at defects in the coating. One of the very first observations and diagnoses of corrosion fatigue, made by B. Haigh in about 1916, involved premature failure of steel towing cables exposed to seawater, and galvanizing

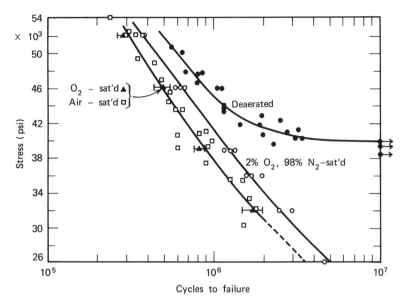

Figure 8.18. Effect of dissolved oxygen concentration in 3% NaCl, 25°C, on fatigue behavior of 0.18% C steel [81]. (Reprinted with permission of ASM International®. All rights reserved. www.asminternational.org.)

provided greatly increase life in this application [84]. Electrodeposits of tin, lead, copper, or silver on steel are also said to be effective without reducing normal fatigue properties; presumably, they owe their effectiveness to exclusion of the environment [85]. Organic coatings are useful if they contain inhibiting pigments in the prime coat. Shot peening the metal surface, or otherwise introducing compressive stresses, is beneficial.

8.6.3 Mechanism of Corrosion Fatigue

The mechanism of fatigue in air proceeds by localized slip within grains of the metal caused by alternating stress, resulting in slip steps at the metal surface. Adsorption of air on the clean metal surface exposed at slip steps probably prevents rewelding on the reverse stress cycle. Continued slip produces displaced clusters of slip bands, which protrude above the metal surface (extrusions); corresponding incipient cracks (intrusions) form elsewhere (Fig. 8.19). Below the fatigue limit, work hardening accompanying each cycle of plastic deformation eventually impedes further slip, which in turn impedes the fatigue process.

The basic effect of the corrosion process is to accelerate plastic deformation accompanied by formation of extrusions and intrusions. For this reason, damage by corrosion fatigue—a conjoint action of corrosion and fatigue—is greater than the damage caused by the sum of both acting separately. In addition, corrosion resistance of a metal is usually more important than high tensile strength in

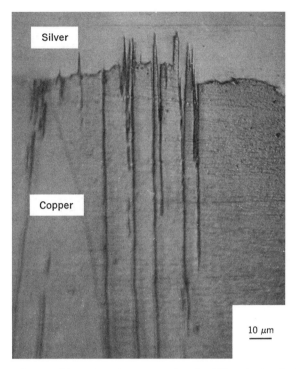

Figure 8.19. Extrusions and intrusions in copper after 6×10^5 cycles in air. Specimen plated with silver after test and mounted at an angle to magnify surface protuberances by about 20×. Overall magnification about 500×. [Figure 3 from W. Wood and H. Bendler, The fatigue process in copper as studied by electron metallography, *Trans. Metall. Soc. AIME* **224**, 182 (1962).]

establishing optimum resistance to corrosion fatigue. Since pure metals are not immune to uniform corrosion, they are also not immune to corrosion fatigue.

The mechanism of the slip dissolution process [86] takes place in the following steps:

- Diffusion of the active species to the crack tip
- Rupture of the protective film at the slip step
- Dissolution of the exposed surface
- Nucleation and growth of the protective film on the bare surface

8.7 FRETTING CORROSION

Fretting corrosion is another phenomenon that occurs because of mechanical stresses, and, in the extreme, it may lead to failure by fatigue or corrosion fatigue.

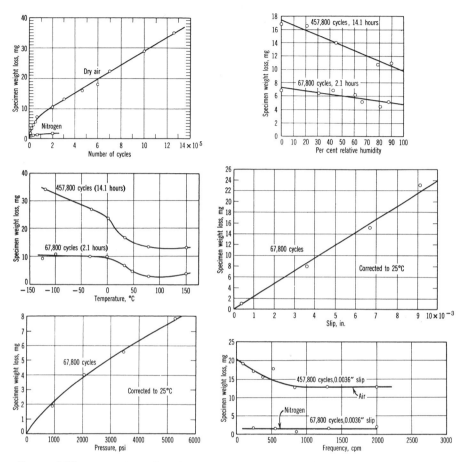

Figure 8.20. Weight loss of mild steel versus mild steel by fretting corrosion in air [87].

It is defined as damage occurring at the interface of two contacting surfaces, one or both being metal, subject to slight relative slip. The slip is usually oscillatory as, for example, that caused by vibration. Continuous slip, as when one roll moves slightly faster than another roll in contact, leads to similar damage. Wear corrosion and friction oxidation are terms that have also been applied to this kind of damage.

Damage by fretting corrosion is characterized by discoloration of the metal surface and, in the case of oscillatory motion, by formation of pits. It is at such pits that fatigue cracks eventually nucleate. The rapid conversion of metal to metal oxide may, in itself, cause malfunction of machines, because dimensional accuracy is lost, or corrosion products may cause clogging or seizing. The corrosion products are exuded from the faying surfaces, which, in the case of steel, are largely composed of αFe_2O_3 plus a small amount of iron powder [87]. In the case of nickel for continuous-slip experiments, the products are NiO and a small amount of nickel; for copper, they are Cu_2O with lesser amounts of CuO and copper [88].

Fretting corrosion is frequently the cause of failure of suspension springs, bolt and rivet heads, king pins of auto steering mechanisms, jewel bearings, variable-pitch propellers, shrink fits, contacts of electrical relays, connecting rods, and many parts of vibrating machinery. It may cause discoloration of stacked metal sheets during shipment. One of the first examples of fretting corrosion was recognized when automobiles were shipped some years ago by railroad from Detroit to the West Coast. Because of vibration, the ball-bearing races of the wheels became badly pitted by fretting corrosion, so that the automobiles were not operable. Damage was worse in winter than in summer, but could be avoided if the load on the wheels was relieved during shipment.

Fretting corrosion occurs on airframe structural surfaces that move against each other in a corrosive environment. High vibration levels and other types of mechanical stress result in small relative movement between parts of operational aircraft systems [89]. Another example of fretting corrosion can occur with railroad car wheels that are shrink-fitted onto their axles. If failure occurs causing the wheel to come off the axle, it can cause derailment and other damage [90]. Electrical contacts made of electroplated gold coatings can fail by fretting corrosion. Relative motion between the contacts removes the gold plate, and atmospheric corrosion of the substrate increases the contact resistance to intolerable levels [91]. Fretting corrosion is also a problem in nuclear reactors, particularly on heat-exchanger tubes and on fuel elements, where fluid flow generates vibrations [92].

Laboratory experiments [87] have shown that fretting corrosion of steel versus steel requires oxygen, but not moisture. Also, damage is less in moist air compared to dry air and is much less in a nitrogen atmosphere. Damage increases as temperature is lowered. The mechanism, therefore, is obviously not electrochemical. Increased load increases damage, accounting for the tendency of pits to develop at contacting surfaces because corrosion products—for example, αFe_2O_3—occupy more volume (2.2 times as much in the case of iron) than the metal from which the oxide forms. Because the oxides are unable to escape during oscillatory slip, their accumulation increases the stress locally, thereby accelerating damage at specific areas of oxide formation. Fretting corrosion is also increased by increased slip, provided that the surface is not lubricated. Increase in frequency for the same number of cycles decreases damage, but in nitrogen no frequency effect is observed. These effects are depicted in Fig. 8.20. Note that the initial rate of metal loss during the run-in period is greater than the steady-state rate.

8.7.1 Mechanism of Fretting Corrosion

When two surfaces touch, contact occurs only at relatively few sites, called asperities, where the surface protudes. Relative slip of the surfaces causes asperities to rub a clean track on the opposite surface, which, in the case of metal, immediately becomes covered with adsorbed gas, or it may oxidize superficially. The next asperity wipes off the oxide; or it may mechanically activate a reaction of adsorbed oxygen with metal to form oxide, which in turn is wiped off, forming another fresh metal track (Fig. 8.21). This is the chemical factor of fretting damage. In addition, asperities plow into the surface, causing a certain amount of wear by welding or shearing action, through which metal particles are dislodged. This is the mechanical factor. Any metal particles eventually are converted partially into oxide by secondary fretting action of particles rubbing against themselves or against adjacent surfaces. Also, the metal surface after an initial run-in period is fretted by oxide particles moving relative to the metal surface rather than by the mating opposite surface originally in contact (hence, electrical resistance between the surfaces is at first low, then becomes high and remains so).

An equation for weight loss W of a metal surface undergoing fretting corrosion by oscillatory motion has been derived [93] (Appendix, Section 29.7) on the basis of the model just described, which accounts reasonably satisfactorily for data of Fig. 8.20:

$$W = (k_0 L^{1/2} - k_1 L)\frac{C}{f} + k_2 lLC$$

where L is the load, C is the number of cycles, f is the frequency, l is the slip, and k_0, k_1, and k_2 are constants. The first two terms of the right-hand side of the equation represent the chemical factor of fretting corrosion. These become smaller the higher the frequency, f, corresponding to less available time for chemical reaction (or adsorption) per cycle. The last term is the mechanical factor independent of frequency, but proportional to slip and load. It is found that either the mechanical or chemical factor may dominate in accounting for damage depending on specific experimental conditions. In nitrogen, the mechanical factor alone is operable, the debris is metallic iron powder, and W is independent of frequency, f.

Fretting corrosion is *not* a high-temperature oxidation phenomenon. This is demonstrated by increased damage at below-room temperatures; by less

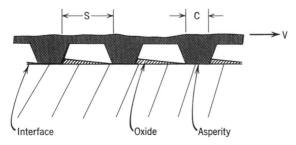

Figure 8.21. Idealized model of fretting action at a metallic surface.

damage at high frequencies, for which surface temperatures are highest; by the fact that oxide produced in fretting corrosion of iron is αFe_2O_3 and not the high-temperature form, Fe_3O_4; and, finally, by steel being badly fretted in contact with polymethacrylate plastic, which melts at 80°C, and hence the surface could have reached temperatures of this order, but not higher [94].

The effect of moisture when adsorbed on the metal surface may be that of a lubricant. In addition, hydrated αFe_2O_3 is probably less abrasive than the anhydrous oxide. At low temperatures, damage is greater presumably because O_2 can adsorb more rapidly or more completely than at high temperatures. More fundamental research is needed to help clarify these details of the general mechanism.

8.7.2 Remedial Measures

1. *Combination of a Soft Metal with a Hard Metal.* At sufficiently high loads, soft metals serve to exclude air at the interface; furthermore, a soft metal may yield by shearing instead of sliding at the interface, thereby reducing damage. Tin-, silver-, lead- indium-, and cadmium-coated metals in contact with steel have been recommended. Brass in contact with steel tends to be less damaging than steel versus steel. Combinations of stainless steels tend to be worst.

2. *Design of Contacting Surfaces to Avoid Slip Completely* (e.g., grit blasting, or otherwise roughening the surface). Intentional design to completely avoid slip is not always easy to accomplish, because damage is presumably caused by relative movement approaching the order of atomic dimensions. Increased load is effective in this direction if it is high enough to prevent slip; otherwise, damage is worse.

3. *Application of Cements* (e.g., rubber cement to the faying surfaces). Cements exclude air from the interface.

4. *Use of Lubricants.* Low-viscosity oils, particularly in combination with a phosphate-treated surface, can be helpful in reducing damage if the load is not too high. Low-viscosity oils diffuse more readily to the clean metal surface produced by oscillatory slip. Molybdenum sulfide is effective as a solid lubricant, particularly if baked onto the surface, but the beneficial effects tend to be temporary because the lubricant is eventually displaced by surface movement.

5. *Use of Elastomer Gaskets or Materials of Low Coefficient of Friction.* Rubber absorbs motion, thereby avoiding slip at the interface. Polytetra-fluoroethylene (Teflon) has a low coefficient of friction and reduces damage. Because of their relatively poor strength, materials of this kind are expected to be effective only at moderate loads.

6. *Use of Cobalt-base Alloys.* These are effective especially in the presence of liquid water or aqueous solutions (see Chapter 24).

REFERENCES

1. Z. Foroulis and H. Uhlig, *J. Electrochem. Soc.* **111**, 522 (1964).
2. H. Copson, in *Corrosion Handbook*, H. H. Uhlig, editor, Wiley, New York, 1948, p. 577.
3. E. A. Charles and R. N. Parkins, *Corrosion* **51**(7), 518 (1995).
4. W. Schroeder and A. Berk, Intercrystalline Cracking of Boiler Steel and Its Prevention, Bureau of Mines Bulletin 443, U.S. GPO, Washington, D.C., 1941.
5. R. Parkins, in *Stress Corrosion Cracking and Embrittlement*, W. Robertson, editor, Wiley, New York, 1956, p. 154.
6. M. A. Streicher, in *Uhlig's Corrosion Handbook*, 2nd edition, R. W. Revie, editor, Wiley, New York, 2000, pp. 1247–1248.
7. M. Holzworth et al., *Mater. Prot.* **7**(1), 36 (1968).
8. R. Pollard, in *Symposium on Stress-Corrosion Cracking of Metals*, ASTM–AIME, Philadelphia, PA, 1944, p. 437.
9. D. A. Jones, C. D. Kim, and B. E. Wilde, *Corrosion* **33**(2), 50 (1977).
10. R. Treseder and A. Wachter, *Corrosion* **5**, 383 (1949).
11. M. Kowaka and S. Nagata, *Corrosion* **24**, 427 (1968); A Brown, J. Harrison, and R. Wilkins, *Corros. Sci.* **10**, 547 (1970).
12. H. Buchholtz and R. Pusch, *Stahl u. Eisen* **62**, 21 (1942).
13. R. N. Parkins, in *Uhlig's Corrosion Handbook*, 2nd edition, R. W. Revie, editor, Wiley, New York, 2000, pp. 191–194.
14. National Energy Board of Canada, *Public Inquiry Concerning Stress Corrosion Cracking on Canadian Oil and Gas Pipelines*, Report of the Inquiry, MH-2–95, National Energy Board, Calgary, Alberta, Canada, November 1996.
15. H. H. Uhlig and J. Sava, *Trans. Am. Soc. Metals* **56**, 361 (1963).
16. M. Long and H. Uhlig, *J. Electrochem. Soc.* **112**, 964 (1965).
17. R. N. Parkins, *J. Iron Steel Inst.* **172**, 149 (1952).
18. K. Bohnenkamp, in *Proceedings, Conference Fundamental Aspects of Stress Corrosion Cracking*, National Association of Corrosion Engineers, Houston, TX, 1969, p. 374.
19. E. Baerlecken and W. Hirsch, *Stahl u. Eisen* **73**, 785 (1953).
20. H. Uhlig, K. Elayaperumal, and M. Talerman, *Corrosion* **30**, 229 (1974).
21. V. Agarwala, M. S. thesis, Department of Metallurgy and Materials Science, M.I.T., Cambridge, MA, 1966.
22. R. N. Parkins, in *Uhlig's Corrosion Handbook*, 2nd edition, R. W. Revie, editor, Wiley, New York, 2000, p. 200.
23. W. Lynes, *Corrosion* **21**, 125 (1965).
24. M. Scheil, in *Symposium on Stress-Corrosion Cracking of Metals*, ASTM–AIME, Philadelphia, PA, 1945, p. 433.
25. K. Sieradski and R. C. Newman, *Philos. Mag. A* **51** (1), 95 (1985).
26. T. Magnin, A. Chambreuil-Paret, J. P. Château, D. Delafosse, and B. Bayle, in *Corrosion–Deformation Interactions*, European Federation of Corrosion Publications, No. 21, The Institute of Materials, London, 1997, p. 12.
27. H. Uhlig and E. Cook, Jr., *J. Electrochem. Soc.* **116**, 173 (1969).

28. T. Beck and M. Blackburn, *Am. Inst. Aeronaut. Astronaut. J.* **6**, 326 (1968).

29. E. Dix, *Min. Metall. Eng.* **137**, 11 (1940).

30. S. P. Pednekar, A. K. Agrawal, H. E. Chaung, and R. W. Staehle, *J. Electrochem. Soc.* **126**, 701 (1979).

31. R. N. Parkins, Stress corrosion cracking, in *Environment-Induced Cracking of Metals*, Proceedings of the First International Conference on Environment-Induced Cracking of Metals, October 1988, R. P. Gangloff and M. B. Ives, editors, National Association of Corrosion Engineers, Houston, TX, 1990, p. 10.

32. I.-H. Lin and R. M. Thomson, *J. Mater. Res.* **1**, 1 (1986).

33. G. J. Dienes, K. Sieradzki, A. Paskin, and B. Massoumzadeh, *Surf. Sci.* **144**, 273 (1984).

34. R. H. Jones, in *ASM Handbook*, Vol. 13A, *Corrosion: Fundamentals, Testing, and Protection*, ASM International, Materials Park, OH, 2003, p. 362.

35. S. P. Lynch, *Hydrogen Effects in Metals*, A. W. Thompson and I. M. Bernstein, editors, The Metallurgical Society, 1981, p. 863.

36. S. P. Lynch, *Met. Sci.* **15** (10), 463 (1981).

37. S. P. Lynch, *J. Mater. Sci.* **20**, 3329 (1985); see also S. P. Lynch, in *Corrosion–Deformation Interactions, CDI '96, Second International Conference on Corrosion–Deformation Interactions in Conjunction with EUROCORR '96*, European Federation of Corrosion Publications, Number 21, The Institute of Metals, London, 1997, p. 206.

38. H. H. Uhlig, in *Physical Metallurgy of Stress Corrosion Fracture*, T. Rhodin, editor, Interscience, New York, 1959, pp. 1–17.

39. E. Coleman, D. Weinstein, and W. Rostoker, *Acta Metall.* **9**, 491 (1961).

40. A. Griffith, *Philos. Trans.* **A220**, 163 (1920).

41. R. Mears, R. Brown, and E. Dix, Jr., in *Symposium on Stress-Corrosion Cracking of Metals*, ASTM–AIME, Philadelphia, 1945, p. 323.

42. F. Fischer, *Kunststoffe* **55**, 453 (1965).

43. N. Petch and P. Stables, *Nature* **169**, 842 (1952).

44. W. Rostoker, J. M. McCaughey, and H. Markus, *Embrittlement by Liquid Metals*, Reinhold, New York, 1960.

45. H. Uhlig, *J. Appl. Electrochem. (Gr. Britain)* **9**, 191 (1979).

46. R. Newberg and H. Uhlig, *J. Electrochem. Soc.* **120**, 1629 (1973).

47. D. Hixson and H. Uhlig, *Corrosion* **32**, 56 (1976).

48. H. Mazille and H. Uhlig, *Corrosion* **28**, 427 (1972).

49. A. Morris, *Trans. Am. Inst. Min. Metall. Eng.* **89**, 256 (1930).

50. M. Elboujdaini, Y.-Z. Wang, R. W. Revie, R. N. Parkins, and M. T. Shehata, Stress corrosion crack initiation processes: Pitting and microcrack coalescence, Paper No. 00379, CORROSION/2000, NACE International, Houston, TX, 2000.

51. D. Broek, *Elementary Engineering Fracture Mechanics*, 4th revised edition, Kluwer Academic Publishers, The Netherlands, 1986, p. 85.

52. M. Speidel, in *Theory of Stress Corrosion Cracking in Alloys*, J. Scully, editor, North Atlantic Treaty Organization, Scientific Affairs Division, Brussels, 1971, pp. 298 and 302.

53. *ibid*, p. 305.

54. M. Elboujdaini, Hydrogen-induced cracking and sulfide stress cracking, in *Uhlig's Corrosion Handbook*, 2nd edition, R. W. Revie, editor, Wiley, New York, 2000, pp. 205–220.

55. A. Schuetz and W. Robertson, *Corrosion* **13**, 437t (1957).

56. P. Bastien, in *Physical Metallurgy of Stress Corrosion Fracture*, T. Rhodin, editor, Interscience, New York, 1959, p. 311.

57. H. H. Uhlig, *Trans. Am. Inst. Min. Metall. Eng.* **158**, 183 (1944).

58. A. Troiano, *Trans. Am. Soc. Metals* **52**, 54 (1960).

59. H. Johnson, J. Morlet, and A. Troiano, *Trans. Am. Inst. Min. Metall. Eng.* **212**, 528 (1958).

60. A. Asphahani and H. Uhlig, *J. Electrochem. Soc.* **122**, 174 (1975).

61. J. Marquez, I. Matsushima, and H. Uhlig, *Corrosion* **26**, 215 (1970).

62. B. Wilde, C. Kim, and E. Phelps, *Corrosion* **36**, 625 (1980).

63. NACE Committee T-1F, *Materials Performance* **12**(3), 41 (1973).

64. C. Carter and M. Hyatt, in *Stress Corrosion Cracking and Hydrogen Embrittlement of Iron-Base Alloys*, R. Staehle et al., editor, National Association of Corrosion Engineers, Houston, TX, 1977, p. 551.

65. *ASM Handbook*, Vol. 13A, *Corrosion: Fundamentals, Testing, and Protection*, ASM International, Materials Park, OH, 2003, p. 1023.

66. R. H. Jones, in *Environment-Induced Cracking of Metals*, Proceedings of the First International Conference on Environment-Induced Cracking of Metals, October 1988, R. P. Gangloff and M. B. Ives, editors, National Association of Corrosion Engineers, Houston, TX, 1990, p. 284.

67. R. H. Jones and S. M. Bruemmer, Environment-induced crack growth processes in nickel-base alloys, in *Environment-Induced Cracking of Metals*, Proceedings of the First International Conference on Environment-Induced Cracking of Metals, October 1988, R. P. Gangloff and M. B. Ives, editors, National Association of Corrosion Engineers, Houston, TX, 1990, pp. 289–290; G. S. Was and P. L. Andresen, Stress corrosion cracking behavior of alloys in aggressive nuclear reactor core environments, *Corrosion* **63** (1), 19 (2007).

68. F. P. Ford, B. M. Gordon, and R. M. Horn, Corrosion in boiling water reactors, in *ASM Handbook*, Vol. 13C, *Corrosion: Environments and Industries*, ASM International, Materials Park, OH, 2006, p. 347.

69. G. S. Was, J. Busby, and P. L. Andresen, Effect of irradiation on stress-corrosion cracking and corrosion in light water reactors, in *ASM Handbook*, Vol. 13C, *Corrosion: Environments and Industries*, ASM International, Materials Park, OH, 2006, pp. 386–404.

70. Ref. 68, p. 354.

71. M. Simnad, in *The Effects of Radiation on Materials*, J. J. Harwood et al., editors, Reinhold, New York, 1958, pp. 129–133.

72. D. Cox, *J. Nucl. Mater.* **28**, 1 (1968). See also R. Asher et al., *Corros. Sci.* **10**, 695 (1970).

73. Y.-Z. Wang, in *Uhlig's Corrosion Handbook*, 2nd edition, R. W. Revie, editor, Wiley, New York, 2000, p. 221.

74. H. Gough and D. Sopwith, *J. Inst. Metals* **49**, 93 (1932).

75. N. Wadsworth, in *Internal Stresses and Fatigue In Metals*, G. Rassweiler and W. Grube, editors, Elsevier, New York, 1959, pp. 382–396.

76. H. Lee and H. Uhlig, Corrosion fatigue of type 4140 high strength steel, Fig. 6, *Metall. Trans.* **3**, 2952 (1972).

77. D. McAdam, Jr., *Proc. Am Soc. Testing Mater.* **27** (II), 102 (1927).

78. D. Duquette and H. Uhlig, *Trans. Am. Soc. Metals* **62**, 839 (1969).

79. A. Asphahani, Ph.D. thesis, Department of Materials Science and Engineering, M.I.T., Cambridge, MA, February 1975.

80. H. Uhlig, in *Corrosion Fatigue*, O. Devereux et al., editors, National Association of Corrosion Engineers, Houston, TX, 1972, p. 270.

81. D. Duquette and H. Uhlig, *Trans. Am. Soc. Metals* **61**, 449 (1968).

82. Y.-Z. Wang, in *Uhlig's Corrosion Handbook*, 2nd edition, R. W. Revie, editor, Wiley, New York, 2000, p. 231.

83. R. L. Martin, in *ASM Handbook*, Vol. 13A, *Corrosion: Fundamentals, Testing, and Protection*, ASM International, Materials Park, OH, 2003, p. 882.

84. H. Gough, *J. Inst. Metals* **49**, 17 (1932).

85. E. Gadd, in *International Conference on Fatigue of Metals*, Institution of Mechanical Engineers, London, 1956, p. 658.

86. S. Suresh, *Fatigue of Materials*, Cambridge Solid State Science Series, Cambridge University Press, Cambridge, U.K., 1991, pp. 363–368.

87. I-Ming Feng and H. Uhlig, *J. Appl. Mech.* (published by ASME) **21**, 395 (1954).

88. M. Fink and U. Hofmann, *Z. Anorg. Allg. Chem.* **210**, 110 (1933); *Z. Metallk.* **24**, 49 (1932).

89. M. L. Bauccio, in *ASM Handbook*, Vol. 13, *Corrosion*, ASM International, Materials Park, OH, 1987, p. 1030.

90. W. Glaeser and I. G. Wright, in *ASM Handbook*, Vol. 13A, *Corrosion: Fundamentals, Testing, and Protection*, ASM International, Materials Park, OH, 2003, p. 324.

91. M. Antler and M. H. Drozdowicz, *Wear* **74**, 27 (1981–1982).

92. P. L. Ko, J. H. Tromp, and M. K. Weckworth, in *Materials Evaluations under Fretting Conditions*, STP 780, American Society for Testing and Materials, Philadelphia, 1981, pp. 86–105.

93. H. H. Uhlig, *J. Appl. Mech.* **21**, 401 (1954).

94. K. H. R. Wright, *Proc. Inst. Mech. Eng.* **1B**, 556 (1952–1953).

GENERAL REFERENCES

Stress-Corrosion Cracking

C. A. Brebbia, V. G. DeGiorgi, and R. A. Adey, editors, Modelling stress corrosion cracking and corrosion fatigue, Section 5 in *Simulation of Electrochemical Processes*, WIT Press, Southampton, U.K., 2005, pp. 215–244.

B. F. Brown, *Stress Corrosion Cracking Control Measures*, National Association of Corrosion Engineers, Houston, TX, 1977.

R. P. Gangloff and M. B. Ives, editors, *Environment-Induced Cracking of Metals*, Proceedings of the First International Conference on Environment-Induced Cracking of Metals, October 1988, National Association of Corrosion Engineers, Houston, TX, 1990.

R. H. Jones, editor, *Chemistry and Electrochemistry of Corrosion and Stress Corrosion Cracking: A Symposium Honoring the Contributions of R. W. Staehle*, TMS, Warrendale, PA, 2001.

T. Magnin, editor, *Corrosion-Deformation Interactions, CDI '96*, Second International Conference on Corrosion-Deformation Interactions, Nice, France, 1996, The Institute of Materials, London, 1997.

R. N. Parkins, Stress corrosion cracking, in *Uhlig's Corrosion Handbook*, 2nd edition, R. W. Revie, editor, Wiley, New York, 2000, pp. 191–204.

Fatigue and Corrosion Fatigue

S. Suresh, *Fatigue of Materials*, 2nd edition, Cambridge University Press, Cambridge, U.K., 1998.

Y.-Z. Wang, Corrosion fatigue, in *Uhlig's Corrosion Handbook*, 2nd edition, R. W. Revie, editor, Wiley, New York, 2000, pp. 221–232.

Fretting Corrosion

David W. Hoeppner, V. Chandrasekaran, and Charles B. Elliott III, editors, *Fretting Fatigue: Current Technology and Practices*, STP 1367, ASTM, West Conshohocken, PA, 2000.

Materials Evaluation under Fretting Conditions, Special Technical Publication 780, American Society for Testing and Materials, Philadelphia, 1981.

R. B. Waterhouse, *Fretting Corrosion*, Pergamon, Oxford, 1972.

R. B. Waterhouse, *Fretting Fatigue*, Applied Science Publishers, London, 1981.

R. B. Waterhouse, Fretting wear, in *ASM Handbook*, Vol. 18, *Friction, Lubrication, and Wear Technology*, ASM International, Materials Park, OH, 1992, pp. 242–256.

Fretting wear failures, in *ASM Handbook*, Vol. 11, *Failure Analysis and Prevention*, ASM International, Materials Park, OH, 2002, pp. 922–940.

Effects of Radiation

T. R. Allen et al., editors, *Proceedings, 22nd Int. Symposium on Effects of Radiation on Materials*, Boston, 2004, STP 1475, ASTM International, West Conshohocken, PA, 2006.

G. S. Was, J. Busby, and P. L. Andresen, Effect of irradiation on stress-corrosion cracking and corrosion in light water reactors, in *ASM Handbook*, Vol. 13C, *Corrosion: Environments and Industries*, ASM International, Materials Park, Ohio, 2006, pp. 386–414.

Hydrogen Damage

B. Craig, Hydrogen damage, in *ASM Handbook*, Vol. 13A, *Corrosion: Fundamentals, Testing, and Protection*, ASM International, Materials Park, OH, 2003, pp. 367–380.

M. Elboujdaini, Hydrogen-induced cracking and sulfide stress cracking, in *Uhlig's Corrosion Handbook*, 2nd edition, R. W. Revie, editor, Wiley, New York, 2000, pp. 205–220.

M. Elboujdaini, Test methods for wet H_2S cracking, in *Uhlig's Corrosion Handbook*, 2nd edition, R. W. Revie, editor, Wiley, New York, 2000, pp. 1129–1137.

R. D. Kane, Corrosion in petroleum refining and petrochemical operations, in *ASM Handbook*, Vol. 13C, *Corrosion: Environments and Industries*, ASM International, Materials Park, OH, 2006, pp. 992–998.

PROBLEMS

1. Engell and Bäumel (T. N. Rhodin, editor, *Physical Metallurgy of Stress Corrosion Fracture*, Interscience, New York, 1959, p. 354) reported that cracks in iron stressed in boiling calcium nitrate solution propagate discontinuously at the rate of 0.2 mm/s. To what corrosion current density does this rate correspond? If this rate is typical, what does your answer indicate with regard to the electrochemical mechanism of crack growth?

2. The residual energy of a cold-worked steel as determined calorimetrically is 21 J/g. Calculate the increased tendency in volts for iron to corrode when cold worked. (Assume that the entropy change is negligible.)

3. The critical potential for stress-corrosion cracking of moderately cold-worked 18–8 stainless steel in deaerated $MgCl_2$ solution at 130°C is –0.145 V (S.H.E.).
   (a) What effect on SCC of 18–8 results from coupling to nickel (ϕ_{corr} = –0.18 V)?
   (d) On adding HCl? Why?
   (c) Under (b), below what value of pH at 130°C does the cathode potential of 18–8 become more noble than $\phi_{critical}$?

Answers to Problems

1. 545 A/cm².
2. 0.006 V.
3. (c) 1.8.

9

ATMOSPHERIC CORROSION

9.1 INTRODUCTION

In the absence of moisture, iron exposed to the atmosphere corrodes at a negligible rate. For example, steel parts abandoned in the desert remain bright and tarnish-free for long periods of time. Also, the corrosion process cannot proceed without an electrolyte; hence, in climates below the freezing point of water or of aqueous condensates on the metal surface, rusting is negligible. Ice is a poor electrolytic conductor. Incidence of corrosion by the atmosphere depends, however, not only on the moisture content of air, but also on the particulate matter content and gaseous impurities that favor condensation of moisture on the metal surface.

Ambient air quality in the United States has improved dramatically since the Clean Air Act was enacted in 1970. The Environmental Protection Agency (EPA) monitors air quality and compares the data with the National Ambient Air Quality Standards (NAAQS), which have been established for ozone (O_3), carbon monoxide (CO), nitrogen dioxide (NO_2), sulfur dioxide (SO_2), lead (Pb), and particulate matter (PM, airborne particles of any composition). Improvements in air quality help to mitigate atmospheric corrosion, as will be discussed in this chapter.

Corrosion and Corrosion Control, by R. Winston Revie and Herbert H. Uhlig
Copyright © 2008 John Wiley & Sons, Inc.

9.2 TYPES OF ATMOSPHERES

Atmospheres vary considerably with respect to moisture, temperature, and contaminants; hence, atmospheric corrosion rates vary markedly around the world. Approaching the seacoast, air is laden with increasing amounts of sea salt, in particular NaCl. At industrial areas, SO_2, H_2S, NH_3, NO_2, and various suspended salts are encountered. Approximate concentration ranges of corrosive gases in urban areas are presented in Table 9.1 [1]. Acids that can form from these gases include sulfuric acid (H_2SO_4), nitric acid (HNO_3), and organic acids, such as formic acid (HCOOH) and acetic acid (CH_3COOH).

A metal that resists corrosion in one atmosphere may lack effective corrosion resistance elsewhere; hence, relative corrosion behavior of metals changes with location; for example, galvanized iron performs well in rural atmospheres, but is relatively less resistant to industrial atmospheres. On the other hand, lead performs in an industrial atmosphere at least as well as, or better than, elsewhere because a protective film of lead sulfate forms on the surface.

Recognition of marked differences in corrosivity has made it convenient to divide atmospheres into types. The major types are marine, industrial, tropical, arctic, urban, and rural. There are also subdivisions, such as wet and dry tropical, with large differences in corrosivity. Also, specimens exposed to a marine atmosphere corrode at greatly differing rates depending on proximity to the ocean. At Kure Beach, North Carolina, specimens of steel located 24 m (80 ft) from the ocean, where salt water spray is frequent, corroded about 12 times more rapidly than similar specimens located 240 m (800 ft) from the ocean [2].

9.3 CORROSION-PRODUCT FILMS

Corrosion-product films formed in the atmosphere tend to be protective; that is, the corrosion rate decreases with time (Fig. 9.1) [3]. This is true to a lesser extent of pure iron, for which the rate is relatively high, compared to the copper-bearing or low-alloy steels, which are more resistant. Rust films on the latter steels tend

TABLE 9.1. Atmospheric Corrosive Gases in Outdoor Urban Environments [1]

Gas	Approximate Concentration Range (ppbv)
H_2S	0.2–700
SO_2	3–1000
NH_3	1–90
HCl	0.5–100
NO_2	0.5–300
O_3	0.9–600
RCOOH	0.5–30

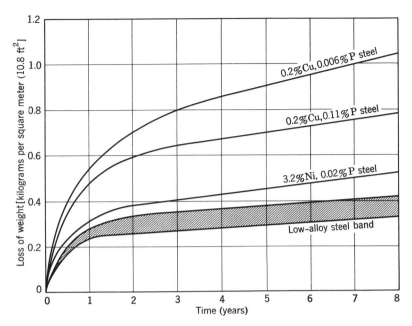

Figure 9.1. Atmospheric corrosion of steels as a function of time in an industrial environment (C. Larrabee, in *Corrosion Handbook*, H. H. Uhlig, editor, Wiley, New York, 1948, p. 124)

to be compact and adherent, whereas on pure iron they are a powdery loose product. The corrosion rate eventually reaches steady state and usually changes very little on further exposure. This is characteristic of other metals as well, as can be seen from data obtained by the American Society for Testing and Materials for various metals exposed 10 or 20 years to several atmospheres (Table 9.2) [4]. Within the experimental error of such determinations, the rate for a 20-year period is about the same as that for a 10-year period. The data in Table 9.2 also show the beneficial effect of alloying elements, primarily 1.1% Cr and 0.4% Cu, in the low-alloy steel compared to the 0.2% C steel.

Data of Fig. 9.1 follow the relation

$$p = kt^n \qquad (9.1)$$

where p can be expressed as specimen weight loss (g/m^2), or as specimen penetration (μm), during time t (years), and k and n are constants that depend on the metal and on the atmospheric conditions (i.e., climatic and pollution factors) at the test site. In addition to carbon steels, this relation is also found to apply to atmospheric test data for galvanized, aluminized, and 55% Al–Zn coatings on steel [5–7]. Values of n typically range from about 0.5 to 1. The constant, $n = 1$, applies for a linear rate law; that is, the corrosion product film is *not* protective. In comparison with carbon steels, weathering steels have very low values of n,

TABLE 9.2. Average Atmospheric Corrosion Rates of Various Metals for 10- and 20-Year Exposure Times, mils/year[a] (American Society for Testing and Materials) [4]

Metal	Atmosphere					
	New York City (Urban Industrial)		La Jolla, CA (Marine)		State College, PA (Rural)	
	10 Years	20 Years	10 Years	20 Years	10 Years	20 Years
Aluminum	0.032	0.029	0.028	0.025	0.001	0.003
Copper	0.047	0.054	0.052	0.050	0.023	0.017
Lead	0.017	0.015	0.016	0.021	0.019	0.013
Tin	0.047	0.052	0.091	0.112	0.018	—
Nickel	0.128	0.144	0.004	0.006	0.006	0.009
65% Ni, 32% Cu, 2% Fe, 1% Mn (Monel)	0.053	0.062	0.007	0.006	0.005	0.007
Zinc (99.9%)	0.202	0.226	0.063	0.069	0.034	0.044
Zinc (99.0%)	0.193	0.218	0.069	0.068	0.042	0.043
0.2% C steel[b] (0.02% P, 0.05% S, 0.05% Cu, 0.02% Ni, 0.02% Cr)	0.48	—	—	—	—	—
Low-alloy steel[b] (0.1% C, 0.2% P, 0.04% S, 0.03% Ni, 1.1% Cr, 0.4% Cu)	0.09	—	—	—	—	—

[a]1 mil/year = 0.001 in./year = 0.0254 mm/year = 25.4 μm/year.
[b]Kearney, New Jersey (near New York City); values cited are one-half reduction of specimen thickness [C. P. Larrabee, *Corrosion* **9**, 259 (1953)].

usually less than $\frac{1}{2}$ [8]. A value of $n = 0.5$ corresponds to a parabolic rate law.

From (9.1), the linear bilogarithmic law may be expressed as

$$\log p = A + B \log t \tag{9.2}$$

where $A = \log k$ and $B = n$. The atmospheric behavior of a specific metal at a specific location can be described using the two parameters A and B. This bilogarithmic law can be very useful in predicting long-term atmospheric corrosion damage based on tests of shorter duration. Extrapolation up to 20–30 years from tests of 4 years is possible with reasonable confidence [9]. Over the longer term, changes in the environment are likely to be more significant than deviations from the model.

The mean corrosion rate, p/t, can be calculated as

$$\log\left(\frac{p}{t}\right) = A + (B - 1)\log t \tag{9.3}$$

and the instantaneous corrosion rate, dp/dt, can be expressed as

$$\log\left(\frac{dp}{dt}\right) = A + \log B + (B-1)\log t \tag{9.4}$$

When there is a linear relationship between $\log p$ and $\log t$, there are also linear relationships between $\log p/t$ and $\log t$, and between $\log dp/dt$ and $\log t$.

Metal surfaces located where they become wet or retain moisture, but where rain cannot wash the surface, may corrode more rapidly than specimens fully exposed. The reason for this is that sulfuric acid, for example, absorbed by rust will continue to accelerate corrosion, perhaps by means of the cycle

$$Fe \xrightarrow{H_2SO_4 + \frac{1}{2}O_2} FeSO_4 \xrightarrow{\frac{1}{4}O_2 + \frac{1}{2}H_2SO_4} \frac{1}{2}Fe_2(SO_4)_3 \xrightarrow{\frac{1}{2}H_2O} \frac{1}{2}Fe_2O_3 + \frac{3}{2}H_2SO_4 \tag{9.5}$$

Intermediate formation of ferric sulfate has not been demonstrated, and so $FeSO_4$ may oxidize directly to Fe_2O_3. Nevertheless, rust contaminated in this way catalyzes the formation of more rust. Direct exposure of a metal to rain may, therefore, be beneficial, compared to a partially sheltered exposure. This advantage presumably would not extend to uncontaminated atmospheres.

9.4 FACTORS INFLUENCING CORROSIVITY OF THE ATMOSPHERE

In all except the most corrosive atmospheres, the average corrosion rates of metals are generally lower when exposed to air than when exposed to natural waters or to soils. This fact is illustrated by the data of Table 9.3 for steel, zinc, and copper in three atmospheres compared to average rates in seawater and a

TABLE 9.3. Comparison of Atmospheric Corrosion Rates with Average Rates in Seawater and in Soils[a]

Environment	Corrosion Rate (gmd)		
	Steel	Zinc	Copper
Rural atmosphere	—	0.017	0.014
Marine atmosphere	0.29	0.031	0.032
Industrial atmosphere	0.15	0.10	0.029
Seawater	2.5	1.0	0.8
Soil	0.5	0.3	0.07

[a]Atmospheric tests on 0.3% copper steel, 7 $\frac{1}{2}$-year exposure, from C. Larrabee, *Corrosion* **9**, 259 (1953). Atmospheric rates for zinc and copper, 10-year exposure, from *Symposium on Atmospheric Exposure Tests on Non-Ferrous Metals*, ASTM, 1946. Seawater data from *Corrosion Handbook*, H. H. Uhlig, editor, Wiley, New York, 1948. Soil data for steel are averaged for 44 soils, 12-year exposure; for zinc, 12 soils, 11-year exposure; for copper, 29 soils, 8-year exposure–from *Underground Corrosion*, M. Romanoff, Circ. 579, National Bureau of Standards, Washington, D.C. 1957.

variety of soils. In addition, atmospheric corrosion of passive metals (e.g., aluminum and stainless steels) tends to be more uniform and with less marked pitting than corrosion in waters or soils.

Specific factors influencing the corrosivity of atmospheres are particulate matter, gases in the atmosphere, and moisture (critical humidity). In the United States, the Environmental Protection Agency (EPA) compiles data on ambient outdoor-air quality from information obtained at air-monitoring stations, where concentrations of the more common gases found in trace amounts in the atmosphere are measured continuously. Because of the importance of trace constituents on atmospheric corrosion behavior, the engineer or architect would be well advised to review air-quality data for a corrosion assessment at the design stage, to avoid disappointment at the operational stage. Microclimatic conditions (e.g., east versus west exposure, sunlight, wind direction, proximity to a highway where deicing salts are used) can be different from macroclimates and should be considered.

9.4.1 Particulate Matter

Particulate matter (PM) is the term for a mixture of solid particles and liquid droplets in air. PM consists of several components, including acids, organic chemicals, metals, and soil or dust particles. The importance of atmospheric dust was established in the early experiments of Vernon [11], who exposed specimens of iron to an indoor atmosphere, some specimens being entirely enclosed by a cage of single-thickness muslin measuring several inches larger in size than the specimen. After several months, the unscreened specimens showed rust and appreciable gain in weight, whereas the muslin-screened specimens showed no rust whatsoever and had gained weight only slightly.

Particulate matter is found near roads and some industries, in smoke and haze; it can be directly emitted from sources such as forest fires, and it can be formed when gases emitted from power plants and automobiles react in the air. The NAAQSs for particulate matter, revised in 2006, are set at $150\,\mu g/m^3$ as the 24-h average for coarse particulates (2.5- to 10-μm diameter) and are set at $35\,\mu g/m^3$ as the 24-h average for fine particulates ($\leq$2.5-μm diameter) [12]. Particle pollution is controlled by reducing directly emitted particles and by reducing emissions of pollutants that are gases when emitted, but form particles in the atmosphere. In the United States in 2006, the average concentration of fine particulates amounted to about $13\,\mu g/m^3$ [13].

Particulate matter can be a primary contaminant of many atmospheres. In the 1950s, it was estimated that the average city air contained about $2\,mg/m^3$, with higher values for an industrial atmosphere, reaching $1000\,mg/m^3$ or more [14]. It was estimated that more than $35,000\,kg$ of dust per km^2 (100 tons per square mile) settled every month over an industrial city [15].

In contact with metallic surfaces, particulate matter influences the corrosion rate in an important way. Industrial atmospheres carry suspended particles of carbon and carbon compounds, metal oxides, H_2SO_4, $(NH_4)_2SO_4$, $NaCl$, and other salts. Marine atmospheres contain salt particles that may be carried many miles inland, depending on magnitude and direction of the prevailing winds. These substances, combined with moisture, initiate corrosion by forming galvanic or differential aeration cells; or, because of their hygroscopic nature, they form an electrolyte on the metal surface. Dust-free air, therefore, is less apt to cause corrosion than is air heavily laden with dust, particularly if the dust consists of water-soluble particles or of particles on which H_2SO_4 is adsorbed.

9.4.2 Gases in the Atmosphere

The carbon dioxide normally present in air neither initiates nor accelerates corrosion. Steel specimens rust in a carbon-dioxide-free atmosphere as readily as in the normal atmosphere. Early experiments by Vernon showed that the normal carbon dioxide content of air actually decreases corrosion [16], probably by favoring a more protective rust film.

A trace amount of hydrogen sulfide in contaminated atmospheres causes the observed tarnish of silver and may also cause tarnish of copper. The tarnish films are composed of Ag_2S and a mixture of $Cu_2S + CuS + Cu_2O$, respectively.

Important atmospheric pollutants are sulfur dioxide (SO_2), nitrogen oxides (NO_x), ozone (O_3), and chloride ions (Cl^-). Sulfur dioxide and nitrogen oxides form corrosive acids, and ozone is a powerful oxidizing agent.

Sulfur dioxide originates predominantly from the burning of coal, oil, and gasoline. In New York City in the 1950s, it was estimated that about 1.5 million tons of sulfur dioxide were produced every year from burning coal and oil [17]. This amount was equivalent to burdening the atmosphere with an average of 6300 tons of H_2SO_4 every day and has been reduced by limiting the allowable sulfur content of fuels burned within city limits. Since fuel consumption is higher in winter, sulfur dioxide contamination is also higher (Fig. 9.2). It is also obvious from this cause that the average sulfur dioxide content of the air (and

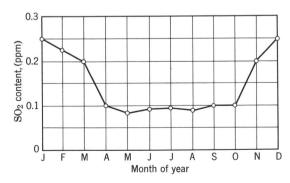

Figure 9.2. Variation of average sulfur dioxide content of New York City air with time of year [17].

TABLE 9.4. Variation of SO_2 Content of Air with Distance from Center of City (H. Meller, J. Alley, and J. Sherrick, quoted in Ref. 17)

City		Parts per Million					
	Distance	0–5	5–10	10–15	15–20	20–25	25–30
	Miles Kilometers	0–8	8–16	16–24	24–32	32–40	40–48
Detroit		0.023	0.012	0.006	0.004	0.004	0.005
Philadelphia–Camden		0.030	0.018	0.016	0.021	0.012	0.012
Pittsburgh		0.060	0.030	0.015	0.018	0.009	0.010
St. Louis		0.111	0.048	0.029	0.020	0.018	0.014
Washington, D.C.		0.003	0.001	0.001	0.001	0.001	0.002

corresponding corrosivity) falls off with distance from the center of an industrial city, and it is clear that this effect is not as pronounced in the case of a residential city, such as Washington, D.C. (Table 9.4).

Although the SO_2 levels in the air of many major centers of population around the world may be similar to the data in Fig. 9.2, the engineering and industrial responses to the U.S. Clean Air Act have resulted in a major decline in the levels of air pollutants. About 10% of SO_2 emissions result from industrial processes, and transportation sources contribute most of the remainder. The EPA estimates that ambient SO_2 levels have decreased by more than 50% since 1983. In comparison with the data in Fig. 9.2, the annual average concentration of SO_2 in New York City air in 2006 was about 0.01 ppm, and the NAAQS for SO_2 is 0.03 ppm (annual arithmetic mean) [18].

A high sulfuric acid content of industrial and urban atmospheres shortens the life of metal structures (see Tables 9.2 and 9.3). The effect is most pronounced for metals that are not particularly resistant to sulfuric acid, such as zinc, cadmium, nickel, and iron. It is less pronounced for metals that are more resistant to dilute sulfuric acid, such as lead, aluminum, and stainless steels. Copper, forming a protective basic copper sulfate film, is more resistant than nickel or 70% Ni–Cu alloy, on which the corresponding films are less protective. In the industrial atmosphere of Altoona, Pennsylvania, galvanized steel sheets [0.381 kg zinc per m^2, 0.028 mm thick (1.25 oz zinc per ft^2, 1.1 mil thick)] began to rust after 2.4 years, whereas in the rural atmosphere of State College, Pennsylvania, rust appeared only after 14.6 years [19].

Copper exposed to industrial atmospheres forms a protective green-colored corrosion product called a *patina*, composed mostly of basic copper sulfate, $CuSO_4 \cdot 3Cu(OH)_2$. A copper-covered church steeple on the outskirts of a town may develop such a green patina on the side facing prevailing winds from the city, but remain reddish-brown on the opposite side, where sulfuric acid is less readily available. Near the seacoast, a similar patina forms, composed in part of basic copper chloride.

Industrial atmospheres can cause S.C.C of copper-base alloys, mostly accounted for by presence of nitrogen oxides (see Section 20.2.3). Copper did not fail, but copper alloys containing >20% Zn failed on exposures for up to 8 years [20].

Nickel is quite resistant to marine atmospheres, but is sensitive to sulfuric acid of industrial atmospheres (Table 9.2), forming a surface tarnish composed of basic nickel sulfate. Corrosion in the industrial atmosphere of New York City is about 30 times higher than in the marine atmosphere of La Jolla, California, and about 20 times higher than in the rural atmosphere of State College, Pennsylvania (Table 9.2).

9.4.3 Moisture (Critical Humidity)

From previous discussions, it is apparent that, in an uncontaminated atmosphere at constant temperature, appreciable corrosion of a pure metal surface would not be expected at any value of relative humidity below 100%. Practically, however, because of normal temperature fluctuations (relative humidity increases on decrease of temperature) and because of hygroscopic impurities in the atmosphere or in the metal itself, the relative humidity must be reduced to values much lower than 100% in order to ensure that no water condenses on the surface. In very early studies, Vernon discovered that a critical relative humidity exists below which corrosion is negligible [21]. Experimental values for the critical relative humidity are found to fall, in general, between 50% and 70% for steel, copper, nickel, and zinc. Typical corrosion behavior of iron as a function of relative humidity of the atmosphere is shown in Fig. 9.3. In a complex or severely polluted atmosphere, a critical humidity may not exist [22].

An important factor determining susceptibility to atmospheric corrosion of a metal in a particular environment is the percentage of time that the critical humidity is exceeded [23]. This period of time is called the "time of wetness." It is determined by measuring the potential between a corroding metal specimen and a platinum electrode [23, 24]. Surface moisture from either precipitation or condensation is the cell electrolyte. To estimate atmospheric corrosion rates, Kucera et al. designed the device shown in Fig. 9.4 [25, 26]. The cell B is placed about 1 m above ground level with the surface inclined at 45°. An electronic integrator automatically integrates cell currents over extended periods of time. Calibration using data from weight-loss experiments at the same site shows that the electrochemical technique is suitable for estimating short-term variations in corrosion rates [26].

Atmospheric corrosion testing is important to the suppliers of metals and to the engineers and architects who use metals under atmospheric conditions. Reviews have been prepared summarizing the atmospheric corrosion standards and testing procedures of the American Society for Testing and Materials [27–30].

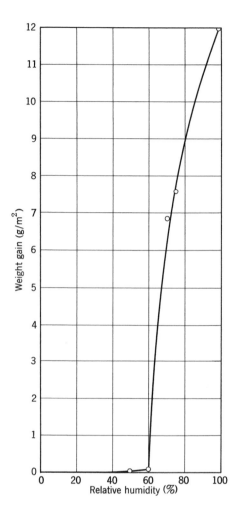

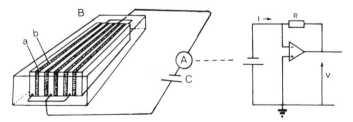

Figure 9.3. Corrosion of iron in air containing 0.01% SO_2, 55 days' exposure, showing critical humidity [21].

Figure 9.4. General arrangement of electrochemical device for measurement of atmospheric corrosion: A is a zero resistance ammeter, the circuit of which is shown on the right; B is an electrochemical cell with (a) electrodes and (b) insulators; C is an external emf. [25, 26]. (Copyright ASTM INTERNATIONAL. Reprinted with permission.)

9.5 REMEDIAL MEASURES

1. *Use of Organic, Inorganic, or Metallic Coatings.* Coatings are discussed in Chapters 14, 15, and 16.

2. *Reduction of Relative Humidity.* Heating air—or, better still, reducing the moisture content—can reduce relative humidity. Lowering the relative humidity to 50% suffices in many cases. If the presence of unusually hygroscopic dust or other surface impurities is suspected, the value should be reduced still further. This protective measure is effective except perhaps when corrosion is caused by acid vapors from nearby unseasoned wood or by certain volatile constituents of adjacent plastics, or paints.

3. *Use of Vapor-Phase Inhibitors and Slushing Compounds.* These are discussed in Chapter 17.

4. *Use of Alloys.* When alloyed with steel in small concentrations, copper, phosphorus, nickel, and chromium are particularly effective in reducing atmospheric corrosion. Copper additions are more effective in temperate climates than in tropical marine regions; chromium and nickel additions combined with copper and phosphorus are effective in both locations (Table 9.5). Corrosion rates of structural steels in tropical atmospheres (e.g., Panama) were found, in general, to be about two or more times higher than in temperate atmospheres (e.g., Kure Beach, North Carolina) mainly because of the higher relative humidity and higher average temperatures.

The usefulness of low-alloy steels to resist atmospheric corrosion through formation of protective rust films has resulted in the development of *weathering steels.* These are used for construction of buildings and bridges and for architectural trim, without the necessity of painting, thereby saving appreciable amounts in maintenance costs over the life of the structure. A typical commercial composition is as follows: 0.09% C, 0.4% Mn, 0.4% Cu, 0.8% Cr, 0.3% Ni, 0.09% P. Such steels do not have any advantage when buried in soil or totally immersed in water, because the corresponding rust films formed under continuously wet conditions are no more protective than those formed on carbon steels. Their more noble corrosion potentials compared to carbon steels may make them useful in certain galvanic couples (see Section 7.3.2.1). It is only by a process of alternate wetting and drying that the rust continues to be protective.

Misawa et al. [32] proposed that an inner cohesive protective rust film is formed on low-alloy steels after long atmospheric exposure (industrial or urban). It consists of amorphous δ-FeOOH, the formation of which is catalyzed by copper and phosphorus on the steel surface; alternate drying and wetting favors its protective qualities. Keiser et al. [33] confirmed that the typical inner adherent rust layer consists mostly of δ-FeOOH.

Stainless steels and aluminum resist tarnish in industrial, urban, and rural atmospheres, as is apparent from their satisfactory use over many years as

TABLE 9.5. Effect of Low-Alloy Components on Atmospheric Corrosion of Commercial Steel Sheet (Eight-Year Exposure)

Steel	Composition, %				Loss of Thickness	
	C	P	Cu	Other	mm	mils
Industrial Atmosphere (Kearney, NJ) [19]						
Carbon	0.2	0.02	0.03		0.20	8.0
Copper bearing	0.2	0.02	0.3		0.11	4.4
Low chromium	0.09	0.2	0.4	1 Cr	0.048	1.9
Low nickel	0.2	0.1	0.7	1.5 Ni	0.051	2.0
Temperate Marine Atmosphere[a](Kure Beach, NC) [19]						
Carbon	0.2	0.02	0.03		0.24	9.5
Copper bearing	0.2	0.01	0.2		0.15	5.8
Low chromium	0.1	0.14	0.4	1 Cr	0.069	2.7
Low nickel	0.1	0.1	0.7	1.5 Ni	0.076	3.0
Tropical Marine Atmosphere (Panama Canal Zone) [31]						
Carbon	0.25	0.08	0.02		0.52	20.4
Copper bearing	0.2	0.004	0.24		0.45	17.6
Low chromium	0.07	0.008	0.1	3.2 Cr	0.23	9.1
Low nickel	0.2	0.04	0.6	2.1 Ni	0.19	7.5

[a]7.5-year exposure.

architectural trim of buildings. Hastelloy C (54% Ni, 17% Mo, 5% Fe, 15% Cr, 4% W) is very resistant to tarnish in marine atmospheres, making it a useful alloy for reflectors on board ship.

REFERENCES

1. C. Leygraf and T. Graedel, *Atmospheric Corrosion*, Wiley, New York, 2000, p. 41.

2. F. L. LaQue, *Am. Soc. Testing Mater. Proc.* **51**, 495 (1951).

3. C. P. Larrabee and S. K. Coburn, in *Proceedings, First International Congress On Metallic Corrosion*, Butterworths, London, 1962, pp. 276–284.

4. *Symposium on Atmospheric Corrosion of Non-Ferrous Metals*, STP 175, American Society for Testing Materials, Philadelphia, PA, 1955. Also previous *Symposium*, 1946.

5. M. Tullmin and P. R. Roberge, Atmospheric corrosion, in *Uhlig's Corrosion Handbook*, 2nd edition, R. W. Revie, editor, Wiley, New York, 2000, pp. 315–317.

6. R. Legault and V. Pearson, *Corrosion* **34**, 433 (1978).

7. H. Townsend and J. Zoccola, *Mater. Perf.* **18**(10), 13 (1979).

8. L. Veleva and R. D. Kane, in *ASM Handbook, Vol. 13A, Corrosion: Fundamentals, Testing, and Protection*, ASM International, Materials Park, OH, 2003, pp. 203–204.

9. T. Kodama, in *ASM Handbook, Vol. 13B, Corrosion: Materials*, ASM International, Materials Park, OH, 2003, p. 7.

10. M. Pourbaix, The linear bilogarithmic law for atmospheric corrosion, in *Atmospheric Corrosion*, W. H. Ailor, editor, Wiley, New York, 1982, pp. 107–121.

11. W. Vernon, Trans. *Faraday Soc.* **23**, 113 (1927).

12. http://www.epa.gov/oar/particlepollution/naaqsrev2006.html

13. http://www.epa.gov/air/airtrends/pm.html

14. A. Kutzelnigg, *Werks. Korros.* **8**, 492 (1957).

15. *Chem. Eng. News*, January 4, 1971, p. 29.

16. W. Vernon, Trans. *Faraday Soc.* **31**, 1668 (1935).

17. L. Greenburg and M. Jacobs, *Eng. Chem.* **48**, 1517 (1956).

18. http://www.epa.gov/air/airtrends/sulfur.html

19. C. P. Larrabee, *Corrosion* **9**, 259 (1953).

20. J. Popplewell and T. Gearing, *Corrosion* **31**, 279 (1975).

21. W. Vernon, Trans. *Faraday Soc.* **23**, 113 (1927); ibid., **27**, 255 (1931) and **31**, 1668 (1935); *Trans. Electrochem. Soc.* **64**, 31 (1933).

22. D. Rice, P. Phipps, and R. Tremoureux, *J. Electrochem. Soc.* **126**, 1459 (1979); **127**, 563 (1980).

23. P. Sereda, in *Corrosion In Natural Environments*, STP 558, American Society for Testing and Materials, Philadelphia, 1974, p. 7.

24. F. Mansfeld, in *Atmospheric Corrosion*, W. Ailor, editor, Wiley, New York, 1982, p. 139.

25. V. Kucera and E. Mattsson, in Ref. 23, p. 239.

26. V. Kucera and M. Collin, In *Eurocorr '77, Sixth European Congress On Metallic Corrosion*, London, 1977, Society of Chemical Industry, London, 1977, p. 189.

27. W. Ailor, in *Atmospheric Factors Affecting the Corrosion of Engineering Metals*, S. Coburn, editor, STP 646, American Society for Testing and Materials, Philadelphia, 1978, p. 129.

28. S. Dean, in Ref. 24, p. 195.

29. H. H. Lawson, Outdoor atmospheres, in *Corrosion Tests and Standards: Application and Interpretation*, 2nd edition, R. Baboian, editor, ASTM International, West Conshohocken, 2005, pp. 343–348.

30. J. D. Sinclair, Indoor atmospheres, in *Corrosion Tests and Standards: Application and Interpretation*, 2nd edition, R. Baboian, editor, ASTM International, West Conshohocken, 2005, pp. 349–361.

31. C. Southwell, B. Forgeson, and A. Alexander, *Corrosion* **14**, 435t (1958).

32. T. Misawa, T. Kyuno, W. Suetaka, and S. Shimodaira, *Corros. Sci.* **11**, 35 (1971).

33. J. Keiser, C. Brown, and R. Heidersbach, *Corros. Sci.* **23**, 251 (1983).

GENERAL REFERENCES

W. Ailor, editor, *Atmospheric Corrosion*, Wiley, New York, 1982.

R. B. Griffin, Corrosion in marine atmospheres, in *ASM Handbook, Vol. 13C, Corrosion: Environments and Industries*, ASM International, Materials Park, OH, 2006, pp. 42–60.

C. Leygraf, Atmospheric corrosion, in *Corrosion Mechanisms in Theory and Practice*, 2nd edition, P. Marcus, editor, Marcel Dekker, New York, 2002.

C. Leygraf and T. Graedel, *Atmospheric Corrosion*, Wiley-Interscience, New York, 2000.

C. Leygraf, J. Hedberg, P. Qiu, H. Gil, J. Henriquez, and C. M. Johnson, Molecular in situ studies of atmospheric corrosion, *Corrosion* **63**(8), 715 (2007).

P. R. Roberge, *Corrosion inspection and Monitoring*, Wiley Series In Corrosion, R. W. Revie, series editor, Wiley, New York, 2007, pp. 301–311.

L. S. Selwyn and P. R. Roberge, Corrosion of metal artifacts displayed in outdoor environments, in *ASM Handbook, Vol. 13C, Corrosion: Environments and Industries*, ASM International, Materials Park, OH, 2006, pp. 289–305.

M. Tullmin and P. R. Roberge, Atmospheric corrosion, in *Uhlig's Corrosion Handbook*, 2nd edition, R. W. Revie, editor, Wiley, New York, 2000, pp. 305–321.

L. Veleva and R. D. Kane, Atmospheric corrosion, in *ASM Handbook, Vol. 13A, Corrosion: Fundamentals, Testing, and Protection*, ASM International, Materials Park, OH, 2003, pp. 196–209.

PROBLEMS

1. A carbon steel is exposed to a highly polluted, industrial-marine environment. After one year of exposure, the penetration is 130 μm, and after 4 years, the penetration is 190 μm. Assuming the bilogarithmic law applies, calculate the following:
   (*a*) The constants A and B in Eq. (9.2).
   (*b*) The average corrosion rate, in terms of penetration per year, over a 40-year period, assuming that environmental conditions do not change significantly during this time period.
   (*c*) The instantaneous corrosion rate, in terms of penetration per year, at 5, 25, and 40 years of exposure.

2. A weathering steel is exposed to the same environment as the carbon steel in Problem 1. After one year of exposure the penetration is 25 μm, and after four years the penetration is 28 μm. Assuming the bilogarithmic law applies, calculate the following:
   (*a*) The constants *A* and *B* in Eq. (9.2).
   (*b*) The average corrosion rate, in terms of penetration per year, over a 40-year period, assuming that environmental conditions do not change significantly during this time period.
   (*c*) The instantaneous corrosion rate, in terms of penetration per year, at 5, 25, and 40 years of exposure.

10

CORROSION IN SOILS

10.1 INTRODUCTION

There are more than 3.7 million kilometers (2.3 million miles) of pipelines crossing the United States, transporting natural gas and hazardous liquids from sources such as wells, refineries, and ports to customers [1]. Underground corrosion is of major importance and results in a significant portion of pipeline failures [1]. Because of corrosion, these pipelines must be regularly inspected, maintained, and sometimes replaced.

In soils, water and gas occupy the spaces between solid particles, and these spaces can constitute as much as half the volume of dry soil. Some of this water is bound to mineral surfaces, whereas bulk water can flow through porous soil. Fluid flow through soil is controlled by the permeability of the soil, which, in turn, depends on the size distribution of the solid particles in the soil. Coarse-grained sand, for example, allows good drainage and access of atmospheric oxygen to a depth greater than, for example, fine-grained soils high in clay. Capillary action in fine-grained soil can draw water up, keeping the soil water-saturated, preventing drainage, retarding evaporation, and restricting oxygen access from the atmosphere to a buried structure, such as a pipeline [2].

Corrosion and Corrosion Control, by R. Winston Revie and Herbert H. Uhlig
Copyright © 2008 John Wiley & Sons, Inc.

The electrochemical corrosion processes that take place on metal surfaces in soils occur in the groundwater that is in contact with the corroding structure. Both the soil and the climate influence the groundwater composition. For example, some clay soils buffer the groundwater pH [2]. Groundwater in desert regions can be high in chloride and very corrosive. On the other hand, groundwaters in tropical climates tend to be very acidic. The influence of soil on groundwater composition is discussed in detail elsewhere [3].

The corrosion behavior of iron and steel buried in the soil approximates, in some respects, the behavior on total immersion in water. Minor composition changes and structure of a steel, for example, are not important to corrosion resistance. Hence, a copper-bearing steel, a low-alloy steel, a mild steel, and wrought iron are found to corrode at approximately the same rate in any given soil. In addition, cold working or heat treatment do not affect the rate. Gray cast iron in soils, as well as in water, is subject to graphitic corrosion. Galvanic effects of coupling iron or steel of one composition to iron or steel of a different composition are important, because they are under conditions of total immersion (Section 7.3.2.1).

In other respects, corrosion in soils resembles atmospheric corrosion in that observed rates, although usually higher than in the atmosphere, vary to a marked degree with the type of soil. A metal may perform satisfactorily in some parts of the country, but not elsewhere, because of specific differences in soil composition, pH, moisture content, and so on. For example, a cast iron water pipe may last 50 years in New England soil, but only 20 years in the more corrosive soil of southern California.

Corrosion rates underground have been measured using the Stern–Geary linear polarization method (Section 5.8) as well as weight loss [4]. The former method has been useful, for example, in assessing the corrosion rates of footings of galvanized-steel towers used to support power lines [5].

10.2 FACTORS AFFECTING THE CORROSIVITY OF SOILS

Among the factors that control corrosivity of a given soil are (1) porosity (aeration), (2) electrical conductivity or resistivity, (3) dissolved salts, including depolarizers or inhibitors, (4) moisture, and (5) pH. Each of these variables may affect the anodic and cathodic polarization characteristics of a metal in a soil. A porous soil may retain moisture over a longer period of time or may allow optimum aeration, and both factors tend to increase the initial corrosion rate. The situation is more complex, however, because corrosion products formed in an aerated soil may be more protective than those formed in an unaerated soil. In most soils, particularly if not well-aerated, observed corrosion takes the form of deep pitting. Localized corrosion of this kind is obviously more damaging to a pipeline than a higher overall corrosion rate occurring more uniformly. Another factor to be considered is that, in poorly aerated soils containing sulfates, sulfate-reducing bacteria may be active; these organisms often produce the highest corrosion rates normally experienced in any soil.

Aeration of soils may affect corrosion not only by the direct action of oxygen in forming protective films, but also indirectly through the influence of oxygen reacting with and decreasing the concentration of the organic complexing agents or depolarizers naturally present in some soils. In this regard, the beneficial effect of aeration extends to soils that harbor sulfate-reducing bacteria because these bacteria become dormant in the presence of dissolved oxygen. Corrosion rates tend to increase with depth of burial near the soil surface, but not invariably. Bureau of Standards tests [6] on steel specimens exposed 6–12 years and buried 30–120 cm (12–48 in.) below the surface showed greater corrosion at greater depth for five soils, but showed the reverse trend for two soils.

A soil containing organic acids derived from humus is relatively corrosive to steel, zinc, lead, and copper. The measured total acidity of such a soil appears to be a better index of its corrosivity than pH alone. High concentrations of sodium chloride and sodium sulfate in poorly drained soils, such as are found in parts of southern California, make the soil very corrosive. Macrogalvanic cells or "long-line" currents established by oxygen concentration differences, by soils of differing composition, or by dissimilar surfaces on the metal become more important when electrical conductivity of the soil is high. Anodes and cathodes may be thousands of feet, or even miles, apart. A poorly conducting soil, whether from lack of moisture or lack of dissolved salts or both, is, in general, less corrosive than a highly conducting soil. But conductivity alone is not a sufficient index of corrosivity; in accordance with the principles of electrode kinetics discussed in Chapter 5, anodic and cathodic polarization characteristics of a corroding metal in the soil are also a factor.

From studies at the National Physical Laboratory in the United Kingdom, it was concluded that the corrosivity of soil towards ferrous metals could be estimated by measuring the resistivity of the soil and the potential of a platinum electrode in the soil with respect to a saturated calomel reference electrode [7]. Soils of low resistivity (<2000 Ω-cm) were considered to be aggressive. Soils in which the potential at pH 7 was low [<0.40 V (S.H.E.), or <0.43 V if the soil is clay] were considered to provide a suitable environment for sulfate-reducing bacteria and to be aggressive. Borderline cases by these two criteria were resolved by considering that a water content of more than 20% causes a soil to be aggressive.

10.3 BUREAU OF STANDARDS TESTS

The most extensive series of field tests on various metals and coatings in almost all types of soils were begun in 1910 by K. H. Logan of the National Bureau of Standards (now the National Institute of Standards and Technology). These tests continued until 1955 and now constitute the most important source of information on soil corrosion available [6]. They showed similarity in corrosion rates in a given soil for various kinds of iron and steels; confirmation was obtained through 5-year tests in England [8]. A few typical corrosion rates averaged for many soils are listed in Table 10.1. In addition, data are listed for two soils relatively corrosive to steel and for one that is relatively noncorrosive, showing the large variations in corrosion rate from one soil to another. For San Diego soil, the symbol > means that the thickness of test specimen was completely penetrated by pitting at the end of the exposure period.

Referring again to Table 10.1, we see that *copper*, on average, corrodes at about one-sixth the rate of iron, but, in tidal marsh, for example, the rate is comparatively higher than in most other soils, being one-half that of iron. The rate for copper is normal in otherwise corrosive San Diego soil. Pitting is not pronounced, with the maximum depth reaching less than 0.15 mm (6 mils).

Lead also corrodes less on the average than does steel. In poorly aerated soils or soils high in organic acids, the corrosion rate may be much higher (four to six times) than the average. Pitting in some of these soils penetrated the test specimen thickness, accounting for an average maximum penetration greater than the average given in Table 10.1. *Zinc* also pitted in some soils to an extent greater than the specimen thickness. In 5-year tests carried out in Great Britain, commercially pure *aluminum* was severely pitted in four soils (0.1 to >1.6 mm, 4 to >63 mils), but was virtually unattacked in a fifth soil [9].

Increase in chromium content of steel decreases observed weight loss in a variety of soils; but, above 6% Cr, depth of pitting increases. In 14-year tests, 12% Cr and 18% Cr steels were severely pitted. The 18% Cr, 8% Ni, Type 304 stainless steel was not pitted or was only slightly pitted (<0.15 mm, <6 mils), nor was weight loss appreciable in 10 out of 13 soils. However, in three soils, at least one specimen was perforated [0.4–0.8 mm (16–32 mils) thick] by pitting. Type 316 stainless steel did not pit in any of the 15 soils to which the alloy was exposed for 14 years. It is expected, however, that pitting would also occur with this alloy in longer time tests since pitting occurs in seawater within about $2\frac{1}{2}$ years.

Zinc coatings are effective in reducing weight loss and pitting rates of steel exposed to soils. In 10-year tests in 45 soils, coatings of 0.85 kg/m² (2.8 oz/ft²) based on one side of the specimen (0.13 mm or 5 mils thick) protected steel against pitting with the exception of one soil (Merced silt loam, Buttonwillow, California) in which some penetration of the base steel could be measured. In later tests extending up to 13 years, a coating of 0.95 kg/m² (3.1 oz/ft²) effectively reduced (but did not prevent) corrosion, even in cinders in which the zinc coating was destroyed within the first two years.

Cinders constitute one of the most corrosive environments. For 4- or 5-year exposures, corrosion rates of steel and zinc in cinders was 5 times, copper 8 times, and lead 20 times higher than the average rates in 13 different soils.

10.3.1 Pitting Characteristics

Because the dimensions of all field test specimens were on the order of inches up to about 1 ft (3–30 cm), the reported pitting rates represent minimum rather than maximum values. Actual depth of pits in a given time is found to increase with size of test specimen, probably because cathodic area per pit increases (i.e.,

TABLE 10.1. Corrosion of Steels, Copper, Lead, and Zinc in Soils (National Bureau of Standards [6])

Maximum Penetration in mils (1 mil = 0.001 in. = 0.025 mm) for Total Exposure Period
Average corrosion rates in g m^{-2} d^{-1} (gmd)

Soil	Open Hearth Iron, 12-Year Exposure		Wrought Iron, 12-Year Exposure		Bessemer Steel, 12-Year Exposure		Copper, 8-Year Exposure		Lead, 12-Year Exposure		Zinc, 11-Year Exposure	
	gmd	mils	gmd	mils	gmd	mils	gmd	mils	gmd	mils	gmd	mils
Average of several soils	0.45 (44 soils)	70	0.47 (44 soils)	59	0.45 (44 soils)	61	0.07 (29 soils)	<6	0.052 (21 soils)	>32	0.3 (12 soils)	>53
Tidal marsh, Elizabeth, New Jersey	1.08	90	1.16	80	1.95	100	0.53	<6	0.02	13	0.19	36
Montezuma clay, Adobe, San Diego	1.37	>145	1.34	>132	1.43	>137	0.07	<6	0.06 (9.6 years)	10	—	—
Merrimac gravelly sandy loam, Norwood, Massachusetts	0.09	28	0.10	23	0.10	21	0.02 (13.2 years)	<6	0.013	19	—	—

anode/cathode area ratio decreases), thus accounting for higher current densities at the pits. In addition to this factor, long-line currents or macrocells, if present, increase pit depth over the values obtained on small specimens where such cells do not operate.

The rate at which pits grow in the soil under a given set of conditions tends to decrease with time and follows a power-law equation $P = kt^n$, where P is the depth of the deepest pit in time t, and k and n are constants. It has been reported [10] that values of n for steels range from about 0.1 for a well-aerated soil, to 0.9 for a poorly aerated soil. The smaller the value of n, the greater the tendency for the pitting rate to fall off with time. As n approaches unity, the pitting rate approaches a constant value, or penetration is proportional to time.

Pits tend to develop more on the bottom side of a pipeline than on the top side. This difference is sometimes sufficiently great to make it worthwhile rotating a pipeline $180°$ after a given period of exposure in order to increase pipe life. Pitting on the bottom side results from constant contact with the soil, whereas the top side, because of the pipe settling, tends to become detached, producing an air space between the pipe and the soil.

10.4 STRESS-CORROSION CRACKING

Stress-corrosion cracking (S.C.C.) is a well-known cause of underground oil and gas transmission pipeline failures [11]. Stress-corrosion cracking of cathodically protected pipelines originates on the outer pipe surface, most commonly at areas where the coating is disbonded (i.e., no longer attached to the surface of the steel pipe).

In most engineering structures where cathodic protection is used to protect steel from general corrosion, the steel is polarized to a potential of $-0.85\,V$ versus $Cu/CuSO_4$, corresponding to $-0.53\,V$ (S.H.E.). Cathodic protection of underground pipelines can result in the accumulation of alkali at the pipe surface [12], consisting of sodium hydroxide and carbonate/sodium bicarbonate solutions. Hydrogen ions, cations (e.g., Na^+), and water containing dissolved oxygen migrate to the cathodic pipe surface (via pores or defects in the coating). The reaction in aerated near-neutral solution at the pipe surface is

$$O_2 + 2H_2O + 4e^- \rightarrow 4OH^- \tag{10.1}$$

Carbon dioxide from either the air, surrounding water, or decaying vegetation may then react with the hydroxide forming bicarbonates and carbonates [13]:

$$CO_2 + OH^- \rightarrow HCO_3^- \tag{10.2}$$

$$HCO_3^- + OH^- \rightarrow CO_3^{2-} + H_2O \tag{10.3}$$

The potential range at which stress-corrosion cracks form in carbonate solutions is reported as noble to the optimum cathodic protection potential and active to the corrosion potential; for example, one high-strength (541 MPa, 78,000 psi yield strength) linepipe steel developed intergranular cracks between −0.31 and −0.46 V (S.H.E.) in 1 N Na_2CO_3 + 1 N $NaHCO_3$ solution at 40 °C [14]. Accumulation of NaOH can also cause S.C.C. in a more active potential range. In 35% NaOH at 85 °C, S.C.C. of steel is maximum in the region of −0.7 V (S.H.E.), whereas in ammonium carbonate solution at 70 °C, the maximum susceptibility occurs near −0.3 V (S.H.E.) (see Figs. 8.6 and 8.7, Section 8.3.5). From these potential relations, one could assume that NaOH is more likely to be the cause of failure in sections of pipeline that are cathodically overprotected, whereas carbonates would be the more likely cause in less active potential regions.

In addition to the intergranular S.C.C. that occurs in high-pH solutions, transgranular cracking has been found to occur in dilute aqueous solutions of near neutral pH [15].

10.5 REMEDIAL MEASURES

1. *Organic and Inorganic Coatings.* Paint coatings of thickness normal to atmospheric protection fail within months when exposed to the soil. On the other hand, thickly applied coatings of coal tar, with reinforcing pigments or inorganic fibers to reduce cold flow, are found to be practical. They provide effective protection at reasonable cost. Bureau of Standards tests showed that soft rubber coatings 6 mm (0.25 in.) thick afforded excellent protection to steel in 11-year exposure tests.

   Coatings based on bituminous compositions offer excellent service but are now being supplemented by others—for example, fusion-bonded epoxy resin typically 1 mm thick applied at the pipe mill.

2. *Metallic Coatings.* The effectiveness of zinc coatings has already been mentioned. Such coatings deteriorate more rapidly when galvanically coupled to large areas of bare iron, steel, or copper, in which case insulating couplings can be useful.

3. *Alteration of Soil.* A soil high in organic acids can be made less corrosive by surrounding the metal structure with limestone chips. A layer of chalk ($CaCO_3$) surrounding buried pipes has been used in some soil formations likely to produce microbiologically influenced corrosion [16].

4. *Cathodic Protection.* In general, buried pipelines and tanks are protected by a combination of cathodic protection and an organic coating. This combination protects steel against corrosion in all soils, both effectively and economically, for as many years as cathodic protection is adequately maintained.

REFERENCES

1. R. E. Ricker, *Analysis of Pipeline Steel Corrosion Data from NBS (NIST) Studies Conducted between 1922–1940 and Relevance to Pipeline Management*, NISTIR 7415, National Institute of Standards and Technology, U.S. Department of Commerce, May 2, 2007, pp. 1–2.

2. M. J. Wilmott and T. R. Jack, in *Uhlig's Corrosion Handbook*, 2nd edition, R. W. Revie, editor, Wiley, New York, 2000, pp. 329–348.

3. W. Stumm and J. J. Morgan, *Aquatic Chemistry: Chemical Equilibria and Rates in Natural Waters*, 3rd edition, Wiley, New York, 1996.

4. D. A. Jones, in *Underground Corrosion*, E. Escalante, editor, Special Technical Publcation No. 741, American Society for Testing and Materials, Philadelphia, 1981, pp. 123–132.

5. P. H. Hewitt and W. B. R. Moore, in *Eurocorr '77, Sixth European Congress On Metallic Corrosion*, London, 1977, Society of Chemical Industry, London, 1977, pp. 205–211.

6. M. Romanoff, *Underground Corrosion*, Circ. 579, U.S. National Bureau of Standards, 1957; reprinted with a Preface by John A. Beavers, NACE, Houston, Texas, 1989; see also the reexamination and analysis of the original data, Richard E. Ricker, *Analysis of Pipeline Steel Corrosion Data from NBS (NIST) Studies Conducted between 1922–1940 and Relevance to Pipeline Management*, NISTIR 7415, National Institute of Standards and Technology, U.S. Department of Commerce, May 2, 2007.

7. G. H. Booth, *Microbiological Corrosion*, M&B Monographs CE/1, Mills & Boon, London, 1971, pp. 46–49.

8. J. C. Hudson and G. Acock, in *Corrosion of Buried Metals*, Iron and Steel Institute Special Report 45, London, 1952.

9. P. Gilbert and F. Porter, in *Corrosion of Buried Metals*, Iron and Steel Institute, Special Report 45, London, 1952.

10. Ref. 6, p. 39.

11. *Public Inquiry Concerning Stress Corrosion Cracking on Canadian Oil and Gas Pipelines, Report of the Inquiry*, Report MH-2-95, National Energy Board, Calgary, Alberta, November 1996.

12. W. J. Schwerdtfeger and R. J. Manuele, *J. Natl. Bur. Std. (U.S.)* **65C**, 171 (1961).

13. W. E. Berry, T. J. Barlo, J. H. Payer, R. R. Fessler, and B. L. McKinney, Paper No. 64, *Corrosion/78*, National Association of Corrosion Engineers, Houston, TX, 1978.

14. R. Revie and R. Ramsingh, *Can. Metall. Q.* **22**, 235 (1983).

15. R. N. Parkins, in *Uhlig's Corrosion Handbook*, 2nd edition, R. W. Revie, editor, Wiley, New York, 2000, pp. 191–194.

16. U. R. Evans, *The Corrosion and Oxidation of Metals*, Arnold, London, 1960, p. 272.

GENERAL REFERENCES

E. Escalante, editor, *Underground Corrosion*, Special Technical Publication 741, ASTM, Philadelphia, 1981.

R. E. Ricker, *Analysis of Pipeline Steel Corrosion Data from NBS (NIST) Studies Conducted between 1922–1940 and Relevance to Pipeline Management*, NISTIR 7415, National Institute of Standards and Technology, U.S. Department of Commerce, May 2, 2007.

W. Stumm and J. J. Morgan, *Aquatic Chemistry: Chemical Equilibria and Rates in Natural Waters*, 3rd edition, Wiley, New York, 1996.

M. J. Wilmott and T. R. Jack, in *Uhlig's Corrosion Handbook*, 2nd edition, R. W. Revie, editor, Wiley, New York, 2000, pp. 329–348.

11

OXIDATION

11.1 INTRODUCTION

In high-temperature corrosion, the protective oxide films—usually either chromia, Cr_2O_3, or alumina, Al_2O_3—that form between the metal and the environment are of critical importance. One of the major advances in corrosion science over the past few decades has been the characterization of the reactive element effect (REE), which identifies the role of small additions of reactive elements, such as yttrium, hafnium, lanthanum, zirconium, and cerium, to improve high-temperature oxidation resistance.

Because increasing the temperature of most high-temperature processes (e.g., energy conversion, waste incineration, chemical synthesis) results in increased efficiency, there is a high priority for designers and engineers to increase process temperatures, not only to increase efficiency, but also to reduce emissions, thereby helping to achieve sustainability and improving economics. For example, partially insulating the combustion chamber on a Diesel engine increased the piston surface temperature from 450 °C to 900 °C, and it also increased the combustion efficiency (decreased fuel consumption) by 20% and reduced the non-burned hydrocarbons emission [1, 2]. Because of the present trend to increase process temperatures, it is more important than ever for engineers to understand

Corrosion and Corrosion Control, by R. Winston Revie and Herbert H. Uhlig
Copyright © 2008 John Wiley & Sons, Inc.

the principles of high-temperature corrosion, the limitations of materials, and the methods that are being used to improve the performance of engineering materials at high temperatures. The basics of these matters will be explored in this chapter.

When a metal is exposed at room or elevated temperatures to an oxidizing gas, such as oxygen, sulfur, or halogens, corrosion may occur in the absence of a liquid electrolyte. This is sometimes called "dry" corrosion, in contrast to "wet" corrosion, which occurs when a metal is exposed to water or to the soil. In dry corrosion, reaction products can be solid films or *scales* (a scale is a thick film), liquids, or volatile compounds. When a protective scale—such as alumina (Al_2O_3), chromia (Cr_2O_3), or silica (SiO_2)—forms, the metal, the environment, or both must diffuse through the scale in order for the reaction to continue. Usually ions, rather than atoms, migrate through solid oxides, sulfides, or halides; hence, the reaction product can be considered an electrolyte. For copper oxidizing in O_2, or silver tarnishing in a contaminated atmosphere, the solid electrolytes are Cu_2O and Ag_2S, respectively. The migrating ions are not hydrated, and they diffuse simultaneously with electrons, but along different paths.

11.2 INITIAL STAGES

The processes that take place at a clean, reactive metal surface exposed to oxygen obey the following sequence: (1) adsorption of oxygen, (2) formation of oxide nuclei that grow laterally, and (3) growth of a continuous oxide film. Because the free energy of adsorption of atomic oxygen exceeds the free energy of dissociation of oxygen, the first-formed adsorbed film consists of atomic oxygen. Low-energy electron-diffraction data show that some metal atoms enter the approximate plane of adsorbed oxygen to form a relatively stable two-dimensional structure of mixed O ions (negative) and metal ions (positive). As was discussed earlier in relation to the passive film (Section 6.5), this initial adsorbed partial monolayer is thermodynamically more stable than the metal oxide; for example, for nickel, the adsorbed film resists decomposition up to the melting point of nickel [3], whereas NiO decomposes accompanied by oxygen dissolving in the metal.* Continued exposure to low-pressure oxygen is followed by adsorption of O_2 molecules on metal atoms exposed through the first adsorbed layer. Since the second layer of oxygen is bonded less energetically than the first layer, it is adsorbed without dissociation to its atoms. The resultant structure is usually more stable on transition than on nontransition metals [4]. Any additional layers of adsorbed oxygen are still less strongly bonded, and the outer layers, at

---

*J. Moreau and J. Bénard [*C. R. Acad. Sci (Paris)* **242**, 1724 (1956)] showed, through observation of the metal surface in $H_2O–H_2$ mixtures at elevated temperatures, that oxygen adsorbed on an 18% Cr stainless steel is thermodynamically more stable than the metal oxide. For analogous data on iron, see A. Pignocco and G. Pellissier, *J. Electrochem. Soc.* **112**, 1118 (1965); E. Hondros, *Acta Metall.* **16**, 1377 (1968).

elevated temperatures, eventually become mobile, with the corresponding diffraction pattern being that of an amorphous structure.

Because the free energy of adsorption per mole of oxygen decreases with amount of oxygen adsorbed (the O–substrate bond becomes weaker), multilayer adsorbed oxygen on metal M eventually favors transformation to a crystalline stoichiometric oxide. In other words, ΔG for $O \cdot M_{ads} + nO_2 \rightarrow (nO_2) \cdot O \cdot M_{ads}$ becomes less negative per mole O_2 as n increases, whereas ΔG for $2nM + (nO_2) \cdot O \cdot M_{ads} \rightarrow (2n + 1)MO$ to form the oxide becomes correspondingly more negative. Therefore, oxide formation nucleates preferably at surface sites where multilayer adsorption is favored, such as at surface vacancies, ledges, or other imperfections. When conditions are favorable, oxide nuclei form at specific sites of the surface, aided by rapid surface diffusion of M and O ions (Fig. 11.1) [5]. Because the amount of adsorbed oxygen on any likely site increases with oxygen pressure, more sites are brought into play, and the density of nuclei increases. Similarly, since multilayer adsorption decreases at elevated temperatures, the density of nuclei decreases with increase of temperature [6, 7]. Nuclei grow rapidly to a certain height on the order of nanometers,* and then they grow more

Figure 11.1. Nuclei of Cu_2O formed on copper surface at 10^{-1} mmHg oxygen pressure (13.3 Pa), 525 °C, 20 s (17,600×). Black lines are bands of imperfections (stacking faults) in the copper lattice (Brockway and Rowe [5]).

*Limited perhaps by the electron tunneling distance [4] below which electrons can penetrate the metal-oxide–interface barrier without first acquiring the usual activation energy.

rapidly in a lateral rather than in a vertical direction. On copper exposed to 1 mm Hg (133 Pa) oxygen pressure at 550 °C, a continuous film of Cu_2O forms within 6 s; at 1 atm O_2, the time required is still less.

Using transmission electron microscopy and computer modeling, it has been shown that the metal–scale interface is very tightly bonded as a result of epitaxial relations between the metal and the scale, at least for thin scales [8]; that is, as the crystalline oxide grows on the metal substrate, the crystal structure of the oxide is aligned with the structure of the metal. Growth of the resulting thin continuous oxide film may be controlled by transfer of electrons from metal to oxide [9], or by migration of metal ions in a strong electric field set up by negatively charged oxygen adsorbed on the oxide surface [10]. When the continuous film reaches a thickness in the order of several hundred nanometers, diffusion of ions through the oxide may become rate controlling instead. The latter situation prevails so long as the oxide remains continuous. Eventually, at a critical thickness, the stresses set up in the oxide may cause it to crack and detach—called *spalling*—and the oxidation rate increases.

11.3 THERMODYNAMICS OF OXIDATION: FREE ENERGY–TEMPERATURE DIAGRAM

Graphical data of the standard Gibbs free energy of formation of oxides versus temperature, shown in Fig. 11.2, can be used to predict the conditions under which a metal is oxidized or a metal oxide is reduced [11]. For every reaction in Fig. 11.2, the reactant is 1 mole of O_2, so that the reactions are

$$\frac{2x}{y} M + O_2 = \frac{2}{y} M_x O_y$$

The standard Gibbs free energy of formation, $\Delta G°$, for any oxide can be read directly off the vertical axis of the graph. For example, at 1000 °C, the standard Gibbs free energy of formation of NiO is approximately −250 kJ for two moles of NiO.

From the diagram, we can estimate the equilibrium values of oxygen partial pressure, as well as CO/CO_2 and H_2/H_2O ratios in contact with metals and their oxides, by drawing straight lines through the points labeled O, H, and C, respectively, and through the point on the metal/metal oxide line at the temperature of interest, extrapolating to the scales of CO/CO_2, H_2/H_2O, and p_{O_2}.

11.4 PROTECTIVE AND NONPROTECTIVE SCALES

The rate of reaction in the later stages of oxidation depends on whether the thick film or scale remains continuous and protective as it grows, or whether it contains cracks and pores and is relatively nonprotective. Because reaction-product films

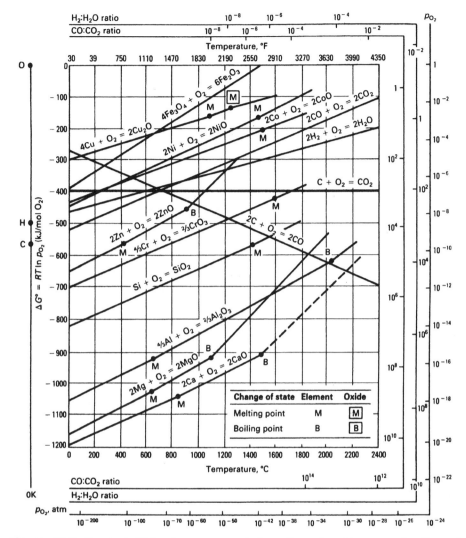

Figure 11.2. Standard Gibbs free energy of formation of oxides as a function of temperature [11]. (Reprinted with permission of ASM International®. All rights reserved. www.asminternational.org.)

are usually brittle and lack ductility, the initiation of cracks depends in some measure on whether the surface film is formed in tension, favoring fracture, or in compression, favoring protection. This situation, in turn, depends on whether the volume of reaction product is greater or less than the volume of metal from which the product forms; that is, it depends on the Pilling–Bedworth ratio, Md/nmD, where M is the molecular weight, D is the density of scale, m and d are the atomic weight and density of metal, respectively, and n is the number of metal

TABLE 11.1. Pilling–Bedworth Ratios

Metal	Oxide	Pilling–Bedworth Ratio
Chromium	Cr_2O_3	2.0
Cobalt	CoO	1.9
Titanium	TiO_2	1.8
Iron	FeO	1.7
Copper	CuO	1.7
Nickel	NiO	1.7
Aluminum	Al_2O_3	1.3
Magnesium	MgO	0.8
Calcium	CaO	0.6
Lithium	Li_2O	0.6

atoms in a molecular formula of scale substance ($n = 2$ for Al_2O_3) [12]. If Md/nmD >1, a protective scale is predicted to form. On the other hand, when this ratio is less than unity, the scale is formed in tension and tends to be nonprotective.

Table 11.1 lists the Pilling–Bedworth ratios for several metals. For calcium and magnesium, which tend to form nonprotective oxides, the ratios are 0.64 and 0.79, respectively, whereas for aluminum and chromium, which tend to form protective oxides, the ratios are 1.3 and 2.0, respectively. Tungsten has a ratio of 3.6, and hence, the oxide, WO_3, is expected to be protective, except at high temperatures (approximately >800 °C), where it volatilizes.

Although the Pilling–Bedworth ratio may be greater than 1, so that a protective scale would be predicted, at high temperatures the rapid growth of oxide may cause compressive stresses to become so great that the scale spalls; for example, Cr_2O_3 scale on pure chromium at 1100 °C, with a Pilling–Bedworth ratio of 2.0 [13]. Similarly, thick scales of TiO_2 on titanium are nonprotective. Typically, oxide is protective during the early stages of oxidation, but it becomes nonprotective above a certain thickness.

In addition to the Pilling–Bedworth ratio, protection by an oxide depends on adherence of the oxide to the substrate, low vapor pressure and high melting temperature of the oxide, slow oxide growth rate, and high thermodynamic stability. The growth mechanism of the oxide is also relevant to oxide protection. For example, if the scale forms at the metal–oxide interface by migration of oxide ions, then compressive stresses can develop because of the constrained space available for the oxide to occupy. On the other hand, scale that forms at the oxide–gas interface by migration of metal ions outward is not constrained to occupy the volume of the metal that was oxidized. Under these conditions, protective scales are expected.

11.4.1 Three Equations of Oxidation

The three main equations that express thickness y of film or scale forming on any metal within time t are (1) the linear, (2) the parabolic, and (3) the logarithmic.

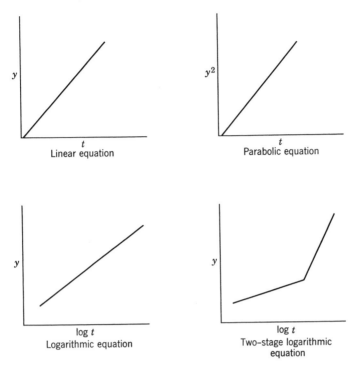

Figure 11.3. Equations expressing growth of film thickness, *y*, as a function of time of oxidation, *t*.

The particular equation that applies depends, in part, on the ratio Md/nmD and the relative thickness of the film or scale.

For the linear equation, the rate of oxidation is constant, or $dy/dt = k$ and $y = kt + \text{const}$, where k is a constant. Hence, the thickness of scale, y, plotted with time, t, is linear (Fig. 11.3). This equation holds whenever the reaction rate is constant at an interface, as, for example, when the environment reaches the metal surface through cracks or pores in the reaction-product scale. Hence, for such metals, the ratio Md/nmD is usually less than unity. In special cases, the linear equation may also hold even though the latter ratio is greater than unity, such as when the controlling reaction rate is constant at an inner or outer phase boundary of the reaction-product scale; for example, tungsten first oxidizes at 700–1000 °C, in accord with the parabolic equation, forming an outer porous WO_3 layer and an inner compact oxide scale [14]. When the rate of formation of the outer scale becomes equal to that of the inner scale, the linear equation is obeyed.

For the parabolic equation, the diffusion of ions or migration of electrons through the scale is controlling, and the rate, therefore, is inversely proportional to scale thickness.

$$\frac{dy}{dt} = \frac{k'}{y} \quad \text{or} \quad y^2 = 2k't + \text{const}$$

Accordingly, if y^2 is plotted with t, a linear relation is obtained (Fig. 11.3). This equation holds for protective scales corresponding to $Md/nmD > 1$ and is applicable to the oxidation of many metals at elevated temperatures, such as copper, nickel, iron, chromium, and cobalt.

For relatively thin protective films, as when metals oxidize initially or when oxidation occurs at low temperatures, it is found that

$$\frac{dy}{dt} = \frac{k''}{t} \quad \text{or} \quad y = k'' \ln\left(\frac{t}{\text{const}} + 1\right)$$

This is called the logarithmic equation. Correspondingly, if y is plotted with log $(t + \text{const})$, or with log t for $t \gg \text{const}$, a linear relation is obtained. First reported by Tammann and Köster [15], the logarithmic equation has been found to express the initial oxidation behavior of many metals, including Cu, Fe, Zn, Ni, Pb, Cd, Sn, Mn, Al, Ti, and Ta. The logarithmic equation can be derived on the condition that the oxidation rate is controlled by transfer of electrons from metal to reaction-product film [9] when the latter is electrically charged—that is, when it contains a space charge throughout its volume. The preponderance of electric charge of usually negative sign in oxides near the metal surface, similar to the electrical double layer in aqueous electrolytes, has been demonstrated experimentally. Therefore, any factor changing the work function of the metal (energy required to extract an electron), such as grain orientation, lattice transition, or magnetic transition (Curie temperature), changes the oxidation rate, as has been observed [16]. When the thickness of the film exceeds the maximum thickness of the space-charge layer, diffusion or migration through the film then usually becomes controlling, the parabolic equation applies, and the factors just mentioned, such as grain orientation and Curie temperature, no longer affect the rate. On this basis, metals forming protective films first follow the logarithmic equation and then the parabolic (or linear) equation.

If, for thin-film behavior, migration of ions controls the rate, and the prevailing electric field within the film is considered to be set up by gaseous ions adsorbed on the outer surface, the rate of ion migration is an exponential function of the field strength, and the inverse logarithmic equation [10] results:

$$\frac{1}{y} = \text{const} - k \ln t$$

This relation has been reported to hold for copper and iron oxidized at low temperatures [17]. It is often difficult to distinguish between the logarithmic and the inverse logarithmic equations because of the limited range of time over which data can be accumulated for thin-film behavior, with either equation apparently applying equally well. This situation also makes it difficult to evaluate other types of equations that have been proposed, such as the cubic equation, $y^3 = kt + \text{const}$. Data obeying this equation can also be represented, in many cases, by a two-stage logarithmic equation where an initial lower rate is followed by a final higher rate

[16] (Fig. 11.3). The higher rate is ascribed to formation of a diffuse space-charge layer overlying an initially constant charge-density layer [9].

The oxidation rate for thin- or thick-film conditions increases with temperature, obeying the Arrhenius equation

$$\text{reaction} - \text{rate constant} = A \exp\left(\frac{-\Delta E'}{RT}\right)$$

where $\Delta E'$ is the activation energy, R is the gas constant, T is the absolute temperature, and A is a constant.

11.5 WAGNER THEORY OF OXIDATION

When certain metals, such as copper, zinc, and nickel, oxidize, it is found that metal ions migrate through the oxide to the outer oxide surface, reacting there with oxygen. For these metals, outward diffusion of metal ions occurs in preference to diffusion of the larger oxygen ions inward. Reaction occurring at the outer rather than the inner oxide surface was first reported by Pfeil [18], who noticed that when an iron surface was painted with a slurry of Cr_2O_3, the green-colored layer, after oxidation, was buried in the middle or lower layers of iron oxide. Iron ions, in other words, diffused through the Cr_2O_3 marker and reacted with oxygen at the gas–oxide interface. Similarly, Wagner [19] showed, by quantitative studies, that Ag^+ ions, and not S^{2-} ions, migrate through Ag_2S. By placing two weighed pellets of Ag_2S over silver, on top of which was molten sulfur (Fig. 11.4), he showed that, after 1 h at 220 °C, the bottom pellet, next to the silver, neither gained nor lost weight, whereas the top pellet, in contact with the sulfur, gained weight chemically equivalent to the loss in weight of metallic silver. Wagner also showed that by assuming independent migration of Ag^+ and electrons, the observed reaction rate could be calculated from independent physical chemical data. He [20] derived an expression for the parabolic rate constant that has the following simplified version [21]:

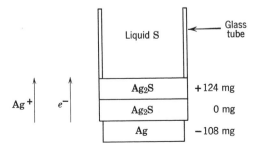

Figure 11.4. Formation of silver sulfide from silver and liquid sulfur, 1-h test, 220 °C (Wagner [19]).

$$k = \frac{En_3(n_1 + n_2)\kappa}{F} \qquad (11.1)$$

where E is the emf of the operating cell derived either from potential measurement or free-energy data; n_1, n_2, and n_3 are the mean cation, anion, and electron transference numbers, respectively, within the reaction-product film; κ is the mean specific conductivity of film substance; and F is the Faraday. The constant, k, appears in the relation

$$\frac{dw}{dt} = k\frac{A}{y}$$

where dw/dt is the rate of formation of film substance in equivalents/second, A is the area, and y is the thickness. The calculated value of k for silver reacting with sulfur at $220\,°C$ is $2\text{–}4 \times 10^{-6}$ compared to the observed value of 1.6×10^{-6} eq/cm-s. For copper oxidizing at $1000\,°C$ and 100-mm oxygen pressure (1.3×10^4 Pa), the calculated and observed values of k are 6×10^{-9} and 7×10^{-9} eq/cm-s, respectively. The excellent agreement between theory and observation confirms the validity of the model suggested by Wagner for the oxidation process within the region for which the parabolic equation applies.

Equation (11.1) holds for diffusion of either cations or anions or both in the reaction-product scale. Some metals—for example, titanium and zirconium—oxidize, in part, by migration of oxygen ions, O^{2-}, via anion vacancies in the corresponding outer oxides.

11.6 OXIDE PROPERTIES AND OXIDATION

Metal oxides are commonly in the class of electrical conductors called semiconductors; that is, their conductivity lies between insulators and metallic conductors. Conductivity increases with a slight shift from stoichiometric proportions of metal and oxygen and with an increase of temperature. There are two types of semiconducting oxides, namely, p- and n-types (p = positive carrier, n = negative carrier). In the p-type, shift of stoichiometric proportions takes the form of a certain number of missing metal ions in the oxide lattice called cation vacancies, represented by $\square$. At the same time, to maintain electrical neutrality, an equivalent number of positive holes, $\oplus$, form—that is, sites where electrons are missing. A cupric ion, Cu^{2+}, in a Cu_2O lattice is an example of a positive hole. Oxides of p-type are Cu_2O, NiO, FeO, CoO, Bi_2O_3, and Cr_2O_3. A model for the Cu_2O lattice is shown in Fig. 11.5. During the oxidation of copper, cation vacancies and positive holes are formed at the outer O_2–oxide surface. These migrate to the metal surface, a process that is equivalent to the reverse migration of Cu^+ and electrons.

For n-type oxides, excess metal ions exist in interstitial positions of the oxide lattice, and it is these, together with electrons, that migrate during oxidation

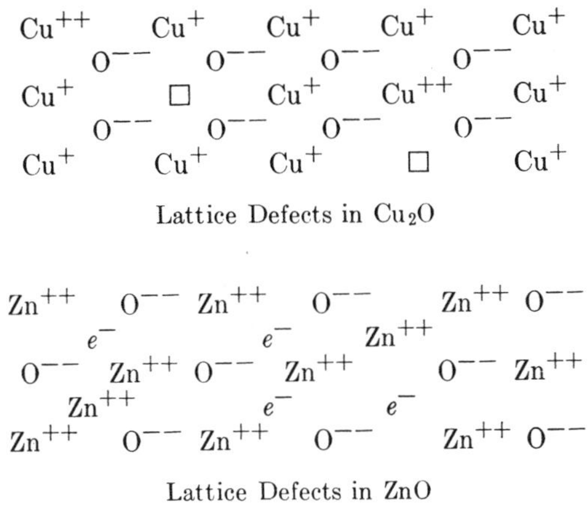

$$
\begin{array}{ccccc}
\text{Cu}^{++} & \text{Cu}^{+} & \text{Cu}^{+} & \text{Cu}^{+} & \text{Cu}^{+} \\
\quad \text{O}^{--} & \text{O}^{--} & \text{O}^{--} & \text{O}^{--} & \\
\text{Cu}^{+} & \square & \text{Cu}^{+} & \text{Cu}^{++} & \text{Cu}^{+} \\
\quad \text{O}^{--} & \text{O}^{--} & \text{O}^{--} & \text{O}^{--} & \\
\text{Cu}^{+} & \text{Cu}^{+} & \text{Cu}^{+} & \square & \text{Cu}^{+}
\end{array}
$$

Lattice Defects in Cu$_2$O

$$
\begin{array}{cccccc}
\text{Zn}^{++} & \text{O}^{--} & \text{Zn}^{++} & \text{O}^{--} & \text{Zn}^{++} & \text{O}^{--} \\
e^- & & e^- & & \text{Zn}^{++} & \\
\text{O}^{--} & \text{Zn}^{++} & \text{O}^{--} & \text{Zn}^{++} & & \text{O}^{--}\ \text{Zn}^{++} \\
& \text{Zn}^{++} & & e^- & e^- & \\
\text{Zn}^{++} & \text{O}^{--} & \text{Zn}^{++} & \text{O}^{--} & & \text{Zn}^{++}\ \text{O}^{--}
\end{array}
$$

Lattice Defects in ZnO

Figure 11.5. Lattice defects.

(Fig. 11.5) to the outer oxide surface. Examples of n-type oxides are ZnO, CdO, TiO$_2$, and Al$_2$O$_3$. Wagner showed that the law of mass action can be applied to the concentration of interstitial ions and electrons, and also to cation vacancies and positive holes. Hence, for Cu$_2$O, the equilibrium relations are

$$
\frac{1}{2}O_2 \rightleftarrows Cu_2O + 2Cu_{\square}^- + 2 \oplus \tag{11.2}
$$

and

$$
C_{Cu_{\square}^-}^2 \cdot C_{\oplus}^2 = \text{const} \cdot p_{O_2}^{1/2} \tag{11.3}
$$

For ZnO, the corresponding equilibrium is

$$
ZnO \rightleftarrows Zn_{int}^{2+} + 2e^- + \frac{1}{2}O_2 \tag{11.4}
$$

and

$$
C_{Zn_{int}^{2+}} \cdot C_{e^-}^2 = \frac{\text{const}}{p_{O_2}^{1/2}} \tag{11.5}
$$

These relations lead to predictions regarding the effect of impurities in the oxide lattice on oxidation rate; for example, if a few singly charged Li$^+$ ions are substituted for doubly charged Ni^{2+} ions in the NiO lattice, the concentration of positive holes must increase in order to maintain electrical neutrality. Hence, to maintain equilibrium for the reaction

$$\frac{1}{2}O_2 \rightleftarrows NiO + Ni_\square^{2-} + 2\oplus$$

as expressed by

$$C_{Ni_\square^{2-}} \cdot C_\oplus^2 = const \cdot p_{O_2}^{1/2}$$

the concentration of cation vacancies must decrease. This is accompanied by a decrease in oxidation rate [22] because the rate is controlled by migration of cation vacancies. On the other hand, if small amounts of Cr^{3+} are added to NiO, the concentration of positive holes decreases, the concentration of cation vacancies correspondingly increases, and the oxidation rate increases. Data of Wagner and Zimens [23] showing the effect of alloyed chromium on oxidation of nickel are given in Table 11.2.

At 10% Cr, the rate decreases again, possibly because a scale composed of Cr_2O_3 forms instead of NiO, which alters the rate of ion migration apart from factors described previously.

In a sulfur atmosphere, for reasons paralleling those pertaining in the situation with O_2, up to 2% Cr alloyed with nickel accelerates reaction at 600–900 °C [24]. Outward diffusion of Ni^{2+} occurs through cation vacancies in $Ni_{1-x}S$, where x connotes a number between 0 and 1, and incorporation of Cr^{3+} increases cation vacancy concentration. In Cr–Ni alloys containing >40% Cr, outward diffusion of Cr^{3+} occurs in a scale composed of Cr_2S_3. Incorporation of Ni^{2+} ions into the Cr_2S_3 scale decreases cation vacancy concentration, thereby decreasing the reaction rate to a value below that for pure chromium. At intermediate chromium compositions, the scale is heterogeneous, consisting of both nickel and chromium sulfides, with the rate of sulfidation being less than that of pure chromium for Cr–Ni alloys containing >20% Cr.

From Eq. (11.3), it is apparent that higher partial pressures of oxygen for p-type semiconductors must be accompanied by higher concentrations of vacancies and holes at the O_2–oxide interface. Hence, copper oxidizes at higher rates the higher the oxygen pressure, in accord with prediction [25].

TABLE 11.2. Oxidation of Nickel Alloyed with Chromium, 1000 °C, 1 atm O_2

Weight Percent Cr	Parabolic Rate Constant, k (g²cm⁻⁴s⁻¹)
0	3.1×10^{-10}
0.3	14×10^{-10}
1.0	26×10^{-10}
3.0	31×10^{-10}
10.0	1.5×10^{-10}

TABLE 11.3. Oxidation of Zinc, 390 °C, 1 atmO_2

	Parabolic Rate Constant, k ($g^2cm^{-4}h^{-1}$)
Zn pure	0.8×10^{-9}
Zn + 0.1% Al	1×10^{-11}
Zn + 0.4% Li	2×10^{-7}

If small amounts of Li^+ are added to ZnO, which is an n-type semiconductor, the electron concentration decreases in order to preserve neutrality, and the concentration of interstitial zinc ions increases in accord with the law of mass action [Eq. (11.5)]. This increase in concentration facilitates the diffusion of ZN_{int}^{2+}; hence, Li^+ increases the oxidation rate of zinc, contrary to its effect in NiO. By similar reasoning, Al^{3+} decreases the rate. Data showing these effects are presented in Table 11.3 [26]. The oxidation rate of zinc is almost independent of oxygen pressure because the concentration of interstitial zinc ions is already low at the O_2–oxide interface, and any further decrease brought about by increasing oxygen pressure has little effect on their concentration gradient referred to the metal surface where ZN_{int}^{2+} concentration is highest.

Therefore, traces of impurities, which play a major role in semiconductor properties, also appreciably affect rates of oxidation of metals protected by semiconductor films. On the other hand, alloying elements present in large percentages (e.g., >10% Cr–Ni) affect the oxidation rate by a gross alteration of the actual composition and structure of the film in addition to any effects on semiconducting properties.

11.7 GALVANIC EFFECTS AND ELECTROLYSIS OF OXIDES

The electrochemical nature of oxidation at elevated temperatures suggests that galvanic coupling of dissimilar metals should affect the rate, and, in fact, such effects do occur [27]. The reaction of silver with gaseous iodine at 174 °C, for example, is accelerated by contact of the silver with tantalum, platinum, or graphite. AgI, which is mainly an ionic conductor, forms on silver at a rate limited by transport of electrons across the AgI layer. When silver is coupled to tantalum, Ag^+ ions diffuse over the surface of the tantalum, with the latter supplying electrons that hasten conversion of silver to AgI. In addition, therefore, to the usual film of AgI on the silver, the compound spreads progressively over the tantalum surface (Fig. 11.6). Analogously, it was found [28] that, when silver coated with a porous gold electroplate is exposed to sulfur vapor at 60 °C, Ag_2S forms a tightly adherent coating over the gold surface.

A second example is given by nickel completely immersed in molten Na–K borax to a depth of about 3 mm at 780 °C in 1 atm O_2 (Fig. 11.7). The rate of

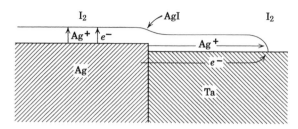

Figure 11.6. Effect of coupling tantalum to silver on reaction of iodine vapor with silver (Ilschner-Gensch and Wagner [27]).

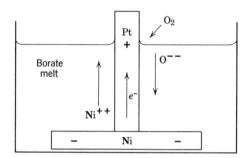

Figure 11.7. Galvanic cell—Pt; O_2, borate melt; Ni—illustrating accelerated oxidation of nickel through contact with platinum.

oxidation under these conditions is low because of limited access of oxygen from the gaseous phase. If nickel is coupled to platinum or silver gauze, the latter reaching above the liquid borax surface, corrosion of the nickel is accelerated by a factor of 35–175 for 1-h exposure. Nickel under these conditions corrodes even more rapidly than if exposed to pure oxygen at the same temperature because a protective NiO scale does not form. Instead, Ni^{2+} dissolves in the borax electrolyte, and the platinum acts as an oxygen electrode; the open-circuit potential difference between platinum and nickel is 0.7 V under the conditions cited previously. Addition of 1% FeO to borax increases the oxidation rate still more, probably by supplying Fe^{2+} ions, which are oxidized to Fe^{3+} by oxygen near the electrolyte surface, with the trivalent ions then being reduced again either at the cathode or by local cell action at the nickel anode.

The dependence of oxidation on ion migration within a reaction-product layer suggests that the oxidation rate should be affected by an applied electric current. This was first shown to be the case by Schein et al. [29] By wrapping a platinum wire around oxidized iron and passing a current of 1.5 A/cm², these investigators decreased the oxidation rate at 880 °C when iron was made cathodic, and they accelerated the rate when iron was made anodic. Similar relations were shown by Jorgensen for the oxidation of zinc in oxygen at 375 °C [30].

11.8 HOT ASH CORROSION

An 8% Al–Cu alloy, oxidizing in air at $750\,°C$ in the presence of MoO_3 vapor from an adjoining molybdenum wire not in contact, was found to react at an abnormally high rate [31]. Similar accelerated oxidation in air was reported for stainless steels containing a few percent molybdenum or vanadium [32], or as little as 0.04% boron [33]. The oxidation products were voluminous and porous.

An analogous accelerated damage is found in boiler tubes or gas turbine blades operating at high temperatures in contact with combustion gases of crude oils high in vanadium [34]. High-vanadium oils are found only in certain wells, some of which are located in South America. The vanadium is present in the oil as a soluble organic complex and, hence, is not easily removed. The solid residual ash of such oils may reach 65% V_2O_5 or higher, and the damage caused by the ash is not different from that observed when vanadium is alloyed with heat-resistant materials.

Damage of this kind is sometimes called *catastrophic* or *accelerated* oxidation, or *hot ash corrosion* [35]. It has been suggested that the cause is a low-melting oxide phase that acts as a flux to dislodge or dissolve the protective scale. The melting points of V_2O_5, MoO_3, and B_2O_3, for example, are $658\,°C$, $795\,°C$, and $294\,°C$, respectively, compared to $1527\,°C$ for Fe_3O_4. Eutectic combinations of oxides reduce the melting point still further. In the case of vanadium-containing oils, the presence of sulfur and sodium compounds in the oil and the catalytic effect of V_2O_5 for converting SO_2 to SO_3 result in a scale containing Na_2SO_4 and various metal oxides, the melting point of which has been found to be as low as $500\,°C$. Additions to oil of substances (e.g., calcium or magnesium soaps, powdered dolomite, or powered Mg) that raise the melting point of the ash by forming CaO (m.p. $2570\,°C$) or MgO (m.p. $2800\,°C$) are found, accordingly, to be beneficial. Also, operating at temperatures below the melting point of the liquid phase avoids catastrophic oxidation. Similarly, alloys high in nickel are more resistant because of the high melting point of NiO ($1990\,°C$).

11.9 HOT CORROSION

Hot corrosion is the accelerated oxidation of a material at elevated temperatures induced by a thin film of fused salt deposit [36]. It is called 'hot corrosion' because, being caused by a thin electrolyte film, it shares some similarities with aqueous atmospheric corrosion, in which corrosion is commonly controlled by the diffusion of oxygen to the metal surface. In hot corrosion, the soluble oxidant is SO_3 ($S_2O_7^{2-}$) in the fused salt.

In the late 1960s, during the Viet Nam conflict, hot corrosion caused severe corrosion of gas turbine engines of military aircraft during operation over seawater. Sulfide formation results from the reaction of the metallic substrate with a thin film of fused salt of sodium sulfate. The condensed liquid film deposits either by chemical deposition (i.e., the vapor pressure of Na_2SO_4 in the vapor phase exceeds its equilibrium partial pressure at the substrate temperature) or by physical deposition (solid or liquid salt detaches from an upstream component and attaches to a hot substrate on impact) [36]. In metallographically examining corroded components, oxide particles were observed to be dispersed in the adherent salt film. As a result, a fluxing mechanism may apply, in which an otherwise protective oxide scale on the substrate surface dissolves at the oxide/salt interface, but precipitates as non-protective particles within the salt film [36]. The fluxing mechanism of hot corrosion has been thoroughly reviewed by Rapp [36].

Hot corrosion may occur in the absence of metallic contaminants in the fuel. It was first identified in marine propulsion gas turbines where service temperatures of first-stage blades and vanes were in the range 650–700 °C (1200–1300 °F). The accelerated attack was repeated in the laboratory [37] by applying a thin coating of Na_2SO_4 to either 30% Cr–Co or 30% Cr–Ni alloys and oxidizing at 600–900 °C (1110–1650 °F) in O_2 at 1 atm containing 0.15% ($SO_2 + SO_3$). The highest oxidation rates occurred at 650–750 °C, in accord with practical experience. Oxidation products consisted mostly of once-liquid mixtures of Na_2SO_4 + $CoSO_4$ (or $NiSO_4$). Metallic sulfides often form inclusions or a network of sulfides along grain boundaries of the alloy; the corresponding attack is then called *sulfidation.*

It has been suggested that lower oxidation rates above 750 °C result from formation of protective Cr_2O_3 films unable to form at lower temperatures; e.g., at 650 °C [37]. In turn, the presence of Cr_2O_3 films may be related to the increased volatility and lesser tendency of Na_2SO_4 to deposit on the alloy surface at higher temperatures [38, 39].

Damaging deposits of Na_2SO_4 may originate alone from sea salt contamination of intake air. Sulfur dioxide and trioxide of oil combustion products also contribute, but hot corrosion of marine turbine blades may occur even with use of very low-sulfur fuels [38]. High-chromium alloys are more resistant to hot corrosion than are low-chromium alloys.

Ilschner-Gensch and Wagner suggested [27] that the mechanism is not one of melting point or fluxing alone, but may also involve galvanic effects. A spongy, porous network of an electronically conducting oxide, such as Fe_3O_4 filled with a liquid electrolyte, reproduces the cell described previously containing a platinum cathode and a nickel anode in contact with molten borax. The Fe_3O_4 sponge acts as an oxygen electrode of large area, and the base metal acts as an anode. Supplied with a liquid electrolyte in which oxygen and metal ions migrate rapidly, such a cell accounts for an accelerated oxidation process far exceeding the rate for a metal reacting directly with gaseous oxygen through a continuous oxide scale.

11.10 OXIDATION OF COPPER

Copper oxidizes in air at low temperatures (<260 °C) in accord with the two-stage logarithmic equation, forming a film of Cu_2O. The rate varies with crystal face, decreasing in the order (100) > (111) > (110). Heat treatment of polycrystalline copper with hydrogen at 300–450 °C decreases the oxidation rate in oxygen at 200 °C because submicroscopic surface facets are formed by adsorbed hydrogen, presumably favoring the (111) orientation. On the other hand, heat treatment with nitrogen or helium increases the rate because adsorbed oxygen (traces from gas or metal) favors submicroscopic facets of predominantly (100) orientation [40].

Between about 260 °C and 1025 °C, the Cu_2O film is overlaid by a superficial film of CuO. Oxidation changes from logarithmic to parabolic behavior above 400–500 °C. Only Cu_2O forms in air above 1025 °C. Copper oxidizes at a rate slightly higher than that for iron, and much more rapidly than that for nickel or the heat-resistant Cr–Fe alloys. This is shown by the following temperatures [41], below which the scaling losses in air are less than approximately $2–4\,g\,m^{-2}h^{-1}$: Cu, 450 °C; Fe, 500 °C; Ni, 800 °C; 8–10% Cr–Fe (0.1% C), 750 °C; 25–30% Cr–Fe (0.1% C), 1050–1100 °C.

Alloying elements that are particularly effective for improving oxidation resistance at high temperatures are aluminum, beryllium, and magnesium; for example, at 256 °C, a 2% Be–Cu alloy oxidizes in 1 h at 1/14 the rate of copper [42]. Maximum improvement by aluminum additions occurs at about 8% [43].

11.10.1 Internal Oxidation

When alloyed with small percentages of certain metals (e.g., aluminum, beryllium, iron, silicon, manganese, tin, titanium, and zinc), copper oxidizes with precipitation of oxide particles within the body of the metal as well as forming an outer oxide scale. Oxidation within the metal is called subscale formation or internal oxidation. Similar behavior is found for many silver alloys, but without formation of an outer scale. Internal oxidation is not observed, in general, with cadmium-, lead-, tin-, or zinc-based alloys. A few exceptions have been noted, such as for alloys of sodium–lead, aluminum–tin, and magnesium–tin [44]. Internal oxidation is usually not pronounced for any of the iron alloys.

The mechanism is apparently one of oxygen diffusing into the alloy and reacting with alloying componens of higher oxygen affinity than that of the base metal before the alloying components can diffuse to the surface. Above a critical concentration of the alloying component, a compact protective layer of the component oxide tends to be formed at the external surface which thereafter suppresses internal oxidation. In accord with a diffusion-controlled mechanism, the depth of subscale grows in accord with the parabolic equation [44]. The subject was reviewed by Rapp [45].

11.10.2 Reaction with Hydrogen ("Hydrogen Disease")

The tendency of copper to dissolve oxygen when the metal is heated in air leads to rupture of the metal along grain boundaries by formation of steam when the metal is subsequently heated in hydrogen. Cast tough-pitch copper containing free Cu_2O is very sensitive to this type of damage. Instances are on record where damage has been caused by hydrogen at temperatures as low as 400 °C (750 °F). Oxygen-free coppers are not susceptible, but they may become moderately so, however, should they be heated at any time in oxygen or in air.

Silver similarly dissolves oxygen when heated at elevated temperatures in air, and it becomes blistered or loses ductility if later heated in hydrogen above 500 °C (925 °F). The mechanism is the same as that applying to copper. Oxygen-free silver heated in hydrogen for 1 h at 850 °C (1550 °F) is not embrittled or damaged. However, when it is heated immediately afterward in air at the same temperature, loss of ductility occurs that is similar to, but not as severe as, the characteristic loss when silver containing oxygen is heated in hydrogen [46]. Some dissolved hydrogen undoubtedly escapes before oxygen can diffuse into the silver, hence diminishing subsequent damage. Gold and platinum dissolve little or no oxygen and, consequently, are not subject to similar damage when heated in hydrogen.

11.11 OXIDATION OF IRON AND IRON ALLOYS

In the low-temperature region (approx 250 °C), the oxidation rate of iron is sensitive to crystal face, decreasing in the order (100) > (111) > (110) [47]. The oxide nuclei, apparently consisting of Fe_3O_4, grow to form a uniform film of oxide. Subsequently, α Fe_2O_3 nucleates and covers the Fe_3O_4 layer [48, 49].

Oxidation of iron in the parabolic range is complicated by formation of as many as three distinct layers of iron oxide, and the proportions of these layers change as the temperature or oxygen partial pressure changes. Data reported by various investigators are not in good agreement, probably because of variations in the purity of iron used for oxidation tests—particularly its carbon content.

At 600 °C in 1 atm O_2 for 100 min, it is reported that the scale is composed of two layers: an inner FeO layer equal in weight to an outer Fe_3O_4 layer [50]. At 900 °C for 100 min, a three-layer scale is composed of 90% FeO, 9% Fe_3O_4, and less than 1% Fe_2O_3. Below 570 °C (1058 °F), FeO is unstable and, if any is formed above this temperature, it decomposes at room temperature into Fe_3O_4 plus Fe.

The most efficient alloying elements for improving oxidation resistance of iron in air are chromium and aluminum. Use of these elements with additional alloyed nickel and silicon is especially effective. An 8% Al–Fe alloy is reported to have the same oxidation resistance as a 20% Cr–80% Ni alloy [51]. Unfortunately, the poor mechanical properties of aluminum–iron alloys, the sensitivity of their protective oxide scales to damage, and the tendency to form aluminum nitride that causes embrittlement have combined to limit their application as oxidation-resistant materials. In combination with chromium, some of these drawbacks of aluminum–iron alloys are overcome.

The good oxidation resistance of the chromium–iron alloys, combined with acceptable mechanical properties and ease of fabrication, accounts for their wide commercial application. Typical oxidation behavior is shown in Fig. 11.8.

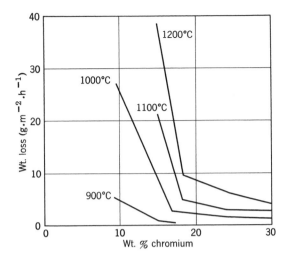

Figure 11.8. Effect of alloyed chromium on oxidation of steels containing 0.5% C, 220 h [E. Houdremont, *Handbuch der Sonderstahlkunde*, Vol. I, *Chromstähle*, Springer, Berlin (1956), p. 815, Fig. 677]. (With kind permission of Springer Science and Business Media.)

Improvement in oxidation resistance of iron by alloying with aluminum or chromium probably results from a marked enrichment of the innermost oxide scale with respect to aluminum or chromium. The middle oxide scales are known, from chemical analysis, to be so enriched, and electron-microprobe analyses confirm marked enrichment of chromium in the oxide adjacent to the metal phase in the case of chromium–iron alloys [52]. These inner oxides resist ion and electron migration better than does FeO. For chromium–iron alloys, the enriched oxide scale is accompanied by depletion of chromium in the alloy surface immediately below the scale. This situation accounts for occasional rusting and otherwise poor corrosion resistance of hot-rolled stainless steels that have not been adequately pickled following high-temperature oxidation.

11.12 LIFE TEST FOR OXIDATION-RESISTANT WIRES

The merit of a particular alloy for resisting high-temperature environments, especially on long exposures, depends not only on the diffusion-barrier properties of reaction-product scales, but also on the continuing adherence of such scales to the metal. Scales that are otherwise protective often spall (become detached) during cooling and heating cycles because their coefficient of expansion differs from that of the metal. Hence, an ASTM accelerated test [53] for oxidation resistance of wires calls for a cyclic heating period of 2 min at a specific temperature followed by a cooling period of 2 min. Alternate heating and cooling results in much shorter life than would occur if the wire is heated continuously. The life of a wire in this test is measured as the time to failure or the time to reach a 10%

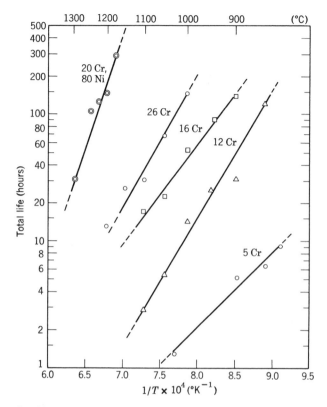

Figure 11.9. Life of heat-resistant alloy wires in ASTM life test as a function of temperature (20% Cr–Ni and 5–26% Cr–Fe), in air of 100% relative humidity at 25 °C (Brasunas and Uhlig [54]). (Copyright ASTM International. Reprinted with permission.)

increase in electrical resistance. An equation related to the Arrhenius reaction-rate equation expresses the dependence of life on temperature [54]:

$$\log \text{life}(h) = \frac{\Delta E'}{2.3RT} + \text{const}$$

Typical test results for life of several alloys are given in Fig. 11.9. The life of the most resistant alloys is more sensitive to temperature change (higher activation energy $\Delta E'$) than that of alloys with inherently shorter life.

11.13 OXIDATION-RESISTANT ALLOYS

11.13.1 Reactive Element Effect (REE)

The *reactive element effect* (REE) is obtained when 1 wt.% or less of a reactive element, such as yttrium, hafnium, lanthanum, zirconium, or cerium, is added to high-temperature alloys containing chromium or aluminum. The effect can be obtained when the reactive element is added as an alloy addition or when a coating of the reactive element oxide is applied to the surface of the alloy. The useful life of high-temperature alloys is typically limited by spalling of the oxide from the metal surface. Reactive elements greatly improve the spalling resistance of both chromia-forming and alumina-forming alloys. Commercial chromia-forming heater alloys now all contain small amounts of one or more reactive elements, and yttrium has become a commonly used reactive element addition to high-temperature alloys [55, 56].

Applying a 4-nm-thick coating of ceria (CeO_2) or yttria (Y_2O_3) has been found to increase the oxidation resistance of chromium and iron–chromium alloys. By using secondary ion mass spectrometry (SIMS) to establish the location of the reactive element in the oxide scale, the reactive element was found in the middle of the oxide scale, suggesting that inward oxygen transport in chromium oxide is the dominant transport mechanism. Addition of the reactive element inhibits the diffusion of Cr and results in growth primarily by the inward diffusion of oxygen [56, 57].

11.13.2 Chromium–Iron Alloys

The 4–9% Cr alloys are widely used for oxidation resistance in oil-refinery construction. The 12% Cr–Fe alloy is used for steam turbine blades because of excellent oxidation resistance and good physical properties. The 9–30% Cr alloys are used for furnace parts and burners; when combined with silicon and nickel, and sometimes other alloying elements, they are used for valves of internal-combustion engines. The approximate upper temperature limits for exposure to air are presented in Table 11.4.

Addition of 1% yttrium to a 25% Cr–Fe alloy was reported to extend the upper limit of oxidation resistance to about 1375 °C (2500 °F) [58, 59]. Rare-earth metal additions, such as yttrium, in general, are beneficial to oxidation resistance of chromium and chromium alloys, including gas turbine alloys [60].

TABLE 11.4. Approximate Upper Temperature Limits for Exposure of Cr–Fe Alloys to Air

% Cr in Cr–Fe Alloys	Maximum Temperature in Air	
	°C	°F
4–6	650	1200
9	750	1375
13	750–800	1375–1475
17	850–900	1550–1650
27	1050–1100	1925–2000

11.13.3 Chromium–Aluminum–Iron Alloys

The chromium–aluminum–iron alloys have exceptional oxidation resistance, combining the oxidation-resistant properties imparted by chromium and aluminum; for example, the 30% Cr, 5% Al, 0.5% Si alloy (Trade name: Megapyr) resists oxidation in air up to at least 1300 °C (2375 °F). Similarly, the 24% Cr, 5.5% Al, 2% Co alloy (Trade name: Kanthal A) is resistant up to 1300 °C. They are used, among other applications, for furnace windings and parts and for electric-resistance elements. Drawbacks in their properties are poor high-temperature strength and tendency toward embrittlement at room temperature after prolonged heating in air, caused in part by aluminum nitride formation. For this reason, furnace windings must be well-supported and are usually corrugated in order to allow for expansion and contraction.

11.13.4 Nickel and Nickel Alloys

The good oxidation resistance of nickel is improved by adding 20% Cr; this alloy (one U.S. trade name: Nichrome V) is resistant in air to a maximum temperature of about 1150 °C (2100 °F). It is a heat-resistant alloy that combines excellent oxidation resistance with good physical properties at both low and elevated temperatures. Oxidation resistance of the commercial alloy is considerably improved by the addition of calcium metal deoxidizer during the melting process, which is said to avoid oxidation of the alloy along grain boundaries. Small amounts of zirconium, thorium, and rare-earth metals (e.g., cerium) are also beneficial, probably in part because they decrease the tendency of protective oxides to spall. This explanation was also mentioned earlier as applying to the beneficial effect of the rare-earth yttrium when alloyed with the chromium–iron alloys.

The 16% Cr, 7% Fe, 76% Ni alloy (Trade name: Inconel 600) is slightly less resistant to oxidation than the 20% Cr–Ni alloy, but similarly has excellent physical properties, is readily fabricated or welded, and can be used in air up to a maximum temperature of about 1100 °C (2000 °F). Electrical heating units for some stoves are fabricated of this alloy in tubular form. An electric heating wire of the 20% Cr–Ni alloy inside the tube is insulated from the outer sheath by powdered MgO. Because of its high nickel content and good strength (nickel does not readily form a carbide or nitride), this alloy is often used for construction of carburizing or nitriding furnaces.

Some heat-resistant thermocouple wires consist of nickel alloys. The 10% Cr–Ni alloy (Chromel P) and 2% Al, 2% Mn, 1% Si, bal. Ni alloy (Alumel) can be exposed to air at a maximum temperature of about 1100 °C (2000 °F).

Nickel and high-nickel alloys tend to oxidize along grain boundaries when subject to alternate oxidation and reduction. Alloying with chromium reduces this tendency. Also, in contact with sulfur or sulfur atmospheres at elevated temperatures, nickel and high-nickel alloys are subject to intergranular attack. Consequently, nickel is not usefully resistant to such atmospheres above about 315 °C

(600 °F). For best resistance to sulfur-containing environments, iron-base alloys should contain high chromium and low nickel.

Gas turbine blades are essentially nickel-base or cobalt-base alloys containing substantial amounts of chromium, several percent aluminum, and a few hundredths percent yttrium. Their susceptibility to hot corrosion and sulfidation has already been discussed. Applied coatings of aluminum or of aluminum–chromium–yttrium increase resistance to attack.

11.13.5 Furnace Windings

Twenty percent chromium–nickel and various chromium–aluminum–iron alloys are in common use as furnace windings. To achieve higher temperatures in air, a 10% Rh–Pt alloy can be used up to at least 1400 °C (2550 °F). This alloy performs better than pure platinum because of higher strength and a lower rate of grain growth. A single crystal of the same dimensions as the resistance wire cross section tends to shear easily and cause failure.

Molybdenum-wound furnaces are operated to at least 1500 °C (2725 °F). Because molybdenum oxidizes in air, such furnace windings are blanketed in a protective atmosphere of hydrogen.

REFERENCES

1. C. Coddet, High temperature corrosion and protection of materials 6, *Proceedings*, 6th International Symposium on High Temperature Corrosion and Protection of Materials, Les Embiez, France, May 2004, Part 1, Trans Tech Publications, Uetikon-Zuerich, Switzerland, 2004, p. 193.

2. T. Hejwowski and A. Weronski, *Vacuum* **65**, 427 (2002).

3. A. MacRae, *Surface Sci.* **1**, 319 (1964).

4. H. H. Uhlig, *Corros. Sci.* **7**, 325 (1967).

5. L. Brockway and A. Rowe, in *Fundamentals of Gas–Surface Interactions*, H. Saltsburg et al., editors, Academic Press, New York, 1967. p. 147.

6. F. Gronlund, *J. Chim. Phys.* **53**, 660 (1956).

7. J. Bénard, F. Gronlund, J. Oudar, and M. Duret, *Z. Elektrochem.* **63**, 799 (1959).

8. Robert A. Rapp, *Mater. Sci. Forum* **154**, 119 (1994).

9. H. H. Uhlig, *Acta Metall.* **4**, 541 (1956).

10. N. Cabrera and N. Mott, *Rep. Progr. Phys.* **12**, 163 (1949).

11. B. Mishra, in *Metals Handbook*, Desk Edition, 2nd edition, ASM International, Materials Park, OH, 1998, p. 718.

12. N. Pilling and N. Bedworth, *J. Inst. Metals* **29**, 534 (1923).

13. ASM Handbook, Vol. 13A, *Corrosion: Fundamentals, Testing, and Protection*, ASM International, Materials Park, OH, 2003, pp. 228–229.

14. Per Kofstad, *High-Temperature Oxidation of Metals*, Wiley, New York, 1966, pp. 252–254.

15. G. Tammann and W. Köster, *Z. Anorg. Allg. Chem.* **123**, 196 (1922).

16. H. Uhlig, J. Pickett, and J. MacNairn, *Acta Metall.* **7**, 111 (1959).

17. D. Gilroy and J. Mayne, *Corros. Sci.* **5**, 55 (1965).

18. L. Pfeil, *J. Iron Steel Inst.* **119**, 520 (1929).

19. C. Wagner, *Z. Phys. Chem.* **21B**, 25 (1933).

20. C. Wagner, *Atom Movements*, American Society for Metals, Cleveland, OH, 1951, pp. 153–173; C. Wagner and K. Grünewald, *Trans. Faraday Soc.* **34**, 851 (1938).

21. Ref. 14, pp. 113–127; T. P. Hoar and L. E. Price, *Trans. Faraday Soc.* **34**, 867 (1938).

22. K. Hauffe and H. Pfeiffer, *Z. Elektrochem.*, **56**, 390 (1952).

23. C. Wagner and K. Zimens, *Acta Chem. Scand.* **1**, 547 (1947).

24. S. Mrowec, T. Werber, and N. Zastawnik, *Corros. Sci.* **6**, 47 (1966).

25. C. Wagner and K. Grünewald, *Z. Phys. Chem.* **40B**, 455 (1938).

26. C. Gensch and K. Hauffe, *Z. Phys. Chem.* **196**, 427 (1950).

27. C. Ilschner-Gensch and C. Wagner, *J. Electrochem. Soc.* **105**, 198, 635 (1958).

28. T. Egan and A. Mendizza, *J. Electrochem. Soc.* **107**, 353 (1960).

29. F. Schein, B. LeBoucher, and P. LaCombe, *C. R. Acad. Sci. (Paris)* **252**, 4157 (1961).

30. P. Jorgensen, *J. Electrochem. Soc.* **110**, 461 (1963).

31. G. Rathenau and J. Meijering, *Metallurgia* **42**, 167 (1950).

32. A. Brasunas and N. Grant, *Trans. Am. Soc. Metals* **44**, 1117 (1952).

33. A. Fry, *Tech. Mitt. Krupp* **1**, 1 (1933).

34. F. Monkman and N. Grant, *Corrosion* **9**, 460 (1953).

35. J. Demo, *Mater. Perf.* **19** (3), 9, (1980).

36. Robert A. Rapp, *Corros. Sci.* **44** (2), 209 (2002).

37. K. Luthra and D. Shores, *J. Electrochem. Soc.* **127**, 2202 (1980).

38. P. Bergman, *Corrosion* **23**, 72 (1967).

39. J. Elliott, *Solid State Chemistry of Energy Conversion and Storage*, J. Goodenough and M. Whittingham, editors, Advances in Chemistry Series, No. 163, American Chemical Society, Washington, D.C., 1977.

40. A. Swanson and H. Uhlig, *J. Electrochem. Soc.* **118**, 1325 (1971).

41. B. Lustman, *Metal Progr.* November, 850 (1946).

42. W. Campbell and U. Thomas, *Trans. Electrochem. Soc.* **91**, 623 (1947).

43. H. Nishimura, *J. Min. Metall, Kyoto* **9**, 655 (1938).

44. F. N. Rhines et al., *Trans. Am. Inst. Min. Metall. Eng.* **147**, 205, 318 (1942)

45. R. Rapp, *Corrosion* **21**, 382 (1965).

46. D. Martin and E. Parker, *Min. Metall. Eng.* **152**, 269 (1943).

47. J. Wagner, Jr., K. Lawless, and A. Gwathmey, *Trans. Metall. Soc. AIME* **221**, 257 (1961).

48. R. Grauer and W. Feitknecht, *Corros. Sci.* **6**, 301 (1966).

49. P. Sewell and M. Cohen, *J. Electrochem. Soc.* **111**, 501 (1964).

50. M. Davies, M. Simnad, and C. Birchenall, *J. Metals* **3**, 889 (1951).

51. N. Ziegler, *Min. Metall. Eng.* **100**, 267 (1932).

52. G. Wood and D. Melford, *J. Iron Steel Inst.* **198**, 142 (1961).

53. ASTM B76-90(2007), *Standard Test Method for Accelerated Life of Nickel–Chromium and Nickel–Chromium–Iron Alloys for Electrical Heating*, ASTM, West Conshohocken, PA.

54. A. deS. Brasunas and H. H. Uhlig, *ASTM Bulletin*, No. 182, ASTM, Philadelphia, May 1952, p. 71.

55. J. Stringer, *Mater. Sci. Engi.* **A120**, 129 (1989).

56. B. A. Pint, in *John Stringer Symposium on High Temperature Corrosion, Proceedings from Materials Solutions Conference 2001*, ASM International, Materials Park, OH, 2003.

57. M. J. Graham, *Corrosion* **59** (6), 475 (2003).

58. J. Fox and J. McGurty, in *Refractory Metals and Alloys*, M. Semchyshen and J. Harwood, editors, Interscience, New York, 1961, p. 207.

59. E. Fellen, *J. Electrochem. Soc.* **108**, 490 (1961).

60. R. Viswanathan, *Corrosion* **24**, 359 (1968).

GENERAL REFERENCES

N. Birks, G. H. Meier, and F. S. Pettit, *Introduction to the High Temperature Oxidation of Metals*, 2nd edition, Cambridge University Press, Cambridge, U.K., 2006.

P. Kofstad, *High Temperature Corrosion*, Elsevier, London, 1988.

P. Kofstad, *High-Temperature Oxidation of Metals*, Wiley, New York, 1966.

R. B. Pond and D. A. Shifler, High-temperature corrosion-related failures, in *ASM Handbook*, Vol. 11, *Failure Analysis and Prevention*, ASM International, Materials Park, OH, 2002, pp. 868–880.

R. Rapp, editor, *High Temperature Corrosion*, National Association of Corrosion Engineers, Houston, TX, 1983.

P. Steinmetz, I. G. Wright, G. Meier, A. Galerie, B. Pieraggi, and R. Podor, editors, *High Temperature Corrosion and Protection of Materials 6*, Proceedings of the 6th International Symposium on High Temperature Corrosion and Protection of Materials, Les Embiez, France, May 2004, Trans Tech Publications, Uetikon-Zuerich, Switzerland, 2004.

P. F. Tortorelli, I. G. Wright, and P. Y. Hou, editors, *John Stringer Symposium on High Temperature Corrosion*, Proceedings from Materials Solutions Conference 2001, ASM International, Materials Park, OH, 2003.

Materials Science and Engineering, A120/121, November 15, 1989, Proceedings, 2nd International Symposium on High Temperature Corrosion of Advanced Materials and Coatings, Les Embiez, France, May 1989.

PROBLEMS

1. Calculate Md/nmD for aluminum forming Al_2O_3 and for sodium forming Na_2O. Indicate whether the oxides are protective.

2. (*a*) Calculate the volume of FeO ($d = 5.95\,\text{g/cm}^3$) that results from the oxidation of $1\,\text{cm}^3$ of iron.

    (*b*) Similarly, calculate the volume of Fe_3O_4 ($d = 5.18\,\text{g/cm}^3$) that results from the oxidation of $1\,\text{cm}^3$ of iron.

    (*c*) Same as above, but, for Fe_2O_3 ($d = 5.24\,\text{g/cm}^3$).

3. Estimate the free energy of formation of NiO at $780\,°C$ from the cell illustrated in Fig. 11.7 (Section 11.7). Assume that NiO in borax is saturated. Compare with reported values as follows:

Temperature (K)	$\Delta G°$ (kJ/mole)
1000	−146.8
1100	−137.7

4. Copper oxidizes within the parabolic range at a higher rate the higher the oxygen pressure.

    (*a*) What is the quantitative relation [use Eq. (11.3), Section 11.6]?

    (*b*) Make the same calculation for nickel.

5. Using the Gibbs free energy of formation versus temperature diagram, Fig. 11.2, at $600\,°C$, estimate

    (*a*) the minimum pressures of oxygen required to oxidize aluminum, chromium, and silicon to Al_2O_3, Cr_2O_3, and SiO_2;

    (*b*) the maximum ratios of H_2/H_2O required to produce the oxides in (*a*); and

    (*c*) the maximum ratios of CO/CO_2 required to produce the oxides in (*a*).

Answers to Problems

1. 1.28; 0.58

2. (*a*) $1.7\,\text{cm}^3$; (*b*) $2.1\,\text{cm}^3$; (*c*) 2.1cm^3

3. Calculated $\Delta G° = -135\,\text{kJ/mole}$

4. (*a*) Rate is proportional to $po_2^{1/8}$; (*b*) Rate is proportional to $po_2^{1/6}$.

12

STRAY-CURRENT CORROSION

12.1 INTRODUCTION

Stray electric currents are those that follow paths other than the intended circuit, or they may be any extraneous currents in the earth. If currents of this kind enter a metal structure, they cause corrosion at areas where the currents leave to enter the soil or water. Usually, natural earth currents are not important from a corrosion standpoint, either because their magnitude is small or because their duration is short. Under some conditions, pipelines can incur considerable corrosion damage as a result of telluric currents—that is, currents induced in the steel pipeline by changes in the geomagnetic field of the earth [1].

Damage by alternating current (ac) is less than that by direct current (dc), with the resultant corrosion usually being greater for lower frequency and less for higher-frequency currents. Jones [2] reported that in $0.1N$ NaCl, the increased corrosion rate of a carbon steel caused by a 60-cycle current density of $300 \, A/m^2$ is negligible in an aerated solution, but severalfold higher (but still low) in a deaerated solution. This probably means that rates of reversible, or partially reversible, anode–cathode reactions in the aerated solution are symmetrical with respect to alternating applied potentials, whereas in the deaerated solution the reactions are not symmetrical, largely because of the hydrogen evolution

Corrosion and Corrosion Control, by R. Winston Revie and Herbert H. Uhlig
Copyright © 2008 John Wiley & Sons, Inc.

reaction. It has been estimated that for metals like steel, lead, and copper in common environments, 60-cycle alternating current (ac) causes less than 1% of the damage caused by an equivalent dc current [3].

The effect of unsymmetrical reactions (Faradaic rectification) is observed especially in the ac corrosion of metals that are passive (mostly by Definition 1, Section 6.1). It has been reported that stainless steels subjected to ac current electrolysis are corroded [4]; similarly, aluminum in dilute salt solutions at 15 A/m^2 (1.4 A/ft^2) suffers 5%, and at 100 A/m^2 (9.3 A/ft^2) as much as 31% of the corrosion damage caused by equivalent dc current densities. In a study of 1-V, 54-cycle ac current superimposed on dc current, Feller and Rückert [4] found that the passive region of potentiostatic polarization curves for nickel in $1N$ H$_2$SO$_4$ completely disappeared and that high anodic current densities persisted throughout the noble potential region. Chin and Fu [5] found analogous behavior of mild steel in $0.5M$ Na$_2$SO$_4$, pH 7. The passive current density increased with increasing superimposed ac current, reaching the same order as the critical current density (no passive region) at a 60-cycle current density of 2000 A/m^2. They also found that the corrosion potential was shifted several tenths of a volt in the active direction by an ac current density of 500 A/m^2. The Flade-potential region simultaneously was shifted in the active direction, but the critical current density remained about the same. These results suggest that passivity lost during the cathodic cycle, if not restored during the anodic cycle, can account for unexpectedly high corrosion rates of otherwise corrosion-resistant metals in aqueous media or in soil.

12.2 SOURCES OF STRAY CURRENTS

Sources of dc stray currents are commonly electric railways, grounded electric dc power lines, electric welding machines, cathodic protection systems, and electroplating plants. Sources of ac stray currents are usually grounded ac power lines or currents induced in a pipeline by parallel power lines. An example of dc stray current from an electric street railway system in which the steel rails are used for current return to the generating station is shown in Fig. 12.1. Because of poor

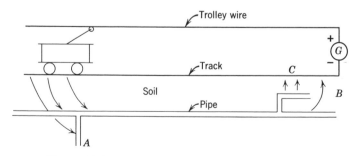

Figure 12.1. Stray-current corrosion of a buried pipe.

bonding between rails, combined with poor insulation of rails to the earth, some of the return current enters the soil and finds a low-resistance path, such as a buried gas or water main. The owner of a household water service pipe at A benefits by cathodic protection and experiences no corrosion difficulty; but owner B, to the contrary, is harassed by corrosion failures because the service pipe of his house is anodic with respect to the rails. If B coats the pipe, which is an understandable layman's reaction to any corrosion problem, matters are made worse because all stray currents now leave the pipe at defects in the insulating coating at high-current densities, accelerating penetration of the pipe. The basic rule: Never coat the anode.

Street railways have now in large part been replaced by other forms of transportation, but the problems of stray-current corrosion originating from metropolitan railway transit systems continue [6]. Also, cathodically protected structures requiring high currents, when located in the neighborhood of an unprotected pipeline, can produce damage similar to that by the railway illustrated in Fig. 12.1.

Another example of stray-current corrosion is illustrated in Fig. 12.2. A welding motor generator located on shore with grounded dc lines to a ship under repair can cause serious damage to the hull of the ship by current returning in part from the welding electrodes through the ship and through the water to the shore installation. In this case, it is better to place the generator on board ship and bring ac power leads to the generator, since ac currents leaking to ground cause less stray-current damage.

Current flowing long a water pipe (e.g., used as an electric ground) usually causes no damage *inside* the pipe because of the high electrical conductivity of steel or copper compared to water; for example, since the resistance of any conductor per unit length equals ρ/A, where ρ is the resistivity and A is the cross-sectional area, then the ratio of current carried by a metal pipe compared to that carried by the water it contains is equal to $\rho_W A_M/\rho_M A_W$, where the subscripts $_W$ and $_M$ refer to water and metal, respectively. For iron, $\rho_M = 10^{-5}\,\Omega$-cm; and for a potable water, ρ_W may be $10^4\,\Omega$-cm. Assuming that the cross-sectional area of water is 10 times that of the steel pipe, it is seen that if current through the pipe is 1 A, only about 10^{-8} A flows through the water. This small current leaving the

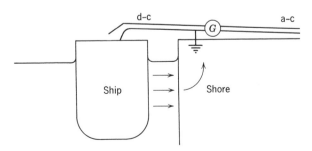

Figure 12.2. Stray-current damage to a ship by a welding generator.

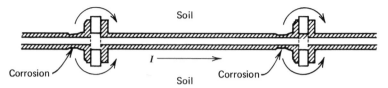

Figure 12.3. Effect of current flowing along a buried pipeline on corrosion near insulated couplings.

TABLE 12.1. Weight Loss of Metals by Stray-Current Corrosion

Metal	Equivalent Weight	Weight of Metal Corroded per Ampere-Year	
		Pounds	Kilograms
Fe	55.85/2	20.1	9.1
Cu	63.57/2	22.8	10.4
Pb	207.2/2	74.5	33.8
Zn	65.38/2	23.6	10.7
Al	26.98/3	6.5	2.9

pipe and entering the water causes negligible corrosion. If seawater is transported instead, with $\rho_W = 20\,\Omega$-cm, the ratio of currents is 2×10^5, indicating that, even in this case, most of the current is carried by the metallic pipe and there is very little stray-current corrosion on the inner surface. However, where such currents leave the pipe and enter the soil, stray-current corrosion of the *outer* surface may be appreciable.

   If insulating joints are installed in the above-mentioned pipe in order to reduce stray-current pickup, corrosion is now focused on the water side of the joint where any current that persists leaves the pipe to enter the water. Or, if a high-resistance joint exists between two sections of a buried pipe, corrosion may be more pronounced on the side where current enters the soil (Fig. 12.3).

12.3 QUANTITATIVE DAMAGE BY STRAY CURRENTS

In general, the amount of metal corroding at anodic areas because of stray current can be calculated using Faraday's law. Weight losses of typical metals for the equivalent of 1 A flowing for one year are listed in Table 12.1.

   At low current densities, local-action corrosion supplements stray-current corrosion. At high current densities in some environments, oxygen may be evolved, reducing the amount of metal corroding per faraday of electricity.

Amphoteric metals (e.g., lead, aluminum, tin, zinc) corrode in alkalies as well as in acids and, hence, may be damaged at cathodic areas where alkalies accumulate by electrolysis. This damage is in addition to damage at anodic areas. The amount of cathodic corrosion is not readily estimated. Plumbites ($NaHPbO_2$), aluminates (NaH_2AlO_3), stannates (Na_2SnO_3), and zincates (Na_2ZnO_2), which are all soluble in excess alkali, form in variable amounts per faraday, depending on diffusion rates, temperature, and other factors. Formulas for such compounds may also vary with conditions of formation. In general, the compounds hydrolyze at lower values of pH some distance away from the cathode to form insoluble metal oxides or hydroxides.

12.4 DETECTION OF STRAY CURRENTS

Stray currents may fluctuate over short or long intervals of time parallel to the varying load of the power source. This is in contrast to galvanic or cathodic protection currents, which are relatively steady. Hence, by recording the potential of a corroding system with respect to a reference electrode over a 24-h period, stray currents can often be detected and their origin can be traced to the generator source for which the load, day or night, varies in a pattern similar to that of the measured potential change. Thus, if stray currents, as indicated by potential measurements, are larger at 7 to 9 A.M. and again at 4 to 6 P.M., a street railway is suspected. If interference by a cathodic protection system is suspected, the protective current can be turned off and on briefly at regular intervals, observing whether the potential of the corroded structure also fluctuates at the same frequency.

The magnitude of current leaving (or entering) a buried pipe from whatever source can be calculated by measuring the potential difference between a position on the soil surface directly over the pipe and a position on the soil surface some distance away and at right angles to the pipe. If $\Delta\phi$ is the measured potential difference, ρ the resistivity of soil, h the depth of pipe below the surface, and y the distance along the soil surface over which the potential difference is measured, we obtain

$$\Delta\phi = \frac{\rho j}{2\pi} \ln \frac{y^2 + h^2}{h^2} \tag{12.1}$$

where j is the total current entering or leaving the pipe surface per unit length (for derivation, see Appendix, Section 29.5). If y is chosen equal to $10h$, we have

$$\Delta\phi = 0.734\rho j \tag{12.2}$$

An outer reference electrode (+) with respect to a (−) reference electrode directly over the pipe corresponds to current entering the pipe.

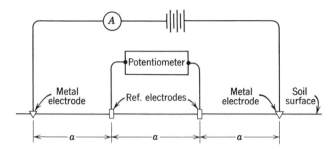

Figure 12.4. Four-electrode method for measuring soil resistivity.

12.5 SOIL-RESISTIVITY MEASUREMENT

The resistivity of a soil can be measured by the four-electrode method with each electrode arranged in a straight line and each separated by the distance a (Fig. 12.4). Steady current I from a battery is passed through the two outer metal electrodes, and the potential difference, $\Delta\phi$, of the two inner reference electrodes (e.g., Cu–CuSO$_4$) is measured simultaneously. The measurement is usually repeated with current direction reversed in order to cancel out any stray currents. Then

$$\rho = \frac{2\pi a\Delta\phi}{I} \tag{12.3}$$

where ρ is the resistivity of a uniform soil measured to an approximate depth equal to a (for derivation, see Appendix, Section 29.6).

12.6 MEANS FOR REDUCING STRAY-CURRENT CORROSION

1. *Bonding.* In Fig. 12.1, stray-current corrosion is completely avoided by placing a low-resistance metallic conductor between service pipe B and the rails at C. This is called bonding the rail and pipe systems. In the event of a cathodic protection system causing the damage, the bond may include a resistor just sufficient to avoid large change of potential of the unprotected system when the cathodic protection current is turned on and off. The resistor avoids major damage to the unprotected system. At the same time, it avoids the necessity of a large increase in protective current caused by current flowing through the bond, which in effect serves to protect the adjoining system in addition to the protected structure. If stray currents for some reason should periodically reverse direction, rectifying units (diodes) are inserted into the bonding connection, thereby ensuring that any current flow is in that direction that is nondamaging to the structure.

2. *Intentional Anodes and Cathodic Protection.* If a bond from B to C in Fig.
12.1 is not feasible, an intentional anode of scrap iron may be buried in
the direction of the rails and attached by a copper conductor to point B.
Stray currents then cause corrosion only of the intentional anode, which
is easily replaced at low cost. If a source of dc current is inserted between
the intentional anode and the pipe such that current flows in the soil in a
direction opposite to that of the stray current, the arrangement is equivalent to cathodically protecting the pipe. Cathodic protection is installed
whenever the intentional anode is not sufficient to overcome all corrosion
caused by stray currents.

3. *Insulating Couplings.* By installing one or more insulating couplings, the
pipeline in Fig. 12.1 becomes a less favorable path for stray currents. Such
couplings are often useful for minimizing stray-current damage. They are
less useful if voltages are so large that current is induced to flow around
the insulating joint, causing corrosion as depicted in Fig. 12.3.

REFERENCES

1. D. H. Boteler and W. H. Seager, *Corrosion* **54**(9), 751 (1998).
2. D. Jones, *Corrosion* **34**, 428 (1978).
3. F. Hewes, *Mater. Prot.* **8**(9), 67 (1969).
4. H. Feller and J. Rückert, *Z. Metallk.* **58**, 635 (1967).
5. D. Chin and T. Fu, *Corrosion* **35**, 514 (1979).
6. G. Kish, *Mater. Perf.* **20** (9), 27 (Sept. 1981).

GENERAL REFERENCES

J. H. Fitzgerald III, Stray current analysis, in *Uhlig's Corrosion Handbook*, 2nd edition,
R. W. Revie, editor, Wiley, New York, 2000, pp. 1079–1087.

Michael J. Szeliga, Stray current corrosion, in *Peabody's Control of Pipeline Corrosion*,
2nd edition, R. L. Bianchetti, editor, NACE International, Houston, Texas, 2001,
pp. 211–236.

M. J. Szeliga, editor, *Stray Current Corrosion: The Past, Present and Future of Rail Transit
Systems*, NACE International, Houston, TX, 1994.

PROBLEMS

1. (*a*) A direct current of 10 A enters and leaves a steel water pipe of 2-in. outside
diameter and 0.25-in. wall thickness, containing water with resistivity of
$10^4\,\Omega$-cm. Calculate the current carried by the steel and that carried by the
water. Assume that the resistivity of the pipe equals $10^{-5}\,\Omega$-cm.

(b) Similarly, calculate the current if the pipe conveys seawater of resistivity 20 Ω-cm.

2. (a) A direct current of 10 A passes through external clamps 2 m apart attached to a copper pipe of 5-cm outside diameter containing water of conductivity $10^{-4} \Omega^{-1} cm^{-1}$. If the pipe wall thickness is 0.35 cm, calculate the total weight of the copper in grams per year that corrodes internally at the positive clamp because of this current.

(b) Similarly, calculate the total weight if the pipe contains seawater of conductivity equal to $0.05 \Omega^{-1} cm^{-1}$.

(c) What weight of copper would in theory corrode in one year if an insulating joint were inserted between the clamps?

3. A long pipeline of 8-in. diameter is buried 6 ft underground. The potential difference between two Cu–CuSO$_4$ reference electrodes located on the soil surface over the pipe and a point at right angles 60 ft distant, is 1.25 V. The electrode over the pipe is negative to the other, and soil resistivity measures 3000 Ω-cm.

(a) Does current flow from or to the pipe? How many amperes per linear foot of pipe?

(b) What is the percent error in calculated current if it is found later that the pipe is buried 7 ft instead of 6 ft below the surface?

4. Assuming that a pipe is buried h meters underground in soil with a resistivity of 3000 Ω-cm, plot ΔE versus y for a constant current, j, entering the pipe per linear length of pipe in meters.

5. In measuring resistivity by the four-electrode method, what distance between electrodes must be specified in order to apply the simplified formula:

$$\rho = \frac{1000 \Delta \phi}{I}$$

6. A stray current of 0.7 A leaves a section of buried steel pipe 2 in. in diameter and 2 ft long. What is the initial corrosion rate in ipy caused by this current?

7. (a) If a spherical electrode of radius a is half-buried in earth of resistivity σ, what is the resistance R to earth as a function of distance D from the electrode center?

(b) At what distance is the resistance within 99% of maximum?

Answers to Problems

2. (*a*) 5.0×10^{-5}; (*b*) 0.025; (*c*) 1.04×10^5 g.
3. (*a*) 0.0173 A/ft.
   (*b*) 6.4%.
5. 1.59 m (5.22 ft).
6. 0.33 ipy.
7. (*a*) $R = \dfrac{\sigma}{2\pi}\left(\dfrac{1}{a} - \dfrac{1}{D}\right)$; (*b*) 100*a*.

13

CATHODIC PROTECTION

13.1 INTRODUCTION

Cathodic protection is probably the most important of all approaches to corrosion control. Using an externally applied electric current, corrosion is reduced essentially to zero. A metal surface that is cathodically protected can be maintained in a corrosive environment without deterioration for an indefinite time. There are two types of cathodic protection: impressed current cathodic protection (ICCP) and sacrificial anode cathodic protection (SACP), also known as galvanic cathodic protection.

As discussed in Section 5.11, the mechanism of cathodic protection depends on external current that polarizes the entire surface to the thermodynamic potential of the anode. The surface becomes equipotential (cathode and anode potentials become equal), and corrosion currents no longer flow. Or, looked at another way, at a high enough value of external current density, a net positive current enters the metal at all regions of the metal surface (including anodic areas); hence, there is no tendency for metal ions to enter into solution.

Cathodic protection can be applied in practice to protect metals, such as steel, copper, lead, and brass, against corrosion in all soils and in almost all aqueous media. Pitting corrosion can be prevented in passive metals, such as the stainless

Corrosion and Corrosion Control, by R. Winston Revie and Herbert H. Uhlig
Copyright © 2008 John Wiley & Sons, Inc.

steels or aluminum. Cathodic protection can be used effectively to eliminate stress-corrosion cracking (e.g., of brass, mild steel, stainless steels, magnesium, aluminum), corrosion fatigue of most metals (but not fatigue), intergranular corrosion (e.g., of Duralumin, 18–8 stainless steel), or dezincification of brass. It can be used to avoid S.C.C. of high-strength steels, but *not hydrogen cracking* of such steels. Corrosion above the water line (e.g., of water tanks) is not affected, because the impressed current cannot reach metal areas that are out of contact with the electrolyte. Nor does the protective current extend into electrically screened areas, such as the interior of water condenser tubes (unless the auxiliary anode enters the tubes), even though the water box may be adequately protected.

13.2 BRIEF HISTORY

As a result of laboratory experiments in salt water, Sir Humphry Davy [1] reported in 1824 that copper could be successfully protected against corrosion by coupling it to iron or zinc. He recommended cathodic protection of copper-sheathed ships, employing sacrificial blocks of iron attached to the hull in the ratio of iron to copper surface of about 1:100. In practice, the corrosion rate of copper sheathing was appreciably reduced, as Davy had predicted; but, unfortunately, cathodically protected copper is subject to fouling by marine organisms, contrary to the behavior of unprotected copper, which supplies a sufficient concentration of copper ions to poison fouling organisms (see Section 6.8.1). Because fouling reduced the speed of ships under sail, the British Admiralty decided against the idea. After Davy's death in 1829, his cousin, Edmund Davy (Professor of Chemistry at the Royal Dublin University), successfully protected the iron work of buoys by attaching zinc blocks, and Robert Mallet, in 1840, produced a zinc alloy particularly suited as a sacrificial anode. When wooden hulls were replaced by steel, the fitting of zinc slabs became traditional on all Admiralty vessels.* These slabs provided localized protection, especially against the galvanic effects of the bronze propeller, but the overall cathodic protection of seagoing ships was not explored again until about 1950, this time by the Canadian Navy [2]. By using antifouling paints in combination with anticorrosion paints, it was shown that cathodic protection of ships is feasible and can make possible appreciable savings in maintenance costs. A cathodically protected, hence smooth, hull also reduces fuel costs of ship operation.

Cathodic protection was incidental to the mechanism of protecting steel sheet coated by dipping into molten zinc (galvanizing)(see Section 14.3.3), a method first patented in France in 1836 and in England in 1837 [3]. However, the practice of zinc coating of steel was apparently described in France as early as 1742 [4]. The first application of impressed electric current for protection of underground structures took place in England and in the United States, about

---

*Communication of Dr. W. H. J. Vernon and Mr. L. Kenworthy with H.H.U.

1910–1912 [3]. Since then, the general use of cathodic protection has spread rapidly, and now tens of thousands of miles of buried pipeline and cables are effectively protected against corrosion by this means. In many parts of the world, cathodic protection of high-pressure oil and gas pipelines is a government-mandated regulatory requirement. Cathodic protection is also applied to canal gates, condensers, submarines, water tanks, marine piling, offshore oil-drilling structures, chemical equipment, bridge decks, parking garages, and other reinforced concrete structures (see Section 7.4).

13.3 HOW APPLIED

Cathodic protection requires a source of direct current and an auxiliary electrode (anode) usually of iron or graphite located some distance away from the protected structure. The direct current (dc) source is connected with its positive terminal to the auxiliary electrode and its negative terminal to the structure to be protected; in this way, current flows from the electrode through the electrolyte to the structure. The applied voltage is not critical—it need only be sufficient to supply an adequate current density to all parts of the protected structure. In soils or waters of high resistivity, the applied voltage must be higher than in environments of low resistivity. Or, when the extremities of a long pipeline are to be protected by a single anode, the applied voltage must be raised. A sketch of a cathodically protected buried pipeline is shown in Fig. 13.1.

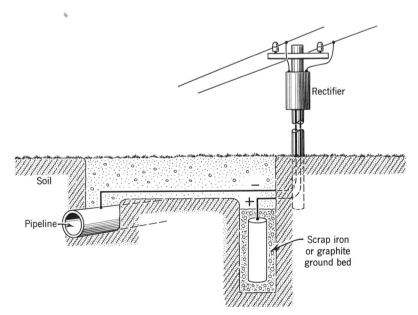

Figure 13.1. Sketch of cathodically protected pipe, auxiliary anode, and rectifier.

The source of current is usually a rectifier supplying low-voltage dc of several amperes. Motor generators have been used, although maintenance is troublesome. Windmill generators are employed in areas where prevailing winds are dependable. Even in periods of calm, some degree of protection of steel persists temporarily because of the inhibiting effect of alkaline electrolysis products at the cathode surface.

13.3.1 Sacrificial Anodes

If the auxiliary anode is composed of a metal more active in the Galvanic Series than the metal to be protected, a galvanic cell is set up with current direction exactly as described in the previous section. The impressed source of current (i.e., the rectifier) can then be omitted, and the electrode is called a *sacrificial anode*, as shown in Fig. 13.2. Sacrificial metals used for cathodic protection consist of magnesium-base and aluminum-base alloys and, to a lesser extent, zinc. Sacrificial anodes serve essentially as sources of portable electrical energy. They are useful particularly when electric power is not readily available, or in situations where it is not convenient or economical to install power lines for the purpose. The open-circuit potential difference of magnesium with respect to steel is about 1 V (ϕ_H for magnesium in seawater = −1.3 V), so that only a limited length of pipeline can be protected by one anode, particularly in high-resistivity soils. This low voltage is sometimes an advantage over higher impressed voltages in that danger of overprotection to some portions of the system is less; and since the total current per anode is limited, the danger of stray-current damage (interference problems) to adjoining metal structures is reduced.

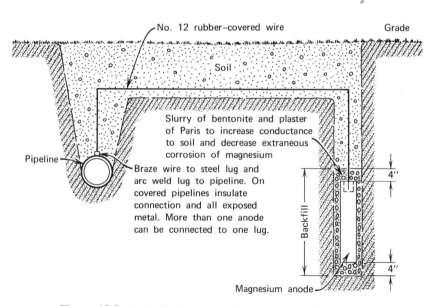

Figure 13.2. Cathodically protected pipe with sacrificial anode.

The potential of zinc is less than that of magnesium (ϕ_H in seawater = -0.8 V); hence, current output per anode is also less. High-purity zinc is usually specified in order to avoid significant anodic polarization with resultant reduction of current output caused by accumulation of adherent insulating zinc reaction products on commercial zinc. This tendency is less pronounced in zinc of high purity.

Aluminum operates theoretically at a voltage between magnesium and zinc. A disadvantage of aluminum as a sacrificial anode is that it tends to become passive in water or in soils with accompanying shift of potential to a value approaching that of steel. A special chemical environment high in chlorides surrounding the electrode can be provided in order to avoid passivity; however, such an environment, called backfill, is only a temporary measure. In seawater, passivity can be avoided by alloying additions, such as tin, indium, antimony, or mercury. For example, alloying aluminum with 0.1% Sn followed by heat treatment at 620 °C for 16 h and water quenching to retain the tin in solid solution very much decreases anodic polarization in chloride solutions [5]. The corrosion potential of the 0.1% Sn alloy in 0.1N NaCl is -1.2 V (S.H.E.) compared to -0.5 V for pure aluminum. Some sacrificial aluminum anodes contain about 0.1% Sn and 5% Zn [6, 7]. Another composition containing 0.6% Zn, 0.04% Hg, and 0.06% Fe, when tested in seawater for 254 days, operated at a current efficiency of 94% (1270 A-h/lb); however, use of mercury has been banned in most locations because of pollution concerns [8]. For cathodic protection of offshore platforms, aluminum anodes, made from aluminum-zinc alloys, are the preferred material [8].

For offshore applications, magnesium anodes have not been popular since the 1980s, because of improvements in aluminum and zinc anodes [8]. Magnesium anodes may be consumed before the structure has reached the end of its lifetime, whereas aluminum anodes are characterized by reliable long-term performance [8]. Magnesium anodes are often alloyed with 6% Al and 3% Zn to reduce pitting-type attack and to increase current efficiency. By using high-purity magnesium containing about 1% Mn, the advantage of a higher potential (with higher current output per anode) is obtained [9]. This alloy operates at a current efficiency in seawater similar to the first-mentioned alloy, but at somewhat lower efficiency in most soils. The observed efficiency of magnesium anodes averages about 500 A-h/lb compared to a theoretical efficiency of 1000 A-h/lb.

A sketch of a magnesium anode rod installed in a steel hot-water tank is shown in Fig. 13.3. Such rods may increase the life of a steel tank by several years, particularly if the rod is renewed as required. The degree of protection is greater in waters of high conductivity, where the currents naturally set up by the magnesium–iron couple reach higher values than in waters of low conductivity (soft waters).

13.4 COMBINED USE WITH COATINGS

The distribution of current in a cathodically protected steel water tank is not ideal—too much current may flow to the sides and not enough to the top and

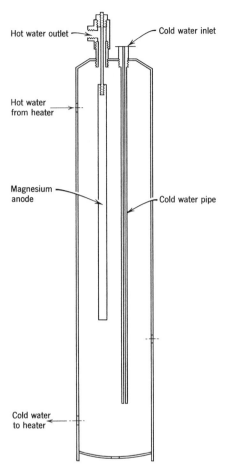

Figure 13.3. Cathodically protected hot-water tank with magnesium anode.

bottom. Better distribution is accomplished by using an insulating coating (e.g., an organic coating at ordinary temperatures or a glass coating at elevated temperatures). The insulating coating need not be pore-free, since the protective current flows preferably to exposed metal areas, wherever located, which are precisely the areas needing protection. Also, since the total required current is less than that for an uncoated tank, the magnesium anode lasts longer.

In hard waters, a partially protective coating may form on steel that consists largely of $CaCO_3$ precipitated by alkalies generated as reaction products at the cathode surface. A similar coating forms gradually on cathodically protected surfaces exposed to seawater (more rapidly at high current densities). Such coatings, if adherent, are also useful in distributing the protective current and in reducing total current requirements.

In the general application of cathodic protection using either impressed current or sacrificial anodes, it is expedient to use an insulating coating, and this combination is the accepted practice today. For example, the distribution of current to a coated pipeline is much improved over that to a bare pipeline; the total current and required number of anodes are less; and the total length of pipeline protected by one anode is much greater. Since the earth, taken as a whole, is a good electrical conductor, and the resistivity of the soil is localized only within the region of the pipeline or the electrodes, one magnesium anode can protect as much as 8 km (5 miles) of a coated pipeline. For a bare pipeline, the corresponding distance might be only 30 m (100 ft). Using an impressed current at higher applied voltages, one anode might protect as much as 80 km (50 miles) of a coated pipeline. The limiting length of pipe protected per anode is imposed not by resistance of the soil, but by the metallic resistance of the pipeline itself.

The potential decay, E_x, along an infinite pipeline measured from the point of attachment to the dc source having potential E_A is expressed as an exponential relation with respect to distance, x, along the pipeline in accord with

$$E_x = E_A \exp\left[-\left(\frac{2\pi r R_L}{kz}\right)^{1/2} x\right] \qquad (13.1)$$

Both E_x and E_A represent differences between polarized potential with current flowing and corrosion potential in absence of current, R_L is the resistance of pipe of radius r per unit length, k is a constant, and z is the resistance of pipe coating per unit area (for derivation, see Appendix, Section 29.4). This equation is derived by assuming that polarization of the cathodically protected surface is a linear function of current density. Note that E_x becomes zero at $x = \infty$.

Considering a finite pipeline for which $a/2$ is half the distance to the next point of bonding, and potential E_x at $a/2 = E_B$,

$$E_x = E_B \cosh\left[\left(\frac{2\pi r R_L}{kz}\right)^{1/2}\left(x - \frac{a}{2}\right)\right] \qquad (13.2)$$

Cathodic protection is optimum within a specific potential range (see Section 13.7), so that the length of pipeline protected by one anode increases as the metallic pipe resistance, R_L, decreases, and coating resistance, z, increases.

13.5 MAGNITUDE OF CURRENT REQUIRED

The current density required for complete protection depends on the metal and on the environment. It can be seen from Fig. 5.15 in Section 5.11 that the applied current density must always exceed the current density equivalent to the measured corrosion rate in the same environment. Hence, the greater the corrosion rate, the higher must be the impressed current density for protection.

When the corrosion rate is cathodically controlled and the corrosion potential approaches the open-circuit anode potential, the required current density is

TABLE 13.1. Orders of Magnitude of Current Density
Required for Cathodic Protection of Steel

Environment	A/m²	A/ft²
Sulfuric acid pickle (hot)	400	35
Soils	0.01–0.5	0.001–0.05
Moving seawater	$\left\{\begin{array}{l}0.03 \text{ (final)} \\ 0.15 \text{ (initial)}\end{array}\right.$	0.003
		0.015
Air-saturated water (hot)	0.15	0.015
Moving fresh water	0.05	0.005

only slightly greater than the equivalent corrosion current. But for mixed control, the required current can be considerably greater than the corrosion current, and it is still greater for corrosion reactions that are anodically controlled.

If the protective current induces precipitation of an inorganic scale on the cathode surface, such as in hard waters or in seawater, the total required current, as described earlier, falls off as the scale builds up. However, at exposed areas of metal, the current density remains the same as before scale formation; only the total current density per apparent unit area is less.

The precise requirements of current density for complete protection can be determined in several ways; the most important is potential measurement of the protected structure (see Section 13.7). In the absence of such measurements, only orders of magnitude can be given. These are listed in Table 13.1 for steel exposed to various environments.

13.6 ANODE MATERIALS AND BACKFILL

Auxiliary anodes for use with impressed current are usually composed of scrap iron or graphite. Scrap iron is consumed at the rate of 15–20 lb/A-year and must be renewed periodically. Graphite anodes are consumed at a lower rate, not exceeding perhaps 2 lb/A-year. Graphite costs more than scrap iron, both initially and in subsequent higher electrical power costs because of the noble potential and accompanying high overpotential of oxygen (or Cl_2) evolution on graphite compared to an active potential and lower overvoltage for the reaction $Fe \rightarrow Fe^{2+} + 2e^-$. Graphite is also fragile compared to scrap iron and must be installed with greater care. The advantages and disadvantages of graphite apply in similar measure to 14% Si–Fe alloy anodes and magnetite anodes, which have also been recommended.

For protection of structures in seawater, platinum-clad copper, 2% Ag–Pb, platinized titanium, or platinized niobium have been recommended as corrosion-resistant anodes using impressed current [10–12]. Whereas sacrificial magnesium anodes require replacement approximately every 2 years, the 2% Ag–Pb anodes are estimated to last more than 10 years, and the 90% Pt–10% Ir anodes still longer [11]. In fresh waters, aluminum anodes are sometimes used for impressed current systems.

Because the effective resistivity of soil surrounding an anode is confined to the immediate region of the electrode, it is common practice to reduce local resistance by backfill. For impressed current systems, this consists of surrounding the anode with a thick bed of coke and adding a mixture of perhaps 3 or 4 parts gypsum ($CaSO_4 \cdot H_2O$) to 1 part NaCl. The coke backfill, being a conductor, carries part of the current, reducing in some measure consumption of the anode itself. Backfill may not be required if the anode is immersed in a river bed, lake, or the ocean.

Whereas auxiliary anodes need not be consumed in order to fulfill their purpose, sacrificial anodes are consumed not less than is required by Faraday's law in order to supply an equivalent electric current. In general, the observed rate of consumption is greater than the theoretical. For zinc the difference is not large, but for magnesium it is appreciable, with the cause being ascribed to local-action currents on the metal surface, to formation of colloidal metal particles [13, 14] or, perhaps more important, to initial formation of univalent magnesium ions [15]. The latter ions are unstable and react in part with water in accord with

$$2Mg^+ + 2H_2O \rightarrow Mg(OH)_2 + Mg^{2+} + H_2$$

Hence, in dilute sodium chloride, about half the magnesium corroding anodically appears as $Mg(OH)_2$ and half as $MgCl_2$, accompanied by hydrogen evolution in about the amount expected according to this reaction [15]. Additional lesser side reactions may also take place at the same time. Accordingly, the observed yield of a magnesium anode is only about one-half the 1000 A-h/lb calculated on the basis of Mg^{2+} formation.

For magnesium anodes, backfill has the advantage of reducing resistance of insulating corrosion-product films, such as $Mg(OH)_2$, as well as increasing conductivity of the immediate environment. A suitable backfill may consist of approximately 20% bentonite (an inorganic colloid used for retention of moisture), 75% gypsum, and 5% Na_2SO_4. Sometimes, the backfill is packaged beforehand in a bag surrounding the anode, so that anode and backfill can be placed simultaneously into position in the soil.

13.6.1 Overprotection

Moderate overprotection of steel structures usually does no harm. The main disadvantages are waste of electric power and increased consumption of auxiliary anodes. In the extreme, additional disadvantages result if so much hydrogen is generated at the protected structure that blistering or disbonding of organic coatings, hydrogen embrittlement of the steel (loss of ductility through absorption of hydrogen), or hydrogen cracking (see Section 8.4) is caused. Damage to steel by hydrogen absorption is particularly apt to occur in environments containing sulfides [16] for reasons discussed in Section 5.5.

In the case of amphoteric metals, such as aluminum, zinc, lead, and tin, excess alkalies generated at the surface of overprotected systems damage the metals by causing increased attack rather than reduction of corrosion. It was shown that cathodic protection of lead continues into the alkaline range of pH, but the critical potential for complete protection (see below) shifts to more active values [17]. Aluminum can be cathodically protected against pitting by coupling it to zinc [18] used as a sacrificial anode; but if coupled to magnesium, overprotection may result with consequent damage to the aluminum.

13.7 CRITERIA OF PROTECTION

The effectiveness of cathodic protection in practice can be established in more than one way, and several criteria have been used in the past to prove whether protection is complete. For example, the observed number of leaks in an old buried pipeline is plotted against time, noting that leaks per year drop to a small number or to zero after cathodic protection is installed. Or the hull of a ship can be inspected at regular intervals for depth of pits.

It is also possible to check effectiveness of protection by short-time tests, including the following measures:

1. *Coupon Tests.* A weighed metal coupon shaped to conform to the outside surface of a buried pipe is attached by a brazed connecting cable, and both the cable and surface between coupon and pipe are overlaid with coal tar. After exposure to the soil for a period of weeks or months, the weight loss, if any, of the cleaned coupon is a measure of whether cathodic protection of the pipeline is complete.

2. *Colorimetric Tests.* A section of buried pipeline is cleaned, exposing bare metal. A piece of absorbent paper soaked in potassium ferricyanide solution is brought into contact, and the soil is returned in place. After a relatively short time, examination of the paper for the blue color of ferrous ferricyanide indicates that cathodic protection is not complete, whereas absence of blue indicates satisfactory protection.

Both the coupon and the colorimetric tests are qualitative and do not provide information about whether just enough or more than enough current is being supplied.

13.7.1 Potential Measurements

A criterion that indicates degree of protection, including overprotection, is obtained through measuring the potential of the protected structure. This measurement is of greatest importance in practice, and it is the criterion generally accepted and used by corrosion engineers. It is based on the fundamental concept that optimum cathodic protection is achieved when the protected structure is polarized to the open-circuit anode potential of local-action cells. This potential for steel, as determined empirically, is equal to $-0.85\,V$ versus the copper-saturated $CuSO_4$ half-cell, or $-0.53\,V$ (S.H.E.).

The theoretical open-circuit anode potential for iron can be calculated assuming that the activity of Fe^{2+} in equilibrium is determined by the solubility of a covering layer of $Fe(OH)_2$, in accord with previous discussions of oxide-film composition on iron exposed to aqueous media (see Section 7.1). Using the Nernst equation, we obtain

$$\phi_H = -0.44 + \frac{0.059}{2}\log(Fe^{2+}), \quad \text{where } (Fe^{2+}) = \frac{\text{solubility product}}{(OH^-)^2}$$

The value of (OH^-) can be estimated assuming that its concentration at equilibrium is twice that of (Fe^{2+}), in accord with $Fe(OH)_2 \rightarrow Fe^{2+} + 2(OH^-)$.* The potential so calculated is $-0.59\,V$ (S.H.E.), equivalent to a potential difference of $-0.91\,V$ versus copper-saturated $CuSO_4$, and is in essential agreement with the empirical value.

The empirical value for lead, known only approximately [19], is about $-0.78\,V$ versus copper-saturated $CuSO_4$ compared to the calculated value for a $Pb(OH)_2$ film on lead equal to $-0.59\,V$. In alkaline media, with formation of plumbites, the calculated value comes closer to the empirical value.† Other calculated values are listed in Table 13.2.

For passive metals, the criterion of protection differs from that just described. Since passive metals corrode uniformly at low rates, but by pitting corrosion at high rates, cathodic protection of metals like aluminum and 18–8 stainless steel depends on polarizing them not to the usual thermodynamic anode potential, but only to a value more active than the critical potential at which pitting initiates (see Section 6.6). The latter potential lies within the passive range and is less noble the higher the Cl^- concentration; in 3% NaCl the value for aluminum is $-0.45\,V$ (S.H.E.). Hence, iron with a corrosion potential in seawater of about $-0.4\,V$ is not suited as a sacrificial anode for cathodically protecting aluminum, unlike zinc, which has a more favorable corrosion potential of about $-0.8\,V$. For 18–8 stainless steel, the critical potential in 3% NaCl is 0.21 V; for nickel it is about 0.23 V. Coupling of the latter metals to a suitable area of either iron or zinc, therefore, can effectively protect them cathodically in seawater against initiation

---

*This relation holds for a saturated solution of $Fe(OH)_2$ in water, the natural pH of which is 9.5. This pH is observed at an iron surface between an external pH of 4–10, as discussed in Section 7.2.3. Values of surface pH for other corroding metals are specific to the metal; but the relation $(2M^{2+}) = (OH^-)$ also applies, provided that the external solution in contact does not alter the natural pH of the surface metal hydroxide.

†A similar situation applies to Cd assumed to be coated with a film of $Cd(OH)_2$ (solubility product $= 2 \times 10^{-14}$). The calculated value of ϕ_H is -0.54, which is noble to iron, contrary to an observed galvanic potential less noble than iron (see Fig. 3.3, Section 3.8). The observed more-active galvanic potential of cadmium is plausibly explained by the known tendency of Cd^{2+} to form complex ions that lower Cd^{2+} activity below the value corresponding to saturated $Cd(OH)_2$.

TABLE 13.2. Calculated Minimum Potential ϕ for Cathodic Protection

Metal	$\phi°$ (volts)	Solubility Product, $M(OH)_2$	ϕ_H (volts)	ϕ_H versus $Cu-CuSO_4$ Reference Electrode (volts)
Iron	−0.440	1.8×10^{-15}	−0.59	−0.91
Copper	0.337	1.6×10^{-19}	0.16	−0.16
Zinc	−0.763	4.5×10^{-17}	−0.93	−1.25
Lead	−0.126	4.2×10^{-15}	−0.27	−0.59

of pitting corrosion. Components of practical structures, such as ships and off-shore oil-drilling platforms, are sometimes designed to take advantage of galvanic couples of this kind.

13.7.2 Doubtful Criteria

Criteria have sometimes been suggested based on empirical rules—for example, polarizing a steel structure 0.3 V more active than the corrosion potential. This criterion is not exact and in many situations leads to under- or overprotection. It has also been suggested that polarization of the structure should proceed to breaks in slope of the voltage versus current plot. Such breaks may, in principle, occur in some environments when the applied current is just equal to or slightly greater than the corrosion current (e.g., oxygen depolarization control); but in other environments, breaks may occur when concentration polarization or *IR* effects through partially protective surface films become appreciable. As Stern and Geary showed [20], breaks of this kind in polarization measurements have varied causes and are of doubtful value as criteria of cathodic protection. Contrary to what is sometimes erroneously supposed, discontinuities in slopes of polarization curves have no general relation to the anode or cathode open-circuit potentials of the corroding system.

13.7.3 Position of Reference Electrode

The potential of a cathodically protected structure is determined ideally by placing the reference electrode as close as possible to the structure to avoid an error caused by *IR* drop through the electrolyte. Any *IR* drop through corrosion-product films or insulating coatings will persist, of course, even where adequate precautions are exercised otherwise, tending to make the measured potential more active than the actual potential at the metal surface. In practice, for buried pipelines, a compromise position is chosen at the soil surface located directly over the buried pipe. This position is chosen because cathodic protection currents flow mostly to the lower surface and are minimal to the upper surface of a pipe buried a few feet below the soil surface.

The reference electrode is sometimes located at a position remote from the pipeline, recommended because currents do not penetrate remote areas, and hence, *IR* drop effects are avoided. Actually, the potential measured at a remote position is a compromise potential at some value between that of the polarized structure and the polarized auxiliary or sacrificial anode. These potentials differ by the *IR* drop through the soil and through coatings. The potential measured at a remote location, therefore, tends to be more active than the true potential of the structure, resulting in a structure that may be underprotected.

13.8 ECONOMICS OF CATHODIC PROTECTION

For buried pipelines, the cost of cathodic protection is far less than for any other means offering equal assurance of protection. The guarantee that no leaks will develop on the soil side of a cathodically protected buried pipeline makes it economically feasible, for example, to transport oil and high-pressure natural gas across entire continents.

In addition, lack of corrosion on the soil side makes it possible to specify thinner wall pipe adequate to withstand internal pressures and to avoid any extra thickness as a safety factor against corrosion. This saving alone is substantial. The Panama Canal gates, for example, are protected by using impressed current, with the initial costs of installation being less than 0.5% the cost of replacing the gates. One important advantage is that the gates can continue to operate without the necessity of periodic long shutdowns for repairs caused by corrosion. Similarly, a ship that is cathodically protected can operate, in principle, for longer periods between dry-docking, thereby saving thousands of dollars annually. The further economic advantages, in other instances, through avoiding stress-corrosion cracking, corrosion fatigue, and pitting of various structural metals are substantial in terms of extended lifetime, enhanced reliability, and assured public safety.

13.9 ANODIC PROTECTION

As mentioned in Section 6.4, some metals, such as iron and stainless steels, can also be protected by making them anodic and shifting their potential into the passive region of the anodic polarization curve (see Fig. 6.1, Section 6.2). The passive potential is automatically maintained, usually electronically, by an instrument called the *potentiostat*. Practical application of anodic protection and use of the potentiostat for this purpose were first suggested by Edeleanu [21].

Anodic protection has found application in handling sulfuric acid [22], but the method is also applicable to other acids (e.g., phosphoric acid) and to alkalies and some salt solutions. It has been shown to be effective for increasing the resistance to corrosion fatigue of various stainless steels in $0.5M$ Na_2SO_4 [23], in 10% H_2SO_4 or 10% NH_4NO_3, and of 0.19% C steel in 10% NH_4NO_3 [24]. Engineering installations have been described for anodically protecting mild steel against uniform corrosion in NH_4NO_3 fertilizer mixtures [25], carbon steel in 86% spent sulfuric acid at temperatures up to 60°C (140°F) [26], and carbon steel in 0.1–0.7M oxalic acid at temperatures up to 50°C (120°F) [27].

Since passivity of iron and the stainless steels is destroyed by halide ions, anodic protection of these metals is not possible in hydrochloric acid or in acid chloride solutions for which the current density in the otherwise passive region is very high. Also, if Cl^- should contaminate the electrolyte, the possible danger of pitting becomes a consideration even if the passive current density remains acceptably low. In the latter case, however, it is only necessary to operate in the potential range below the critical pitting potential for the mixed electrolyte.* Titanium, which has a very noble critical pitting potential over a wide range of Cl^- concentration and temperature, is passive in the presence of Cl^- (low $i_{passive}$) and can be anodically protected without danger of pitting even in solutions of hydrochloric acid.

Anodic protection is applicable only to metals and alloys (mostly transition metals) which are readily passivated when anodically polarized and for which $i_{passive}$ is very low. It is not applicable, for example, to zinc, magnesium, cadmium, silver, copper, or copper-base alloys. Anodic protection of aluminum exposed to high-temperature water has been shown to be feasible (see Section 21.1.2).

Current densities to initiate passivity, $i_{critical}$, are relatively high, with $6\,A/m^2$ being required for Type 316 stainless steel in 67% H_2SO_4 at 24°C (75°F). But currents for maintaining passivity are usually low, with the orders of magnitude being $10^{-3}\,A/m^2$ ($0.1\,\mu A/cm^2$) for Type 316 stainless steel to $0.15\,A/m^2$ ($15\,\mu A/cm^2$) for mild steel [22], both in 67% H_2SO_4. Corrosion rates corresponding to these current densities are 0.02–2.5 gmd.

It is typical of anodic protection that corrosion rates, although small, are not reduced to zero, contrary to the situation for cathodic protection of steel. On the other hand, the required current densities in corrosive acids are usually much lower than those for cathodic protection, since for cathodic protection the current cannot be less than the normal equivalent corrosion current in the same environment. For stainless steels, this value of current density corresponds to the rather high corrosion rate for the active state of the alloys.

For anodic protection, it has been reported [22] that unusual throwing power (protection at distances remote from the cathode or in electrically screened areas) is obtained, far exceeding similar throwing power obtained in cathodic protection. The cause has been ascribed to high electrical resistance of the passive film, but this is probably not correct, because measurements have shown that such resistances are typically low. The cause instead may be related to the corrosion-inhibiting properties of anodic corrosion products released by stainless steels in small amounts (e.g., $S_2O_8^{2-}$, $Cr_2O_7^{2-}$, Fe^{3+}), which shift the corrosion potential into the passive region in the absence of an applied current.

---

*Stress-corrosion cracking of Type 304 stainless steel, which is reported to occur at room temperature in 10N H_2SO_4 + 0.5N NaCl, is prevented by anodically polarizing the alloy to 0.7 V (S.H.E.). See S. Acello and N. Greene, *Corrosion* **18**, 286t (1962); J. Harston and J. Scully, *Corrosion* **25**, 493 (1969).

REFERENCES

1. H. Davy, *Philos. Trans. Roy. Soc.* **114**, 151–158, 242–246, 328–346 (1824–1825).
2. K. Barnard (with G. Christie), *Corrosion* **6**, 232 (1950); ibid. **7**, 114 (1951); (with G. Christie and J. Greenblatt), ibid. **9**, 246 (1953).
3. W. Lynes, *J. Electrochem. Soc.* **98**, 3c (1951).
4. R. Burns and W. Bradley, *Protective Coatings for Metals*, 3rd edition, Reinhold, New York, 1967, pp. 104–106.
5. D. Keir, M. Pryor, and P. Sperry, *J. Electrochem. Soc.* **114**, 777 (1967); ibid. **116**, 319 (1969).
6. J. Burgbacher, *Mater. Prot.* **7**, 26 (1968).
7. T. Lennox, M. Peterson, and R. Groover, *Mater. Prot.* **7**, 33 (1968).
8. R. H. Heidersbach, J. Brandt, D. Johnson, and J. S. Smart III, Marine corrosion protection, in *ASM Handbook*, Vol. 13C, *Corrosion: Environments and Industries*, ASM International, Materials Park, OH, 2006, p. 73.
9. P. George, J. Newport, and J. Nichols, *Corrosion* **12**, 627t (1956).
10. H. Preiser and B. Tytell, *Corrosion* **15**, 596t (1959).
11. K. Barnard, G. Christie, and D. Gage, *Corrosion* **15**, 581t (1959).
12. R. Benedict, *Mater. Prot.* **4**, 36 (Dec. 1965).
13. G. Marsh and E. Schashl, *J. Electrochem. Soc.* **107**, 960 (1960).
14. G. Hoey and M. Cohen, *J. Electrochem. Soc.* **105**, 245 (1958).
15. J. Greenblatt, *J. Electrochem. Soc.* **103**, 539 (1956); (with E. Zinck) *Corrosion* **15**, 76t (1959); ibid. **18**, 125t (1962).
16. W. Bruckner and K. Myles, *Corrosion* **15**, 591t (1959).
17. W. Bruckner and O. Jansson, *Corrosion* **15**, 389t (1959).
18. R. Mears and H. Fahrney, *Trans. Am Inst. Chem. Eng.* **37**, 911 (1941).
19. D. Werner, *Corrosion* **13**, 68 (1957).
20. M. Stern and A. Geary, *J. Electrochem. Soc.* **104**, 56 (1957).
21. C. Edeleanu, *Nature* **173**, 739 (1954); *Metallurgia* **50**, 113 (1954).
22. O. Riggs, M. Hutchison, and N. Conger, *Corrosion* **16**, 58t (1960).
23. H. Spähn, *Z. Phys. Chem. (Leipzig)* **234**, 1 (1967).
24. W. Cowley, F. Robinson, and J. Kerrich, *Br. Corros. J.* **3**, 223 (1968).
25. W. Banks and M. Hutchison, *Mater. Prot.* **8**, 31 (February 1969).
26. L. Hays, *Mater. Prot.* **5**, 46 (1966).
27. L. Perrigo, *Mater. Prot.* **5**, 73 (1966).

GENERAL REFERENCES

C. A. Brebbia, V. G. DeGiorgi, and R. A. Adey, editors, Cathodic protection systems, Section 1 in *Simulation of Electrochemical Processes*, WIT Press, Southampton, U.K., 2005, pp. 1–56.

J. H. Fitzgerald III, Engineering of cathodic protection systems, in *Uhlig's Corrosion Handbook*, 2nd edition, Wiley, New York, 2000, pp. 1061–1078.

R. H. Heidersbach, Cathodic protection, in *ASM Handbook*, Vol. 13A, *Corrosion: Fundamentals, Testing, and Protection*, ASM International, Materials Park, OH, 2003, pp. 855–870.

R. H. Heidersbach, J. Brandt, D. Johnson, and J. S. Smart III, Marine corrosion protection, in *ASM Handbook*, Vol. 13C, *Corrosion: Environments and Industries*, ASM International, Materials Park, OH, 2006, pp. 73–78.

C. E. Locke, Anodic protection, in *ASM Handbook*, Vol. 13A, *Corrosion: Fundamentals, Testing, and Protection*, ASM International, Materials Park, OH, 2003, pp. 851–854.

Peabody's Control of Pipeline Corrosion, 2nd edition, R. L. Bianchetti, editor, NACE International, Houston, TX, 2001.

O. Riggs, Jr. and C. Locke, *Anodic Protection, Theory and Practice in the Prevention of Corrosion*, Plenum Press, New York, 1981.

PROBLEMS

1. Sketch the polarization diagram of a buried steel pipe connected to a sacrificial magnesium anode. Indicate the following on the diagram:

   (*a*) The potential near the pipe surface with respect to the open-circuit potentials of anodic and cathodic areas of the pipe.

   (*b*) The potential of the pipe with the reference electrode located far from and at right angles to the pipe.

   (*c*) The change in potential of the pipe proceeding toward the vicinity of the magnesium anode.

2. Calculate the minimum potential versus $Cu–CuSO_4$ reference electrode to which zinc must be polarized for complete cathodic protection, assuming $Zn(OH)_2$ as corrosion product on the surface. (Solubility product of $Zn(OH)_2 = 4.5 \times 10^{-17}$.)

3. From sketches of polarization diagrams of a corroding metal, compare applied current, as required for complete cathodic protection, to the normal corrosion current when:

   (*a*) The corrosion rate is anodically controlled.

   (*b*) The corrosion rate is cathodically controlled.

4. Current between a magnesium anode and a 50-gal steel tank filled with air-saturated hot water is 100 mA. Disregarding local-action currents, what interval of time is required between filling and emptying the tank to ensure minimum corrosion of outlet steel water piping? (Solubility of oxygen in inlet water, 25 °C, = 6 mL/liter.)

5. Iron corrodes in seawater at a rate of 2.5 gmd. Assuming that all corrosion is by oxygen depolarization, calculate the minimum initial current density (A/m^2) necessary for complete cathodic protection.

6. Calculate the minimum potential versus saturated calomel electrode to which copper must be polarized in $0.1M$ $CuSO_4$ for complete cathodic protection.

7. A copper bar of 300-cm² total exposed area coupled to an iron bar 50-cm² area is immersed in seawater. What minimum current must be applied to the couple in order to avoid corrosion of both iron and copper? (Corrosion rate of uncoupled iron in seawater is 0.13 mm/y.)

8. To what minimum potential versus saturated calomel electrode must indium be polarized in $0.01M$ $In_2(SO_4)_3$ solution for complete cathodic protection? [*Data:* $In^{3+} + 3e^- \rightarrow In$, $\phi° = -0.342\,V$; $\gamma 0.01M$ $In_2(SO_4)_3 = 0.142$.]

9. (*a*) How many milliliters H_2 (S.T.P.) are evolved per square centimeter per day from a steel surface maintained at the critical potential for cathodic protection, $\phi_H = -0.59\,V$?
   (b) Similarly, how many milliliters H_2 are evolved if cathodic protection is increased to a potential 0.1 V more active than the critical value? [*Data:* Assume hydrogen overpotential (volts) on steel $= 0.105 \log i/(1 \times 10^{-7})$, where $i = A/cm^2$, pH of steel surface $= 9.5$.]

10. You have been asked to review a proposal to cathodically protect a structure in an aqueous environment with a pH of approximately 7. The structure will be constructed of a high-strength steel that is very susceptible to hydrogen cracking. Discuss the factors to be considered in deciding on how to protect the structure from corrosion without causing failure by hydrogen cracking.

11. Under what circumstances can cathodic protection (either sacrificial protection or impressed current protection) be used to protect an automobile from corrosion?

Answers to Problems

2. $-1.25\,V$
4. 49.9 h.
5. $0.10\,A/m^2$.
6. 0.042 V.
7. $3.8 \times 10^{-3}\,A$.
8. $-0.634\,V$.
9. (*a*) $0.0019\,mL/cm^2$-day; (*b*) $0.018\,mL/cm^2$-day.

14

METALLIC COATINGS

14.1 METHODS OF APPLICATION

Metal coatings are applied by dipping, electroplating, spraying, cementation, and diffusion. The selection of a coating process for a specific application depends on several factors, including the corrosion resistance that is required, the anticipated lifetime of the coated material, the number of parts being produced, the production rate that is required, and environmental considerations.

Hot dipping is carried out by immersing the metal on which the coating is to be applied, usually steel, in a bath of the molten metal that is to constitute the coating, most commonly zinc, but also aluminum and aluminum–zinc alloys. Hot dipping can be either a continuous process, as in galvanizing steel sheet, or a batch process—for example, galvanizing fabricated parts, nuts, bolts, and fasteners [1].

In *electroplating*, the substrate, or base, metal is made the cathode in an aqueous electrolyte from which the coating is deposited. Although the primary purpose of electroplating coatings is to achieve corrosion resistance, these coatings can also be decorative, with a metallic luster after polishing [2]. A wide of variety of coatings can be applied by electroplating—for example, zinc, cadmium, chromium, copper, gold, nickel, tin, and silver, as well as alloys, such as tin–zinc,

Corrosion and Corrosion Control, by R. Winston Revie and Herbert H. Uhlig
Copyright © 2008 John Wiley & Sons, Inc.

zinc–nickel, brass, bronze, gold alloys, and nickel alloys [2]. *Electrogalvanizing* is the electroplating of zinc on either iron or steel. Electroplated zinc-coated sheet is widely used for exposed automobile body panels because of its uniform coating thickness and surface characteristics compared to hot-dip zinc-coated. Coatings range in thickness from 4 to 14 μm [3].

Coatings are also produced by *electroless plating*—that is, by chemical reduction of metal–salt solutions, with the precipitated metal forming an adherent overlay on the base metal. Nickel coatings of this kind are called *electroless* nickel plate.

In *thermal spraying* of metal coatings, a gun is used that simultaneously melts and propels small droplets of metal onto the surface to be coated. There are several types of thermal spraying, with the three main variables in each type being the temperature of the flame, the velocity of the particles that are sprayed onto the substrate to form the coating, and the nature of the material that is to form the coating (i.e., powder, rod, wire, or liquid) [4, 5]. The material that is to form the coating is called the "feedstock." In all cases, the feedstock is rapidly heated and propelled toward the substrate where, on impact, it consolidates forming an adherent coating. In flame-powder spraying, powder feedstock is melted and carried by the flame onto the workpiece. In flame-wire spraying, the flame melts the wire, and a stream of air propels the molten material onto the workpiece. In plasma spraying, a plasma at a temperature of about 12,000 °C is formed, and the plasma stream carries the powder feedstock to the workpiece [4].

Thermally sprayed coatings tend to be porous, although porosity can be controlled by optimizing the process variables [4]. These coatings can be made adherent and of almost any desired thickness, and they can be applied on already fabricated structures. Sometimes, pores are filled with a thermoplastic resin in order to increase corrosion protection.

Cementation consists of tumbling the work in a mixture of metal powder and a flux at elevated temperatures, allowing the metal to diffuse into the base metal. Aluminum and zinc coatings can be prepared in this way. Diffusion coatings of chromium, nickel, titanium, aluminum, and so on, can also be prepared by immersing metal parts, under an inert atmosphere, in a bath of molten calcium containing some of the coating metal in solution [6].

Coatings are also sometimes produced by *gas-phase reaction*. For example, $CrCl_2$, when volatilized and passed over steel at about 1000 °C (1800 °F), results in formation of a chromium–iron alloy surface containing up to 30% Cr in accord with the reaction

$$\frac{3}{2}CrCl_2 + Fe \rightarrow FeCl_3 + \frac{3}{2}Cr \text{ (alloyed with Fe)}$$

Similar surface alloys of silicon–iron containing up to 19% Si can be prepared by reaction of iron with $SiCl_4$ at 800–900 °C (1475–1650 °F).

Ion implantation is a process of producing thin surface alloy coatings by bombarding the metal with ions in vacuum. Such coatings of, for example, Ti, B,

Cr, or Y have specialized applications for wear and high-temperature oxidation resistance [7, 8].

14.2 CLASSIFICATION OF COATINGS

All coatings provide barrier protection; that is, they provide a barrier between the corrosive environment and the metal substrate; however, all commercially prepared metal coatings are porous to some degree. Furthermore, coatings tend to become damaged during shipment or in use. Therefore, galvanic action at the base of a pore or scratch becomes an important factor in determining coating performance. From the corrosion standpoint, metal coatings can be divided into two classes, namely, *noble coatings*, which provide only barrier protection, and *sacrificial coatings*, which, in addition to barrier protection, also provide cathodic protection.

As the names imply, noble coatings (e.g., nickel, silver, copper, lead, or chromium) on steel are noble in the Galvanic Series with respect to the base metal. At exposed pores, the direction of galvanic current accelerates attack of the base metal and eventually undermines the coating (Figs. 14.1 and 14.2); consequently, it is important that noble coatings always be prepared with a minimum number of pores and that any existing pores be as small as possible, to delay access of water to the underlying metal. This usually means increasing the thickness of coating. Sometimes, the pores are filled with an organic lacquer, or a second metal, of lower melting point, is diffused into the coating at elevated temperatures (e.g., zinc or tin into nickel).

For sacrificial coatings (e.g., zinc, cadmium) and, in certain environments, also aluminum and tin on steel, the direction of galvanic current through the electrolyte is from coating to base metal; as a result, the base metal is cathodically protected (Fig. 14.1). As long as adequate current flows and the coating remains in electrical contact, corrosion of the base metal does not occur. The degree of

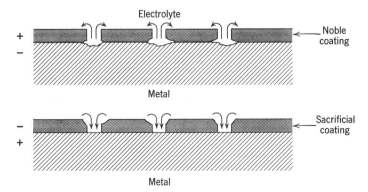

Figure 14.1. Sketch of current flow at defects in noble and sacrificial coatings.

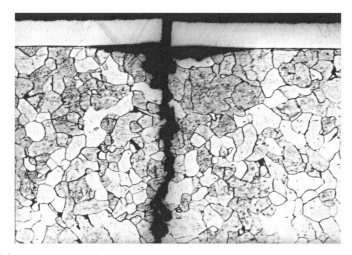

Figure 14.2. Undermining of nickel electrodeposit on steel by galvanic corrosion in 3% NaCl solution (100×). Crack resulted from cyclic stressing in a corrosion-fatigue test [H. Spähn and K. Fässler, *Werkst Korros.* **17**, 321 (1966)].

porosity of sacrificial coatings, therefore, is not of great importance, contrary to the situation for noble coatings. In general, of course, the thicker the coating, the longer cathodic protection continues.

The area of base metal over which cathodic protection extends depends on the conductivity of the environment. For zinc coatings on steel in waters of low conductivity, such as distilled or soft waters, a coating defect about 3 mm (1/8 in.) wide may begin to show rust at the center. However, in seawater, which is a good conductor, zinc protects steel several decimeters or feet removed from the zinc. This difference in behavior results from adequate current densities for cathodic protection extending over a considerable distance in waters of high conductivity, whereas cathodic current densities in waters of low conductivity fall off rapidly with distance from the anode.

14.3 SPECIFIC METAL COATINGS

14.3.1 Nickel Coatings

Nickel coatings are usually prepared by electroplating. The metal is plated either directly on steel or sometimes over an intermediate coating of copper. The copper underlayer is used to facilitate buffing of the surface on which nickel is plated, because copper is softer than steel, and also to reduce the required thickness of nickel (which costs more than copper) for obtaining a coating of minimum porosity. The automotive industry uses nickel as an underlayer for microcracked chromium to protect steel [9].

Because porosity is an important factor in the life and appearance of nickel coatings, a certain minimum thickness is recommended, depending on exposure conditions. For indoor exposures, a coating 0.008–0.013 mm (0.3–0.5 mil) is considered adequate for many applications. For outdoor exposures, a coating 0.02–0.04 mm (0.8–1.5 mil) may be specified. The thicker coatings in this range (or still thicker) are desirable (e.g., for auto bumpers) for use near the sea coast or in industrial environments, but thinner coatings are usually adequate for dry or unpolluted atmospheres. For use in the chemical industry, nickel coatings commonly range from 0.025–0.25 mm (1–10 mils) thick.

Nickel is sensitive to attack, particularly by industrial atmospheres. In the phenomenon called *fogging*, coatings tend to lose their specular reflectivity because a film of basic nickel sulfate forms that decreases surface brightness [10]. To minimize fogging, a very thin (0.0003–0.0008 mm; 0.01–0.03 mil) coating of chromium is electrodeposited over the nickel. This thin chromium overlayer has led to the term "chrome plate," although, in reality, such coatings are composed mostly of nickel.

Optimum protection is obtained with "microcracked" chromium, electrodeposited with many inherent microscopic cracks. The many cracks in chromium favor initiation of corrosion at numerous sites, thereby decreasing depth of penetration that would otherwise result from attack at fewer sites. The density of cracks in chromium deposits varies from 0 to more than 120 cracks/mm, depending on bath chemistry, current density, and temperature [11]. The chromium deposit usually overlies a thin nickel coating formed from a plating bath containing additives (usually sulfur compounds) that improve surface brightness (bright nickel). This thin coating, in turn, overlies a matte coating two to three times thicker, electrodeposited from a conventional nickel plating bath. The bright nickel, which includes a small amount of sulfur, is anodic to the underlying nickel coating of lower sulfur content, so that the bright nickel acts as a sacrificial coating. Any pit beginning under the chromium coating tends to grow sidewise rather than through the nickel layers. In this way, appearance of rust is delayed and the multiple coating system is more protective for a given thickness of chromium and nickel [12].

Electroless nickel plate is produced primarily for the chemical industry. Nickel salts are reduced to the metal by sodium hypophosphite solutions at or near the boiling point. A typical solution is the following:

$NiCl_2 \cdot 6H_2O$	30 g/liter
Sodium hypophosphite	10 g/liter
Sodium hydroxyacetate	50 g/liter
pH	4–6

This particular formulation deposits nickel at the rate of about 0.015 mm (0.6 mil)/h in the form of a nickel–phosphorus alloy [13]. The usual range of phosphorus content in coatings of this kind is 7–9%. Various metal surfaces, including nickel, act as catalysts for the reaction so that deposits can be built up to relatively heavy thicknesses. Specific additions are made to the commercial solution in order to increase the rate of nickel deposition and also for coating glass and plastics, should this be desirable. Metals on which the coating does not deposit include lead, tin solder, cadmium, bismuth, and antimony. The phosphorus content makes it possible to harden the coating appreciably by low-temperature heat treatment—for example, 400 °C (750 °F). Corrosion resistance of the nickel–phosphorus alloy is stated to be comparable to electrolytic nickel in many environments. Electroless nickel plating is useful for uniformly coating, with controlled thickness, the inside surfaces of tubes and other recessed areas [14].

14.3.2 Lead Coatings

Lead coatings on steel are usually formed either by hot dipping or by electrodeposition. In the hot-dipping operation, a few percent tin are usually incorporated to improve bonding with the underlying steel. When 3–15% tin is incorporated, the resulting lead–tin alloy is called *terne*. "Terne," which means "dull," was originally used to differentiate such coatings from bright tin plate. Coatings of lead or lead–tin alloy are resistant to atmospheric attack, the pores tending to fill with corrosion product, most likely lead sulfate ($PbSO_4$) which stifles further reaction [15]. Lead coatings are not very protective in the soil. Lead coatings must not be used in contact with drinking water or food products, because of the poisonous nature of small quantities of lead salts (see Section 1.2). Lead and lead–tin coatings are used to enhance the corrosion resistance of iron castings to H_2SO_3 and H_2SO_4 [16].

14.3.3 Zinc Coatings

Coatings of zinc, whether hot-dipped or electroplated, are called *galvanized coatings*, and the term *electrogalvanized* is used to differentiate electroplated from hot-dipped coatings. The electroplating process has been used to explore new zinc alloy coatings, such as Zn–Fe, Zn–Co, and Zn–Ni [17].

The coating produced by hot dipping is bonded to the underlying steel by a series of Zn–Fe alloys, with a layer of almost pure zinc on the outside surface. The coating produced in the batch process is thicker than that produced in the continuous process, and it has clearly distinguishable alloy layers. The continuous process results in a thinner coating with a very thin alloy layer at the coating–steel interface [17].

Zinc coatings are relatively resistant to rural atmospheres and also to marine atmospheres, except when seawater spray comes into direct contact with the surface. Table 14.1 lists the ranges of typical atmospheric corrosion rates in each of the three types of atmospheres, rural, marine, and urban/industrial [18].

In aqueous environments at room temperature, the overall corrosion rate in short time tests is lowest within the pH range 7–12 (Fig. 14.3). In acid or very alkaline environments, the major form of attack results in hydrogen evolution.

TABLE 14.1. Atmospheric Corrosion Rates of Zinc [18]

Type of Atmosphere	Range of Corrosion Rate (μm/year)
Rural	0.2–3
Marine (outside the splash zone)	0.5–8
Urban and industrial	2–16

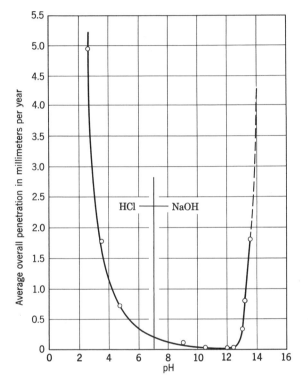

Figure 14.3. Effect of pH on corrosion of zinc, aerated solutions, 30 °C [B. Roetheli, G. Cox, and W. Littreal, *Metals and Alloys* **3**, 73 (1932)].

Above about pH 12.5, zinc reacts rapidly to form soluble zincates in accord with

$$Zn + OH^- + H_2O \rightarrow HZnO_2^- + H_2$$

In seawater, zinc coatings are effectively resistant for protecting steel against rusting, with each mil (0.03 mm) of zinc equaling about one year of life; thus, a coating 5 mils (0.13 mm) thick will protect steel against the appearance of rust for about five years.

In many aerated hot waters, *reversal of polarity* between zinc and iron occurs at temperatures of about 60 °C (140 °F) or above [19–21]. This reversal of polarity leads to zinc having the characteristics of a noble coating instead of a sacrificial coating; hence, a galvanized coating under these circumstances can induce pitting of the base steel.

A 15-year service test on piping carrying Baltimore City water at a mean temperature of 46 °C (115 °F) and maximum of 80 °C (176 °F) confirmed that pitting of galvanized pipe was 1.2–2 times deeper than that in black iron pipe (not galvanized) of the same type, corresponding to shorter life of the galvanized pipe. In cold water, however, pits in galvanized pipe were only 0.4–0.7 as deep as those in black iron pipe, indicating, in this case, a beneficial effect of galvanizing [22]. It was found that waters high in carbonates and nitrates favor the reversal in polarity, whereas those high in chlorides and sulfates decrease the reversal tendency [23].

The potential of zinc relative to steel is apparently related to the formation of porous $Zn(OH)_2$ or basic zinc salts, which are insulators, under those conditions for which zinc is anodic to steel, but is related to formation of ZnO instead under conditions where the reverse polarity occurs [24]. The latter compound conducts electronically, being a semiconductor. It can, therefore, perform in aerated waters as an oxygen electrode, the potential of which, like mill scale on steel, is noble to both zinc and iron. Accordingly, in deaerated hot or cold waters in which an oxygen electrode does not function, because oxygen is absent, zinc is always anodic to iron, but this is not necessarily true in aerated waters. Apparently, the presence of HCO_3^- and NO_3^-, aided by elevated temperatures, stimulates formation of ZnO, whereas Cl^- and SO_4^{2-} favor formation of hydrated reaction products instead.

At room temperature, in water or dilute sodium chloride, the current output of zinc as anode decreases gradually because of insulating corrosion products that form on its surface. In one series of tests, the current in a zinc–iron couple decreased to zero after 60–80 days, and a slight reversal of polarity was reported [25]. This trend is less pronounced with high-purity zinc, on which insulating coatings have less tendency to form.

14.3.4 Cadmium Coatings

Cadmium coatings are produced almost exclusively by electrodeposition. The difference in potential between cadmium and iron is not as large as that between zinc and iron; hence, cathodic protection of steel by an overlayer of cadmium falls off more rapidly with size of coating defects. The lower potential difference provides an important advantage to cadmium coatings applied to high-strength steels (Rockwell C > 40, see Section 8.4.1). By maintaining a potential below the critical potential for S.C.C., but not overlapping the still more active hydrogen cracking potential range, cadmium coatings, compared to zinc coatings, more reliably protect the steel against failure by cracking in moist atmospheres. Cadmium is more expensive than is zinc, but cadmium retains a bright metallic appearance longer, ensures better electrical contact and solderability, and, consequently, finds use in electronic equipment. Furthermore, it is more resistant to attack by aqueous condensate and by salt spray. Otherwise, however, cadmium coatings exposed to the atmosphere are not quite as resistant as zinc coatings of equal thickness. Because the coefficient of friction of cadmium is less than that of zinc, cadmium is preferred for fastening hardware and connectors that are repeatedly taken on and off.

In aqueous media, cadmium, unlike zinc, resists attack by strong alkalies. Like zinc, it is corroded by dilute acids and by aqueous ammonia. Cadmium salts are toxic, and, for this reason, cadmium coatings must not come into contact with food products. Zinc salts are less toxic in this respect, and galvanized coatings are tolerated for drinking water, but they also are not recommended for contact with foods. Applications in which corrosion products of cadmium can enter the environment must be avoided. Because solutions used for plating cadmium are toxic, disposal is difficult. Cadmium should be specified only after thorough investigation of the implications and alternatives [26]. Because of toxicity considerations and regulations concerning use of cadmium and disposal of wastes containing cadmium, effective cadmium replacements are being identified wherever possible [27].

14.3.5 Tin Coatings

Tinplate is a low-carbon steel strip product coated on both sides with a thin layer of tin. Several million tons of tinplate are produced each year, and most of it is used to manufacture many billions of food containers. The nontoxic nature of tin salts makes tinplate ideal for use with beverages and foods.

Because electrodeposited tin is more uniform than is hot-dipped tin and can be produced as thinner coatings, tinplate is now produced universally by electrodeposition. After electroplating the tin on the steel substrate, the tinplate is given a heating cycle that melts the tin coating, creating a layer of tin–iron intermetallic compound at the tin–steel interface. The tinplate is then passivated, either in chromic acid (H_2CrO_4) or in sodium dichromate ($Na_2Cr_2O_7$), and, finally, a thin layer of lubricant is applied. As supplied to the can maker, the typical tinplate product consists of five layers [28]:

1. Steel sheet approximately 200–300 μm (7.8–11.7 mils) thick
2. On each side, a layer of tin–iron intermetallic compound about 0.08 μm (0.004 mil) thick
3. Tin layer, 0.3 μm (0.012 mil) thick
4. Passivation film about 0.002 μm (0.00008 mil) thick
5. Oil film about 0.002 μm (0.00008 mil) thick

Tin coatings so thin are, naturally, very porous; hence, it is essential that the tin act as a sacrificial coating in order to avoid perforation by pitting of the thin-gage steel on which the tin is applied. This condition usually applies.

On the outside of a tinned container, tin is cathodic to iron, in accord with the standard potential of tin equal to $-0.136\,V$ compared to $-0.440\,V$ for iron. On the inside, however, tin is almost always anodic to iron, and hence, tin cathodically protects the base steel. This fortunate reversal of potential occurs because stannous ions, Sn^{2+}, are complexed by many food products that greatly reduce Sn^{2+} activity, resulting in a change in corrosion potential of tin in the active direction (see Section 3.8). On the inside of the can, the corrosion reaction for tin is

$$Sn \rightarrow Sn^{2+} + 2e^{-} \tag{14.1}$$

Two possible reduction reactions are

$$O_2 + 2H_2O + 4e^{-} \rightarrow 4OH^{-} \tag{14.2}$$

and

$$2H^{+} + 2e^{-} \rightarrow H_2 \tag{14.3}$$

Oxygen inside the can is usually limited to that in the headspace (i.e., the space between the top of the contents of the can and the lid), so that oxygen reduction takes place, by reaction (14.2), with accompanying corrosion of the tin, reaction (14.1), until the oxygen is consumed. Any further corrosion of tin is limited by the hydrogen evolution reaction (HER). Because tin has high hydrogen overpotential, reduction of hydrogen ions on tin is not significant, where as hydrogen reduction can occur on the steel substrate. In the absence of other depolarizers, it is this reaction—reduction of hydrogen on the steel substrate at pores in the tin coating—that controls detinning. Because the tin coating is the anode, loss of tin by reaction (14.1) takes place. Corrosion of tin usually proceeds slowly over the years of useful life of the can. As tin dissolves, more steel may become exposed, so that both reactions—reduction of hydrogen on the steel and dissolution of tin—accelerate [29].

In addition to acids or alkalies as incidental and natural components, foods usually contain various organic substances, some of which are complexing agents, as mentioned previously, and others act either as corrosion inhibitors or as cathodic depolarizers. Foods low in inhibitor and high in depolarizer substances may cause more rapid corrosion of food containers than, for example, highly acid foods. Because of the presence of organic depolarizers, corrosion of the inside tin coating of food containers may occur with little or no hydrogen evolution. When the tin coating has all corroded, however, it is observed that subsequent corrosion usually occurs by hydrogen evolution. The reason for this behavior is not firmly established, but it may be related to the fact that Sn^{2+} ions, which are known to inhibit corrosion of iron in acids, increase hydrogen overpotential, thereby favoring reduction of organic substances at the iron cathode. Stannous ions formed continuously at the iron surface during corrosion of the tin layer do not maintain sufficient concentration once the tin layer has dissolved. It is also possible that the potential difference of the iron–tin couple favors adsorption and reduction of organic depolarizers at the cathode, whereas these processes do not take place at lower potential differences. Eventual failure of containers is by so-called hydrogen swells, in which an appreciable pressure of hydrogen builds up within the can, making the contents suspect, since bacterial decomposition also causes gas accumulation.

The amount of hydrogen accumulated within the lifetime of the container is determined not only by the tin coating thickness, the temperature, and the chemical nature of food in contact, but most often by the composition and structure of the base steel. The rate of hydrogen evolution is increased by cold working of the steel (see Section 8.1), which is standard procedure for strengthening the container walls. Subsequent low-temperature heat treatment, incidental or intentional, may increase or decrease the rate (see Fig. 8.1, Section 8.1).

14.3.5.1 Properties of Tin. Tin is an amphoteric metal, reacting with both acids and alkalies, but it is relatively resistant to neutral or near-neutral media. It does not corrode in soft waters and has been used for many years as piping for distilled water. Only the high price of tin accounts for its replacement for distilled water piping by other materials, such as aluminum.

Corrosion in dilute nonoxidizing acids, such as HCl and H_2SO_4, is determined largely by concentration of dissolved oxygen in the acid. The hydrogen overpotential of tin is so high that corrosion accompanied by hydrogen evolution is limited. Deaerated acetic acid, for example, either hot or cold, corrodes tin slowly (1.5–6.0 gmd); NH_4OH and Na_2CO_3 have little effect, but, in concentrated NaOH, the rate of attack is rapid with formation of sodium stannate. As with acids, the rate is increased by aeration. In aerated fruit juices at room temperature, rates are only 0.1–2.5 gmd, but they become more than 10 times higher at the boiling point, indicating appreciable increase of attack with temperature [30].

14.3.6 Chromium-Plated Steel for Containers

Because of price advantage, steel may be coated with a combined thin chromium layer (0.3–0.4 μin; 0.008–0.01 μm), followed by chromium oxide (5–40 mg Cr/m^2) and an organic top coat. The advantages include good storage ability, resistance to sulfide staining, good adhesion, freedom from undercutting of the organic coating on the inside, and resistance to filiform corrosion on the outside of the container. However, it is not readily soldered, limiting its use in some application to can ends.

Chromium ($\phi° = -0.74$ V) is more active in the Emf Series than is iron ($\phi° = -0.44$ V), but chromium has a strong tendency to become passive ($\phi_F° = 0.2$ V). Hence, the potential of chromium in aqueous media is usually noble to that of steel. However, in galvanic couples of the two metals, especially in acid media, chromium is polarized below its Flade potential, so that it exhibits an active potential. Hence, the corrosion potential of chromium-plated steel, which is always porous to some degree, is more active than that of either passive chromium or the underlying steel [31]. Under these conditions, chromium acts as a sacrificial coating in the same manner as does a tin coating, but with dependence of its active potential on passive–active behavior rather than on formation of metal complexes. The metallic chromium layer contributes resistance to undercutting of the organic coating.

A chromium oxide coating enhances adhesion of the organic top coating, with optimum adhesion occurring for a hydrated oxide thickness [32] equivalent to 20mg/m^2 (2mg/ft^2). The organic coating, in turn, seals pores in the metal coating, increases resistance to the flow of electric current produced by galvanic cells, and reduces the amount of iron salts entering the contents of the container and affecting flavor or color [33].

14.3.7 Aluminum Coatings

Steel is aluminized (i.e., coated with aluminum) by hot dipping or spraying and, to a lesser extent, by cementation. Molten baths of aluminum for hot dipping usually contain dissolved silicon in order to retard formation of a brittle alloy layer. Hot-dipped coatings are used for oxidation resistance at moderately elevated temperatures, such as for oven construction and for automobile mufflers. They are unaffected by temperatures up to $480\,^\circ\text{C}$ ($900\,^\circ\text{F}$). At still higher temperatures, the coatings become refractory, but continue to be protective up to about $680\,^\circ\text{C}$ ($1250\,^\circ\text{F}$) [34].

The use, since 1975, of catalytic converters in automobiles has resulted in increased exhaust gas temperatures, up to $870\,^\circ\text{C}$ ($1600\,^\circ\text{F}$), with metal temperatures up to $760\,^\circ\text{C}$ ($1400\,^\circ\text{F}$), as well as increased concentrations of corrosive chemicals, such as sulfuric acid, in the exhaust gas stream. As a result, materials resistant to high-temperature oxidizing conditions, such as hot dip aluminum-coated type 409 stainless steel, are being used in automotive exhaust systems [35–37].

Aluminum coatings are also used for protecting against atmospheric corrosion, an application that is limited by higher costs of aluminum compared to zinc coatings and by variable performance. In soft waters, aluminum exhibits a potential that is positive to steel; hence, it acts as a noble coating. In seawater and in some fresh waters, especially those containing Cl^- or SO_4^{2-}, the potential of aluminum becomes more active and the polarity of the aluminum–iron couple may reverse. An aluminum coating under these conditions is sacrificial and cathodically protects steel. An Al–Zn alloy coating consisting of 44% Zn, 1.5% Si, bal. Al has been reported to have excellent resistance to marine and industrial atmospheres. It also protects against oxidation at elevated temperatures.

Sprayed coatings are commonly sealed with organic lacquers or paints to delay eventual formation of visible surface rust. The usual thickness of sprayed aluminum is 3–8 mils (0.08–0.2 mm). In one series of tests conducted in an industrial atmosphere, sprayed coatings 3 mils (0.08 mm) thick showed an average life of 12 years in comparison with 7 years for zinc, with the latter coatings being sprayed, electrodeposited, or hot-dipped [38].

Cemented coatings ("Calorizing") are produced by tumbling the work in a mixture of aluminum powder, Al_2O_3, and a small amount of NH_4Cl as flux in a hydrogen atmosphere at about 1000 °C (1800 °F). An aluminum–iron surface alloy forms that imparts useful resistance to high-temperature oxidation in air up to 850–950 °C (1550–1750 °F), as well as resistance to sulfur-containing atmospheres, such as those encountered in oil refining. Cemented aluminum coatings on steel are not useful for resisting aqueous environments. They are sometimes used to protect gas turbine blades (nickel-base alloys) from oxidation at elevated temperatures.

REFERENCES

1. T. J. Langill, in *ASM Handbook*, Vol. 13A, *Corrosion: Fundamentals, Testing, and Protection*, ASM International, Materials Park, OH, 2003, p. 801.
2. T. Mooney, ibid., pp. 772–785.
3. H. E. Townsend, in *Automotive Corrosion and Protection, Proceedings of the CORROSION/91 Symposium, NACE International, Houston, TX, 1991*, p. 31-1.
4. M. L. Berndt and C. C. Berndt, in *ASM Handbook*, Vol. 13A, *Corrosion: Fundamentals, Testing, and Protection*, ASM International, Materials Park, OH, 2003, pp. 803–813.
5. M. R. Dorfman, in *Handbook of Environmental Degradation of Materials*, M. Kutz, editor, William Andrew Publishing, Norwich, NY, 2005, pp. 405–422.
6. G. Carter, *Metal Progr.* **93**, 117 (1968).
7. G. Dearnaley, in *Ion Implantation Metallurgy*, C. Preece and J. Hirvonen, editor, The Metallurgical Society of AIME, Warrendale, PA, 1980, pp. 1–20.
8. E. McCafferty, P. G. Moore, J. D. Ayers, and G. K. Hubler, in *Corrosion of Metals Processed by Directed Energy Beams*, C. R Clayton and C. M. Preece, editors, The Metallurgical Society, Warrendale, PA, 1982, pp. 1–21.
9. J. Mazia, D. S. Lashmore, and T. Mooney, in *ASM Handbook*, Vol. 13A, *Corrosion: Fundamentals, Testing, and Protection*, ASM International, Materials Park, OH, 2003, p. 780.
10. W. Vernon, *J. Inst. Metals* **48**, 121 (1932).
11. A. R. Jones, in *ASM Handbook*, Vol. 13, *Corrosion*, ASM International, Materials Park, OH, 1987, p. 871.
12. F. LaQue, *Trans. Inst. Metal Finishing* **41**, 127 (1964).
13. A. Brenner and G. Riddell, *J. Natl. Bur. Std. (U.S.)* **39**, 385 (1947).
14. T. Mooney, in *ASM Handbook*, Vol. 13A, *Corrosion: Fundamentals, Testing, and Protection*, ASM International, Materials Park, OH, 2003, p. 777.
15. Ibid., p. 779.
16. T. C. Spence, in *ASM Handbook*, Vol. 13B, *Corrosion: Materials*, ASM International, Materials Park, OH, 2005, p. 49.
17. X. G. Zhang, in *Uhlig's Corrosion Handbook*, 2nd edition, R. W. Revie, editor, Wiley, New York, 2000, pp. 888–889.
18. E. Mattsson, *Mater. Performance* **21**(7), 9 (1982).

19. G. Schikorr, *Trans. Electrochem. Soc.* **76**, 247 (1939).

20. G. K. Glass and V. Ashworth, *Corros. Sci.* **25**, 971 (1985).

21. X. G. Zhang, in *ASM Handbook*, Vol. 13B, *Corrosion: Materials*, ASM International, Materials Park, OH, 2005, p. 413.

22. C. Bonilla, *Trans. Electrochem. Soc.* **87**, 237 (1945).

23. R. Hoxeng and C. Prutton, *Corrosion* **5**, 330 (1949).

24. P. Gilbert, *J. Electrochem. Soc.* **99**, 16 (1952).

25. H. Roters and F. Eisenstecken, *Arch. Eisenhüttenw* **15**, 59 (1941).

26. Ref. 14, p. 780.

27. M. W. Ingle, Cadmium elimination, in *ASM Handbook*, Vol. 5, *Surface Engineering*, ASM International, Materials Park, OH, 1994, pp. 918–924.

28. ASM Handbook, Vol. 13B, *Corrosion: Materials*, ASM International, Materials Park, OH, 2005, p. 186.

29. T. P. Murphy, in *Uhlig's Corrosion Handbook*, 2nd edition, R. W. Revie, editor, Wiley, New York, 2000, pp. 858–860.

30. B. Gonser and J. Strader, in *Corrosion Handbook*, H. H. Uhlig, editor, Wiley, New York, 1948, p. 327.

31. G. Kamm, A. Willey, and N. Linde, *J. Electrochem. Soc.* **116**, 1299 (1969).

32. M. Vucich, National Steel Corporation, private communication with HHU, 1983.

33. R. McKirahan and R. Ludwigsen, *Mater. Prot.* **7**(12), 29 (1968).

34. Prevention of Corrosion of Motor Vehicle Body and Chassis Components, Society of Automotive Engineers, Warrendale, PA, 1981.

35. *ASM Handbook*, Vol. 13, *Corrosion*, ASM International, Materials Park, OH, 1987, p. 1016.

36. W. R. Patterson, Materials, design, and corrosion effects on exhaust system life, SAE Paper 780921, in *Designing for Automotive Corrosion Prevention*, P-78, Society of Automotive Engineers, 1978, pp. 71–106.

37. J. Douthett, Automotive Exhaust System Corrosion, in *ASM Handbook*, Vol. 13C, *Corrosion: Environments and Industries*, ASM International, Materials Park, OH, 2006, pp. 519–520.

38. J. Hudson, *Chem. Ind. (London)* 1961, p. 3 (January 7).

GENERAL REFERENCES

F. E. Goodwin, Lead and lead alloys, in *Uhlig's Corrosion Handbook*, 2nd edition, R. W. Revie, editor, Wiley, New York, 2000, pp. 767–792.

S. Grainger and J. Blunt, editors, *Engineering Coatings—Design and Application*, 2nd edition, William Andrew Publishing, Cambridge, England, 1998.

G. Krauss and D. K. Matlock, editors, *Zinc-Based Steel Coating Systems: Metallurgy and Performance*, TMS, Warrendale, PA, 1990.

Metal coatings, in *ASM Handbook*, Vol. 13A, *Corrosion: Fundamentals, Testing, and Protection*, ASM International, Materials Park, OH, pp. 772–813.

T. P. Murphy, Tin and tinplate, in *Uhlig's Corrosion Handbook*, 2nd edition, R. W. Revie, editor, Wiley, New York, 2000, pp. 853–862.

L. Pawlowski, *The Science and Engineering of Thermal Spray Coatings*, Wiley, Chichester, England, 1995.

X. G. Zhang, Zinc, in *Uhlig's Corrosion Handbook*, 2nd edition, R. W. Revie, editor, Wiley, New York, 2000, pp. 887–904.

15

INORGANIC COATINGS

15.1 VITREOUS ENAMELS

Vitreous enamels, glass linings, and porcelain enamels are all essentially glass coatings of suitable coefficient of expansion fused on metals. Glass in powdered form (as glass frits) is applied to a pickled or otherwise prepared metal surface, then heated in a furnace at a temperature that softens the glass and allows it to bond to the metal. Several coats may be applied. Vitreous enamel coatings are used mostly on steel, but some coatings are also possible on copper, brass, and aluminum.

In addition to decorative utility, vitreous enamels protect base metals against corrosion by many environments. The glasses are composed essentially of alkali borosilicates and can be formulated to resist strong acids, mild alkalies, or both. Their highly protective quality results from virtual impenetrability to water and oxygen over relatively long exposure times, and from their durability at ambient and above-room temperatures. Their use in cathodically protected hot-water tanks has already been mentioned in Section 13.4. Although pores in the coating are permissible when cathodic protection supplements the glass coating, for other applications the coating must be perfect and without a single defect. This means that glass-lined vessels for the food and chemical industries must be maintained

Corrosion and Corrosion Control, by R. Winston Revie and Herbert H. Uhlig
Copyright © 2008 John Wiley & Sons, Inc.

free of cracks and other defects. Susceptibility to mechanical damage and cracking by thermal shock are the main weaknesses of glass coatings. If damage occurs, repairs can sometimes be made by tamping gold or tantalum foil into the voids.

Enameled steels exposed to the atmosphere last many years when used as gasoline pump casings, advertising signs, decorative building panels, plumbing fixtures, appliances, and so on. Failure occurs eventually by formation of a network of cracks in the coating—called *crazing*—through which rust appears. Vitreous enamels are also used to protect against high-temperature gases (e.g., in airplane exhaust tubes), and they have long life when exposed to soils.

15.2 PORTLAND CEMENT COATINGS

Portland cement coatings have the advantages of low cost, a coefficient of expansion ($1.0 \times 10^{-5}/°C$) approximating that of steel ($1.2 \times 10^{-5}/°C$), and ease of application or repair. The coatings can be applied by centrifugal casting (as for the interior of piping), by troweling, or by spraying. Usual thickness ranges from 5 to 25 mm (0.2–1 in.); thick coatings are usually reinforced with wire mesh.

Portland cement coatings are used to protect cast iron and steel water pipe on the water or soil side or both, with an excellent record of performance. In addition, Portland cement coatings are used on the interior of hot- and cold-water tanks, oil tanks, and chemical storage tanks. They are also used to protect against seawater and mine waters. The coatings are usually cured for 8–10 days before exposure to nonaqueous media, such as oils.

A disadvantage of Portland cement coatings is the sensitivity to damage by mechanical or thermal shock. Open tanks are easily repaired, however, by troweling fresh cement into cracked areas. There is evidence that small cracks in cold-water piping are automatically plugged with a protective reaction product of rust combining with alkaline products leached from the cement. In sulfate-rich waters, Portland cement may be attacked, but cement compositions are now available with improved resistance to such waters.

15.3 CHEMICAL CONVERSION COATINGS

Chemical conversion coatings are protective coatings formed *in situ* by chemical reaction with the metal surface. They include special coatings, such as $PbSO_4$, which forms when lead is exposed to sulfuric acid, and iron fluoride, which forms when steel containers are filled with hydrofluoric acid (>65% HF).

Phosphate coatings on steel ("Parkerizing," "Bonderizing") are produced by brushing or spraying, onto a clean surface of steel, a cold or hot dilute manganese or zinc acid orthophosphate solution (e.g., ZnH_2PO_4 plus H_3PO_4). The ensuing reaction produces a network of porous metal phosphate crystals firmly bonded to the steel surface (Fig. 15.1). Accelerators are sometimes added to the phosphating solution (e.g., Cu^{2+}, ClO_3^-, or NO_3^-) to speed the reaction.

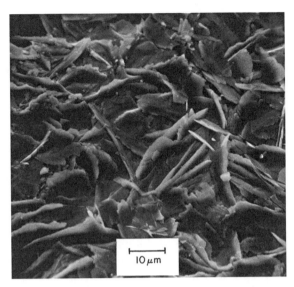

Figure 15.1. Scanning electron photomicrograph of phosphated type 1010 mild steel; acidic zinc phosphate + sodium nitrate accelerator applied at 65°C for 1 min (*Symposium on Interface Conversion for Polymer Coatings*, edited by P. Weiss and G. Cheever, Elsevier, New York, 1968).

Phosphate coatings do not provide appreciable corrosion protection in themselves. They are useful mainly as a base for paints, ensuring good adherence of paint to steel and decreasing the tendency for corrosion to undercut the paint film at scratches or other defects. Sometimes, phosphate coatings are impregnated with oils or waxes, which may provide some protection against rusting, especially if they contain corrosion inhibitors.

In corrosion-resistant automobile bodies, the first—and probably the most important—layer of the paint system is the phosphate coating. Although it is the thinnest coating in the paint system [about 3 μm (0.1 mil) thick], it is the anchor for the subsequent layers [1].

Oxide coatings on steel can be prepared by controlled high-temperature oxidation in air or, for example, by immersion in hot concentrated alkali solutions containing persulfates, nitrates, or chlorates. Such coatings, blue, brown, or black in color, consist mostly of Fe_3O_4 and, like phosphate coatings, are not protective against corrosion. When rubbed with inhibiting oils or waxes, as is often done with oxidized gun barrels, some protection is obtained.

Oxide coatings on aluminum are produced at room temperature by anodic oxidation of aluminum (called *anodizing*) in a suitable electrolyte (e.g., dilute sulfuric acid) at current densities of 100 or more A/m^2. The resultant coating of Al_2O_3 may be 0.0025–0.025 mm (0.1–1 mil) thick. The oxide so formed must be hydrated to improve its protective qualities by exposing anodized articles to steam or hot water for several minutes, a process called *sealing*. Improved corrosion resistance is obtained if sealing is done in a hot dilute chromate solution. The oxide coating may be dyed various colors, either in the anodizing bath or afterward.

Anodizing provides aluminum with some degree of improved corrosion resistance, but the additional protection is not spectacular, and it is certainly less than proportional to oxide thickness. Anodized coatings provide a good base for paints on aluminum, which is otherwise difficult to paint without special surface preparation.

Coatings of MgF_2 on magnesium can be formed by anodizing the metal at 90–120 V in 10–30% NH_4HF_2 solution at room temperature. This coating has been recommended for surface cleaning or as a base for finishing treatments [2].

Chromate coatings are produced on zinc by immersing the cleaned metal for a few seconds in sodium dichromate solution (e.g., 200 g/liter) acidified with sulfuric acid (e.g., 8 mL/liter) at room temperature and then rinsing and drying. A zinc chromate surface is produced that imparts a slight yellow color and that protects the metal against spotting and staining by condensed moisture. It also increases the life of zinc to a modest degree on exposure to the atmosphere. Similar coatings have been recommended for Zn–Al coatings [3] and for cadmium coatings on steel.

REFERENCES

1. L. L. Piepho, L. Singer, and M. R. Ostermiller, in *Automotive Corrosion and Protection*, R. Baboian, editor, NACE International, Houston, TX, 1992, pp. 20–8.
2. E. Emley, *Principles of Magnesium Technology*, Pergamon Press, New York, 1966.
3. H. Townsend and J. Zoccola, *J. Electrochem. Soc.* **125**, 1290 (1978).

GENERAL REFERENCES

Automotive Corrosion and Protection, R. Baboian, editor, NACE International, Houston, TX, 1992.

B. Grambow, Corrosion of Glass, in *Uhlig's Corrosion Handbook*, 2nd edition, R. W. Revie, editor, Wiley, New York, 2000, pp. 411–437.

Porcelain enamels, in *ASM Handbook*, Vol. 13A, *Corrosion: Fundamentals, Testing, and Protection*, ASM International, Materials Park, OH, 2003, pp. 750–754.

16

ORGANIC COATINGS

16.1 INTRODUCTION

In the United States, the Clean Air Act has had a major influence on the formulation and application of organic coatings. Solvent vapors from coating operations undergo photochemical oxidation in the atmosphere and contribute to smog. In order to reduce this pollution, new coating products, termed *compliant coatings*, have been developed with reduced volatile organic compounds. In addition, alternative processes are now being used in place of the traditional vapor degreasing in which chlorofluorocarbons and 1,1,l-trichloroethane, both being class 1 ozone-depleting chemicals, were used.

16.2 PAINTS

The total value of paints, varnishes, and lacquers produced in the United States in 2006 amounted to about 21 billion (21×10^9) dollars [1], half of which is estimated to be used for corrosion protection. Painted systems require continued maintenance to ensure integrity of corrosion protection. For example, in 1995, the Port Authority of New York and New Jersey began a 20-year, $500 million

Corrosion and Corrosion Control, by R. Winston Revie and Herbert H. Uhlig
Copyright © 2008 John Wiley & Sons, Inc.

program to repaint all bridges. The existing, lead-based paint was removed by abrasive blasting down to the bare metal. The new paint system, consisting of three coats, zinc primer, epoxy intermediate, and urethane finish, is expected to last 25 years with "moderate maintenance" [2].

Paints are a mixture of insoluble particles of pigment suspended in a continuous organic or aqueous vehicle. Pigments usually consist of metallic oxides (e.g., TiO_2, Pb_3O_4, Fe_2O_3) or other compounds, such as $ZnCrO_4$, $PbCO_3$, $BaSO_4$, clays, and so on. The vehicle may be a natural oil—linseed or tung oil, for example. When these drying oils are exposed to air, they oxidize and polymerize to solids, a process that can be hastened by small amounts of catalysts, such as lead, manganese, or cobalt soaps. Synthetic resins are now more often used as vehicles or components of vehicles, particularly for continuous contact with water or where resistance to acids, alkalies, or higher temperatures is required. These resins may dry by evaporation of the solvent in which they are dissolved, or they may polymerize through application of heat or by addition of catalysts. *Varnishes* consist generally of a mixture of drying oil, dissolved resins, and a volatile thinner. *Lacquers* consist of resins dissolved in a volatile thinner; they sometimes contain pigments as well. Because of the need to reduce atmospheric pollution, use of aqueous vehicle paints is preferred, and in some cities required, over use of paints employing volatile organic solvents.

Examples of synthetic *resins* include phenol-formaldehyde formulations, which withstand boiling water or slightly higher temperatures and are used in the chemical industry in the form of multiple coats, baked on, for resisting a variety of corrosive media. Silicone and polyimide resins are useful at still higher temperatures. Alkyd resins, because of favorable cost, fast-drying properties, and durability, have found wide application for protecting the metal surfaces of machinery and home appliances.

Vinyl resins have good resistance to penetration by water. Their resistance to alkalies makes them useful for painting structures that are to be protected cathodically. Linseed- and tung-oil paints, by comparison, are quickly saponified and disintegrated by alkaline reaction products formed at a cathode, whether in waters or in soil.

Epoxy resins are also resistant to alkalies and to many other chemical media and have the distinguishing property of adhering well to metal surfaces. The latter property presumably derives from many available polar groups in the molecule. These resins are the basis of plastic mixtures that, on addition of a suitable catalyst, solidify in place within a short time. They are useful, for example, to seal leaks temporarily in ferrous or nonferrous piping.

Paints, in general, are not useful for protecting buried structures, one reason being that mechanical damage to thin coatings by contact with the soil is difficult to avoid. Tests have shown that, for this purpose, their life is relatively short. Coatings based on coal tar or on fusion-bonded epoxy have been found to be far more practical. Similarly, the usual linseed-tung oil paints are not durable for metal structures totally immersed in water, except possibly for short periods of time on the order of one year or less. In hot water, life is still shorter. More adequate protection extending up to several years at ordinary temperatures can be obtained by applying four or five coats of a synthetic vehicle paint, as is done in the chemical industry. Because of the expense of a multiple-coat system of this kind, many applications for fresh water or seawater make use, instead, of thick coal-tar coatings. By far the majority of paints perform best for protecting metals against atmospheric corrosion, and this is their main function.

16.3 REQUIREMENTS FOR CORROSION PROTECTION

To protect against corrosion, a good paint should meet the following requirements:

1. *Provide a Good Vapor Barrier.* All paints are permeable, in some degree, to water and oxygen. Some vehicles are less permeable than others, but their better performance as a diffusion barrier applies only to well-adhering multiple-coat applications that effectively seal pores and other defects. The diffusion path through a paint film is normally increased by incorporating pigments. Particularly effective in this regard are pigments having the shape of flakes oriented parallel (e.g., by brushing) to the metal surface (e.g., micaceous or flaky hematite, aluminum powder). Diffusion, on the other hand, tends to be accelerated by electro-osmosis whenever the coated metal is made the cathode of a galvanic couple or is cathodically protected.

2. *Inhibit against Corrosion.* Pigments incorporated into the prime coat (the coat immediately adjacent to the metal) should be effective corrosion inhibitors. Water reaching the metal surface then dissolves a certain amount of pigment, making the water less corrosive. Corrosion-inhibiting pigments must be soluble enough to supply the minimum concentration of inhibiting ions necessary to reduce the corrosion rate, yet not soluble to a degree that they are soon leached out of the paint.

Among pigments that have been recommended for prime coats, only relatively few actually do the job that is required. Effective pigments for which performance has been established in many service tests include (1) red lead (Pb_3O_4) having the structure of plumbous ortho-plumbate (Pb_2PbO_4) and (2) zinc chromate ($ZnCrO_4$) and basic zinc chromate or zinc tetroxychromate. The inhibiting ion in the case of red lead is probably PbO_4^{4-}, which is released in just sufficient amounts to passivate steel, protecting it against rusting by water reaching the metal surface. It is likely that lead oxides and hydroxides of other compositions also have inhibiting properties in this regard, but red lead appears to be best of the lead compounds.

For zinc chromate, the inhibiting ion is CrO_4^{2-}, solubility relations being just right to release at least the minimum concentration of the ion ($>10^{-4}$ mole/liter) for optimum inhibition of steel. The solubility of zinc tetroxychromate is reported to be 2×10^{-4} mole/liter [3]. Lead chromate, on the other hand, is not nearly soluble enough (solubility $= 1.4 \times 10^{-8}$ mole/liter) and acts only as an inert pigment. Commercial formulations of lead chromate sometimes contain lead oxides, present either inadvertently or intentionally, which may impart a degree of inhibition.

Zinc molybdate has been suggested as an inhibiting pigment for paints [4], being white instead of the characteristic yellow of chromates. It is less toxic than chromates. The sulfate and chloride content of a commercial $ZnCrO_4$ (or $ZnMoO_4$) pigment must be low so that it can passivate the metal surface. Because inhibiting pigments passivate steel, they are relatively ineffective for this purpose in the presence of high concentrations of chlorides, such as in seawater.

Paints pigmented with zinc dust using essentially an aqueous sodium silicate vehicle (called inorganic zinc-rich paint), or an organic vehicle, are also useful as prime coats, the function of the zinc being to cathodically protect the steel in the same manner as galvanized coatings. Such paints are sometimes used over partly rusted galvanized surfaces because they also adhere well to zinc, but rust should first be removed. Sacrificial protection by zinc-rich coatings requires intimate contact with the substrate, and so these coatings are always used as primers. It has been reported [5] that in order to ensure good electrical contact between zinc particles and with the base metal, the amount of pigment in the dried paint film should account for 95% of its weight. A coating of this kind protected steel in seawater against rusting at a scratch for 1–2 years, whereas with 86% and 91% Zn, rust appeared after 1–2 days and 10–20 days, respectively. Vehicles for zinc-rich paints to accommodate such a large fraction of pigment include chlorinated rubber, polystyrene, epoxy, and polyurethane.

The minimum amount of zinc dust pigment required to provide cathodic protection depends on several factors, including Zn particle size, nature of the vehicle, and the amount of ZnO and other pigments that may be present [6]. It probably also depends on the extent to which insulating coatings form on zinc particles before the paint is applied (age of paint).

3. *Provide Long Life at Low Cost.* A reasonable cost of paint should be gauged by its performance. A paint system lasting 5 years justifies double the cost of paint if the more expensive paint provides 35% longer life or lasts short of 7 years (labor to paint cost ratio of 2:1).

The rate of deterioration of a paint depends on the particular atmosphere to which it is exposed, which, in turn, depends on the amount of atmospheric pollution and on the amount of rain and sunshine. The color of the top coat (i.e., its ability to reflect infrared and ultraviolet radiation) and the type of vehicle used play some part. Other things being equal, the performance of good-quality paints used for corrosion protection is largely determined by the thickness of the final paint film. In achieving a given coating thickness, it is advantageous to apply several coats rather than one, probably because pores are better covered by several applications and also because evaporation or dimensional changes during polymerization are better accommodated by thin films.

16.4 METAL SURFACE PREPARATION

Many test results and a great deal of engineering data over many years have shown that the most important single factor influencing the life of a paint is the proper preparation of the metal surface. This factor is, generally, more important than the quality of the paint that is applied. In other words, a poor paint system on a properly prepared metal surface usually outperforms a better paint system on a poorly prepared surface. A well-prepared surface is the foundation on which the paint system is built.

Adequate surface preparation consists of two main processes.

16.4.1 Cleaning All Dirt, Oils, and Greases from the Surface

Initial cleaning can be accomplished by using solvents or alkaline solutions.

Solvents. Mineral spirits, naphtha, alcohols, ethers, chlorinated solvents, and so on, are applied by dipping, brushing, or spraying. One such solvent is Stoddard solvent, a petroleum base mineral spirit with a flash point of 40–55°C (100–130°F), sufficiently high to minimize fire hazard, and it is not particularly toxic.

Chlorinated solvents, on the other hand, although nonflammable, are relatively toxic and contribute to pollution. In addition, they may leave chloride residues on the metal surface that can later initiate corrosive attack. They are used largely for vapor degreasing (tri- or perchlorethylene), in which the work is suspended in the vapor of the boiling solvent. Care must be exercised in the vapor degreasing of aluminum, ensuring that adequate chemical inhibitors are added and maintained in the chlorinated solvent in order to avoid catastrophic corrosion (see Section 21.1.4.1) or, in the extreme, to avoid an explosive reaction.

Alkaline Solutions. Aqueous solutions of certain alkalies provide a method of removing oily surface contamination that is cheaper and less hazardous than the use of solvents. They are more efficient in this particular function than solvents, but perhaps less effective for removing heavy or carbonized oils. Suitable solutions contain one or more of the following substances: Na_3PO_4, $NaOH$, $Na_2O \cdot nSiO_2$, Na_2CO_3, borax, and sometimes sodium pyro- or metaphosphate and a wetting agent. Cleaning may be done by immersing the work in the hot solution [80°C (180°F) to b.p.] containing about 30–75 g/liter (4–10 oz/gal) alkali. The hot solution, in somewhat more dilute concentration, may also be sprayed onto the surface. Electrolytic cleaning in alkaline solutions is also sometimes used; this process makes use of the mechanical action of hydrogen gas and the detergent effect of released OH^- at the surface of the work connected as cathode to a source of electric current.

If the metal is free of mill scale and rust, a final rinse in water and in a dilute chromic–phosphoric acid ensures both removal of alkali from the metal surface, which would otherwise interfere with good bonding of paint, and also temporary protection against rusting.

16.4.2 Complete Removal of Rust and Mill Scale

Rust and mill scale are best removed by either pickling or sandblasting.

Pickling. The metal, cleaned as described previously, is dipped into an acid (e.g., 3–10% H_2SO_4 by weight) containing a pickling inhibitor (see Section 17.3) at a temperature of 65–90°C (150–190°F) for an average of 5–20 min. Oxide next to the metal surface is dissolved, loosening the upper Fe_3O_4 scale. Sometimes, sodium chloride is added to the sulfuric acid, or HCl alone is used at lower temperatures, or a 10–20% H_3PO_4 is used at temperatures up to 90°C (190°F). The latter acid is more costly, but has the advantage of producing a phosphate film on the steel surface that is beneficial to paint adherence. Some pickling procedures, in fact, call for a final rinse in dilute H_3PO_4 in order to ensure removal from the metal surface of residual chlorides and sulfates that are damaging to the life of a paint coating.

Sometimes, the final dip is a dilute solution of chromic (30–45 g/liter; 4–6 oz/gal) or chromic–phosphoric acid, which serves to prevent rusting of the surface before the prime coat is applied. Use of chromates is less popular than formerly because their toxicity places restrictions on disposal of spent solutions. Hexavalent chromium (Cr^{6+}) is a potent human toxin and known cancer-causing agent [7]. A relatively nontoxic pickle for steel consists of hot 3–10% ammoniated citric acid followed by a dilute alkaline sodium nitrite solution to minimize superficial rusting before application of paint [8].

Blasting. Using this procedure, scale is removed by high-velocity particles impelled by an air blast or by a high-velocity wheel. Blast materials usually consist of sand, or sometimes of steel grit, silicon carbide, alumina, refractory slag, or rock wool byproducts.

Other methods of removing mill scale include *flame cleaning*, by which scale spalls off the surface through sudden heating of the surface with an oxyacetylene torch. *Weathering* for several weeks or months is also possible; the natural rusting of the surface dislodges scale, which can then be further removed by wire brushing. But these procedures are less satisfactory than complete removal of scale and rust by pickling or blasting.

The poor performance of paints on weathered steel exposed to an industrial atmosphere was shown in tests reported by Hudson [9] (Table 16.1). The relatively long life of paint shown for intact mill scale would probably not be achieved in actual service. It would, for example, be difficult to keep large areas and various shapes of mill scale from cracking before or after painting. Fragmentation of mill scale allows paint to become dislodged, particularly after electrolytic action has occurred between metal and scale by aqueous solutions that have penetrated to the metal surface.

TABLE 16.1. Effect of Surface Preparation of Steel on Life of Paint Coatings [9]

Surface Preparation	Durability of Paint (years) in Sheffield, England	
	2 Coats Red Lead + 2 Coats Red Iron Oxide Paint	2 Coats Red Iron Oxide Paint
Intact mill scale	8.2	3.0
Weathered and wire-brushed	2.3	1.2
Pickled	9.5	4.6
Sandblasted	10.4	6.3

16.5 APPLYING PAINT COATINGS

The prime coat should be applied to the dry metal surface as soon as possible after the metal is cleaned in order to achieve a good bond. Better still, the metal should first be given a phosphate coat (see Section 15.3), in which case the prime coat, if necessary, can be delayed for a short while. The advantages of a phosphate coat are a better bond of paint to metal and good resistance to undercutting of the paint film at scratches or other defects in the paint at which rust forms and progresses beneath the organic coating. For many years it has been standard practice to coat automobile bodies and electric appliances with phosphate before painting.

Only in unusual cases should paint be applied over a damp or wet surface because poor bonding of paint to steel results under these conditions. A second prime coat can be applied after the first has dried, or a sequence of top coats can follow. A total of four coats with combined thickness of not less than about 0.13 mm (5 mils) is considered by some authorities to be the recommended minimum for steel that will be exposed to corrosive atmospheres [10].

During the past 30 years, there have been vast improvements in the corrosion performance of automobiles, driven not only by the demands of the public for increased corrosion resistance, but also by the Clean Air Act of 1970 [with regulations regarding volatile organic content (VOC)] and subsequent revisions. Tougher and thicker materials have been developed, and the application processes have been improved. One of the most important advances has been the development of the cathodic electrodeposition priming process, introduced in 1976 and, by 1985, fully implemented by most vehicle manufacturers in their assembly plants. Today, nearly 100% of the cars and trucks manufactured in the world are primed using the cathodic electrodeposition process [11]. Following phosphate pretreatment, the vehicle is immersed in the electrodeposition bath for typically 2–3 minutes [11]. Electric current is applied, and a film is deposited.

One of the advantages of this process, compared to spraying, for example, is that it results in a uniform, thin coating [about 25 µm (1 mil) thick] with coverage on both exterior and interior cavity surfaces. In addition, this process is controllable, automated, efficient, and environmentally acceptable [12].

16.5.1 Wash Primer

A wash primer, WP1, was developed during World War II in order to facilitate the painting of aluminum. Subsequently, it was also found advantageous as a prime coat for steel and several other metals. One wash primer solution consists of approximately 9% polyvinyl butyral and 9% zinc tetroxychromate by weight in a mixture of isopropanol and butanol as one solution, which, just before using, is mixed with a solution of 18 wt.% H_3PO_4 in isopropanol and water in the weight ratio of four of the pigmented solution to one of the latter [13]. The mixture must be used within 8–24 h after mixing. It has the advantage of providing in one operation, instead of two, a phosphating treatment of the metal and the application of the prime coat. It has proved to be an effective prime coat on steel, zinc (galvanized steel), and aluminum.

16.5.2 Painting of Aluminum and Zinc

Paints do not adhere well to aluminum without special surface treatment, unless the wash primer is used, which provides its own surface treatment. Otherwise, phosphating or anodizing is suitable. The prime coat should, in general, contain zinc chromate as an inhibiting pigment. Red lead is not recommended, because of the galvanic interaction of aluminum with metallic lead deposited by replacement of lead compounds. Paints pigmented with zinc dust plus ZnO (zinc-rich paints) can also be used satisfactorily as a prime coat, forming a good bond with the metal. In this case, Zn and ZnO apparently react beforehand with organic acids of the vehicle, ensuring that aluminum soaps and other compounds do not form at the paint–metal interface to weaken the attachment of paint to metal.

Zinc or galvanized surfaces are also difficult to paint and should be phosphated beforehand, or a wash primer should be used as a prime coat. As with aluminum, zinc chromate, but not red lead, is a suitable inhibiting pigment in the prime coat, or zinc-rich paints can be used. A coating system consisting of painted galvanized steel provides a protective service life up to 1.5 times that predicted by adding the expected lifetimes of the paint and the galvanized coating in a severe atmosphere, and the synergistic improvement is greater for mild environments [14].

16.6 FILIFORM CORROSION

Metals with organic coatings may undergo a type of corrosion resulting in numerous meandering thread-like filaments of corrosion product. This is sometimes known as underfilm corrosion, and it was called *filiform corrosion* by Sharmon [15] (Fig. 16.1). It has been described by several investigators and reproduced in the laboratory [16]. According to reported descriptions, the filaments, or threads, on steel are typically 0.1–0.5 mm wide. The thread itself is red in color, characteristic of Fe_2O_3, and the head is green or blue, corresponding to the presence of

(a)

(b)

Figure 16.1. Filiform corrosion. (a) Lacquered tin can. 1×. (b) Clear varnish on steel, 10× (86% R.H., 840 h). [Reprinted with permission, M. Van Loo, D. Laiderman, and R. Bruhn, *Corrosion* **9**, 279 (1953). Copyright NACE International 1953.]

ferrous ions. Each thread grows at a constant rate of about 0.4 mm/day in random directions, but threads never cross each other. If a head approaches another thread, it either glances off at an angle or stops growing.

Filiform corrosion occurs independent of light, metallurgical factors in the steel, and bacteria. Although threads are visible only under clear lacquers or varnishes, they probably also occur under opaque paint films. They have been observed under various types of paint vehicles and on various metals, including steel, zinc, aluminum, magnesium, and chromium-plated nickel. This type of corrosion takes place on steel only in air of high relative humidity (e.g., 65–95%). At 100% relative humidity, the threads may broaden to form blisters. They may not form at all if the film is relatively impermeable to water, as is stated to be the case for paraffin [17]. The mechanism appears to be a straightforward example of a differential aeration cell.

16.6.1 Theory of Filiform Corrosion

Various schematic views of a filiform thread or filament are shown in Fig. 16.2. The head is made up of a relatively concentrated solution of ferrous salts, as was shown by analysis [17]. Hence, water tends to be absorbed from the atmosphere in this region of the thread. Oxygen also diffuses through the film and reaches higher concentrations at the interface of head and body and at the periphery of the head, compared to lower concentrations in the center of the head. This sets up a differential aeration cell with the cathode and accumulation of OH^- ions in all regions where the film makes contact with the metal, as well as at the rear of the head. The anode is located in the central and forward portions of the head attended by formation of Fe^{2+}.* The liberated OH^- ions probably play an important role in undermining the film in view of the well-known ability of alkalies to destroy the bond between paints and metals (cathodic delamination). In addition, they diffuse toward the center of the head, reacting with Fe^{2+} to form $FeO \cdot nH_2O$, which, in turn, is oxidized by O_2 to $Fe_2O_3 \cdot nH_2O$. The precipitated oxide assumes a typical V shape because more alkali is produced in the region between the head and the body (more O_2), compared to the periphery of the head. Behind the V-shaped interface, Fe_2O_3 exists predominantly, and, since it is less hygroscopic than ferrous salt solutions, water again diffuses out through the film, leaving this portion relatively dry. Oxygen continues to diffuse through the film, serving to keep the main portion of the filament cathodic to the head.

If a head should approach another filament body, the previous participation of the film in filiform growth will have depleted the film of organic and inorganic anions necessary to the accumulation of high concentrations of ferrous salts in the head and also of cations necessary to build up a high pH at the periphery.

---

*The accumulation of alkali at the periphery can be demonstrated by placing a large drop of dilute sodium chloride solution (1–5%), preferably deaerated, containing a few drops of phenolphthalein and about 0.1% $K_3Fe(CN)_6$ on an abraded surface of iron. Within a few minutes, the periphery turns pink and the center turns blue

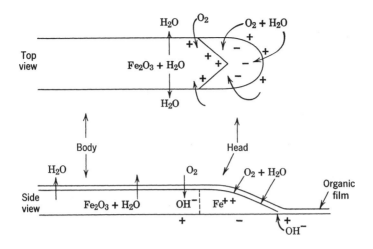

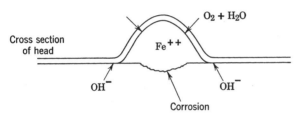

Figure 16.2. Schematic views of filiform filament on iron showing details of differential aeration cell causing attack.

This serves to discourage further growth of the filament in the direction of the old filament. Furthermore, and perhaps more important, the previous accumulation of OH⁻ added to that being formed, plus still greater abundance of oxygen, tends to ensure that the old filament body remains cathodic, encouraging the approaching anode to veer off in another direction. If the head should lose its electrolyte because of delaminated film at the old thread which it approaches, the filament would stop growing, a situation sometimes observed.

Phosphate surface treatments and chromate prime coats of paint serve to limit filiform corrosion, but apparently do not prevent its occurrence. Wholly adequate remedies have not yet been found.

16.7 PLASTIC LININGS

Protection against acids, alkalies, and corrosive liquids and gases in general can be obtained by bonding a thick sheet of plastic or rubber to a steel surface.

Rubber, Neoprene, and vinylidene chloride (Saran) are examples of materials that are so used. A thickness of 3 mm ($\frac{1}{8}$ in.) or more ensures a relatively good diffusion barrier and protects the base metal against attack for a long time. The expense of such coatings usually excludes their use for any but severely corrosive environments, such as are common in the chemical industry.

Plastic coatings of vinyl or polyethylene can be applied as an adhesive tape, particularly to protect buried metal structures. Such tape finds practical use for coating pipe and auxiliary equipment, including pipe connections and valves, exposed to the soil. However, tape can become disbonded with time, leaving the pipe uncoated, not cathodically protected since the tape material is nonconductive, and in contact with the groundwater that leaks into such disbondments. This sequence of events has resulted in failure by S.C.C. of some underground high-pressure oil and natural gas pipelines [18].

One of the most stable plastics in terms of resisting a wide variety of chemical media is tetrafluorethylene (Teflon). It successfully resists aqua regia and boiling concentrated acids, including HF, H_2SO_4, and HNO_3. It resists concentrated boiling alkalies, gaseous chlorine, and all organic solvents up to about 250°C (480°F). It is said to react only with elementary fluorine and with molten sodium. Slow decomposition into HF and fluorinated hydrocarbons begins at temperatures above 200°C (400°F), with the mixture of gases being highly toxic [19]. Toxic gases may also be liberated by heat generated during machining of the plastic.

As a material, Teflon is not strong, tending to creep readily under stress. Its extreme inertness makes bonding to any surface difficult. This compound is used, typically, for linings, gasket material, and diaphragm valves.

REFERENCES

1. Bureau of Census, U.S. Dept. Of Commerce, Paint and Allied Products: 2006, Table 1, Current Industrial Report MA325F(06)-1, Washington, D.C., issued June 2007.

2. Port Authority of New York and New Jersey website: http://www.panynj.gov/CommutingTravel/bridges/html/painting.html

3. P. Nylen and E. Sunderland, *Modern Surface Coatings*, Interscience, New York, 1965, (p. 689.)

4. *Chem. Eng. News*, November 14, 1960, p. 58.

5. J. Mayne, *Br. Corros. J.* **5**, 106 (1970).

6. T. Ross et al., *Corros. Sci.* **18**, 505 (1978).

7. P. Lay and A. Levina, *J. Am. Chem. Soc.* **120**, 6704 (1998).

8. W. Blume, *Mater. Perf.* **16**(3), 18 (1977).

9. J. Hudson, *J. Iron Steel Inst.* **168**, 153 (1951); (with W. Johnson), ibid. **168**, 165 (1951).

10. J. Hudson, *Chem. Ind. (London)* **55**, 3 (1961).

11. L. L. Piepho, L. Singer, and M. R. Ostermiller, In *Automotive Corrosion and Protection*, R. Baboian, editor, NACE International, Houston, TX, 1992, pp. 20–9, 20–20.

12. John J. Vincent, in *Automotive Corrosion and Protection*, R. Baboian, editor, NACE International, Houston, TX, 1992, pp. 21-2–21-3.

13. L. Whiting, *Corrosion* **15**, 311t (1959).

14. T. J. Langill, in *ASM Handbook*, Vol. 13A, *Corrosion: Fundamentals, Testing, and Protection*, ASM International, Materials Park, OH, 2003, p. 800.

15. C. Sharmon, *Nature*, **153**, 621 (1944); *Chem. Ind. (London)* **46**, 1126 (1952).

16. C. Hahin and R. G. Buchheit, In *ASM Handbook*, Vol. 13A, *Corrosion: Fundamentals, Testing, and Protection*, ASM International, Materials Park, OH, 2003, pp. 248–256.

17. W. Slabaugh and M. Grotheer, *Eng. Chem.* **46**, 1014 (1954).

18. *Public Inquiry Concerning Stress Corrosion Cracking on Canadian Oil and Gas Pipelines, Report of the Inquiry*, Report MH-2-95, National Energy Board, Calgary, Alberta, November 1996.

19. *Bulletin*, M.I.T. Occupational Medical Service, December 1961.

GENERAL REFERENCES

A. Forsgren, *Corrosion Control through Organic Coatings*, CRC Press, Taylor & Francis, Boca Raton, FL, 2006.

C. H. Hare, Corrosion control of steel by organic coatings, in *Uhlig's Corrosion Handbook*, 2nd edition, R. W. Revie, editor, Wiley, New York, 2000, pp. 1023–1039.

P. R. Khaladkar, Using plastics, elastomers, and composites for corrosion control, in *Uhlig's Corrosion Handbook*, 2nd edition, R. W. Revie, editor, Wiley, New York, 2000, pp. 965–1022.

R. Norsworthy, Selection and use of coatings for underground or submersion service, in *Uhlig's Corrosion Handbook*, 2nd edition, R. W. Revie, editor, Wiley, New York, 2000, pp. 1041–1059.

D. J. Smukowski, Environmental regulation of surface engineering, in *ASM Handbook*, Vol. 5, *Surface Engineering*, ASM International, Materials Park, OH, 1994, pp. 911–917.

K. B. Tator, Organic coatings and linings, in *ASM Handbook*, Vol. 13A, *Corrosion: Fundamentals, Testing, and Protection*, ASM International, Materials Park, OH, 2003, pp. 817–833.

K. B. Tator, Paint systems, in *ASM Handbook*, Vol. 13A, *Corrosion: Fundamentals, Testing, and Protection*, ASM International, Materials Park, OH, 2003, pp. 837–844.

B. Thomson and R. P. Campion, Testing of polymeric materials for corrosion control, in *Uhlig's Corrosion Handbook*, 2nd edition, R. W. Revie, editor, Wiley, New York, 2000, pp. 1151–1164.

17

INHIBITORS AND PASSIVATORS

17.1 INTRODUCTION

An *inhibitor* is a chemical substance that, when added in small concentration to an environment, effectively decreases the corrosion rate. There are several classes of inhibitors, conveniently designated as follows: (1) passivators, (2) organic inhibitors, including slushing compounds and pickling inhibitors, and (3) vapor-phase inhibitors.

Passivators are usually inorganic oxidizing substances (such as chromates, nitrites, and molybdates) that passivate the metal and shift the corrosion potential several tenths volt in the noble direction. Nonpassivating inhibitors, such as the pickling inhibitors, are usually organic substances that have only slight effect on the corrosion potential, changing it either in the noble or active direction usually by not more than a few milli- or centivolts. In general, the passivating-type inhibitors reduce corrosion rates to very low values, being more efficient in this regard than most of the nonpassivating types. They represent, therefore, the best inhibitors available for certain metal–environment combinations. Except where noted, the discussion in this chapter pertains to steel.

The practice of corrosion inhibition is greatly influenced by new regulations that have been developed because of toxicity and environmental effects resulting

Corrosion and Corrosion Control, by R. Winston Revie and Herbert H. Uhlig
Copyright © 2008 John Wiley & Sons, Inc.

from industrial effluents. For example, there is a trend to replace some widely used inhibitors, especially those that contain hexavalent chromium, Cr^{6+}, a known carcinogen, in applications where toxicity, environmental damage, and pollution caused by these chemicals are important considerations.

17.2 PASSIVATORS

17.2.1 Mechanism of Passivation

The theory of passivators has already been discussed in part, and should be referred to again, in Section 6.3. Passivators in contact with a metal surface act as depolarizers, initiating high current densities that exceed $i_{critical}$ for passivation at residual anodic areas. Only those ions can serve as passivators that have both an oxidizing capacity in the thermodynamic sense (noble oxidation–reduction potential) and that are readily reduced (shallow cathodic polarization curve), as illustrated in Fig. 17.1. Hence, SO_4^{2-} or ClO_4^- ions are not passivators for iron, because they are not readily reduced, nor are NO_3^- ions compared to NO_2^-, because nitrates are reduced less rapidly than are nitrites, with the former reducing too sluggishly to achieve the required high value of $i_{critical}$. The extent of chemical reduction on initial contact of a passivator with metal, according to this viewpoint, must be at least chemically equivalent to the amount of passive film

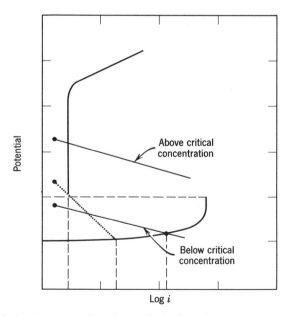

Figure 17.1. Polarization curves that show effect of passivator concentration on corrosion of iron. An oxidizing substance that reduces sluggishly does not induce passivity (dotted cathodic polarization curve) (schematic).

formed as a result of such reduction. For the passive film on iron, as has been discussed earlier, this is on the order of $0.01\,C/cm^2$ $(100\,C/m^2)$ of apparent surface area. The total quantity of equivalents of chemically reduced chromate is found to be of this order, and it is probably also the same for other passivators acting on iron. The amount of chromate reduced in the passivation process was arrived at from measurements [1] of residual radioactivity of a washed iron surface after exposure to a chromate solution containing radioactive ^{51}Cr. The following reaction applies, assuming that all reduced chromate (or dichromate) remains on the metal surface as adsorbed Cr^{3+} or as hydrated Cr_2O_3:

$$Cr_2O_7^{2-} + 8H^+ + Fe_{surface} \rightarrow 2Cr^{3+} + 4H_2O + O_2 \cdot O_{ads\,on\,Fe} \qquad \Delta G^\circ = -406\,kJ$$

Residual radioactivity accounts for 3×10^{16} Cr atoms/cm^2 (1.5×10^{-7} eq or 0.015 C passive-film substance/cm^2). The equation assumes an adsorbed passive-film structure, but the same reasoning applies whatever the structure.

Reduction of passivator continues at a low rate after passivity is achieved, equivalent in the absence of dissolved oxygen to the value of $i_{passive}$ [$<0.3\,\mu A/cm^2$ ($<3\,mA/m^2$)] based on observed corrosion rates of iron in chromate solutions. Iron oxide and chromate reduction products slowly accumulate. The rate of reduction increases with factors that increase $i_{passive}$, such as higher H^+ activity, higher temperatures, and the presence of Cl^-. It is found, in practice, that less chromate is consumed as exposure time continues, consistent with $i_{passive}$ that also decreases with time.

For optimum inhibition, the concentration of passivator must exceed a certain critical value. Below this concentration, passivators behave as active depolarizers and increase the corrosion rate at localized areas, such as pits. Lower concentrations of passivator correspond to more active values of the oxidation–reduction potential, and eventually the cathodic polarization curve intersects the anodic curve in the active region instead of in the passive region alone (Fig. 17.1).

The critical concentration for CrO_4^{2-}, NO_2^-, MoO_4^{2-}, or WO_4^{2-} is about 10^{-3} to $10^{-4}\,M$ [2–5]. A concentration of $10^{-3}\,M$ Na_2CrO_4 is equivalent to 0.016% or 160 ppm. Chloride ions and elevated temperatures increase $i_{critical}$ as well as $i_{passive}$, thereby raising the critical passivator concentration to higher values as well. At 70–90°C, for example, the critical concentration of CrO_4^{2-} and NO_2^- is about $10^{-2}\,M$ [4, 5]. Should the passivator concentration fall below the critical value in stagnant areas (e.g., at threads of a pipe or in crevices), the active potential at such areas in galvanic contact with passive areas elsewhere of noble potential promotes localized corrosion, or pitting, at the active areas through the action of passive–active cells. For this reason, it is important to maintain the concentration of passivators above the critical value at all portions of the inhibited system by stirring, rapid flow rates, and avoiding crevices and surface films of grease and dirt. Since consumption of passivators increases in the presence of chloride and sulfate ions, it is also essential to maintain as low a concentration of these ions as possible.

Some substances indirectly facilitate passivation of iron (and probably of some other metals as well) by making conditions more favorable for adsorption of oxygen. In this category are alkaline compounds (e.g., NaOH, Na_3PO_4, $Na_2O \cdot nSiO_2$, $Na_2B_4O_7$). These are all nonoxidizing substances requiring dissolved oxygen in order to inhibit corrosion; hence, oxygen is properly considered the passivating substance. The mechanism of passivation is similar to that described in Sections 7.2.1.3 and 7.2.3. High concentrations of OH^- displace H adsorbed on the metal surface, thereby decreasing the probability of a reaction between dissolved O_2 and adsorbed H. The excess oxygen is then available to adsorb instead, producing passivity. In addition to passive films of this kind, protection is supplemented by diffusion-barrier films of iron silicate, iron phosphate, and so on.

Passivation of iron by molybdates and tungstates, both of which inhibit in the near-neutral pH range, requires dissolved oxygen [6], contrary to the situation for chromates and nitrites. In this case, dissolved oxygen may help create just enough additional cathodic area to ensure anodic passivation of the remaining restricted anode surface at the prevailing rate of reduction of MoO_4^{2-} or of WO_4^{2-}, whereas in the absence of O_2, $i_{critical}$ is not achieved.

Sodium benzoate [6, 7] (C_6H_5COONa), sodium cinnamate [8] ($C_6H_5 \cdot CH \cdot CH \cdot COONa$), and sodium polyphosphate [9, 10] $(NaPO_3)_n$ (Fig. 17.2) are further examples of nonoxidizing compounds that effectively passivate iron in the near-neutral range, apparently through facilitating the adsorption of dissolved oxygen. As little as $5 \times 10^{-4} M$ sodium benzoate (0.007%) effectively inhibits steel in aerated distilled water [11], but inhibition is not observed in deaerated water. The steady-state corrosion rate of iron in aerated $0.01 M$ (0.14%) sodium benzoate, pH 6.8, is only 0.001 gmd, whereas in deaerated solution the rate is 0.073 gmd. Inhibition occurs only above about pH 5.5; below this value the hydrogen evolution reaction, for which benzoate ions have no inhibiting effect, presumably becomes dominant, and the passive film of oxygen is destroyed.

Gatos [12] found that optimum inhibition of steel in water of pH 7.5 containing 17 ppm NaCl occurred at and above 0.05% sodium benzoate or 0.2% sodium cinnamate. By using radioactive ^{14}C as tracer, only 0.07, 0.12, and 0.16 monomolecular layer of benzoate [$0.25 nm^2$, roughness factor 3] was found on a steel surface exposed 24 h to 0.1, 0.3, and 0.5% sodium benzoate solutions, respectively, and inhibition was observed in all the solutions. Clearly, a very small amount of benzoate on the metal surface increases adsorption of oxygen, or, alternatively, decreases reduction of oxygen at cathodic areas. This effect is specific to the cathodic areas of iron because iron continues to corrode in 0.5% sodium benzoate when coupled to gold, on which the reduction of oxygen is apparently not retarded.

The mechanism of inhibition in the case of sodium polyphosphate solutions may depend in part on the ability of polyphosphates to interfere with oxygen reduction on iron surfaces, making it easier for dissolved oxygen to adsorb and, thereby, to induce passivity. Other factors enter as well; there is, for example, evidence of protective film formation of the diffusion-barrier type on cathodic

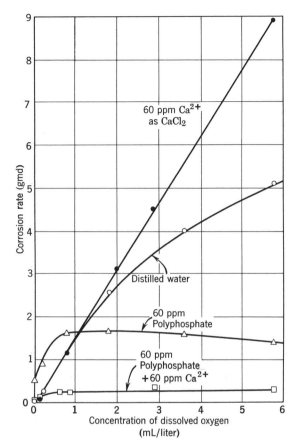

Figure 17.2. Effect of oxygen concentration on sodium polyphosphate as a corrosion inhibitor of iron showing beneficial effect of dissolved O_2 and Ca^{2+}, 48-h test, 25°C [9]. (Reproduced with permission. Copyright 1955, The Electrochemical Society.)

areas, and such diffusion-barrier films probably account for the observed inhibition of steel exposed to as high as 2.5% NaCl solutions containing several hundred parts per million calcium polyphosphate [13]. At low concentration of dissolved oxygen, corrosion of iron is accelerated by sodium polyphosphate because of its metal-ion-complexing properties (Fig. 17.2). Calcium, iron, and zinc polyphosphates are better inhibitors than the sodium compound.

In line with the theory of passivators just described, transition metals are those expected and found to be inhibited best by passivators; their anodic polarization curves have the shape shown in Fig. 17.1, allowing passivity to be established and then maintained at low current densities. A lesser degree of inhibition can be obtained with the nontransition metals, such as Mg, Cu, Zn, and Pb, using, for example, chromates. Protection of these metals apparently results largely from formation of relatively thick diffusion-barrier films of insoluble metal

TABLE 17.1. Effect of Chromate Concentration, Chlorides, and Temperature on Corrosion of Mild Steel [14]

	Velocity of Specimen: 0.37 m/s; 14-day tests			
$Na_2Cr_2O_7 \cdot 2H_2O$ (g/liter) →	0	0.1	0.5	1.0
% NaCl Temperature (°C)	Corrosion Rate (mm/y)			
0 20	0.539	0.002	0.002	0.001
75	0.905	0.356[a]	0.010	0.004
95	0.430	0.266[a]	0.011	0.000
0.002 20	0.668	0.015	0.001	0.001
75	1.705	0.120[a]	0.004	0.000
95	0.529	0.441[a]	0.131[a]	0.007
0.05 20	0.785	0.030	0.037	0.020
75	2.158	0.055	0.066	0.039
95	0.579	0.178[a]	0.120[a]	0.062
3.5 20	0.606	0.042	0.040	0.039
22.0 20	0.185	0.022	0.014	0.034

[a]Pitted.

chromates mixed with oxides. There is also the possibility that adsorption of CrO_4^{2-} on the metal surface contributes in some degree to the lower reaction rate by decreasing the exchange current density for the reaction $M \rightarrow M^{2+} + 2e^-$.

An inhibiting mechanism similar to that for nontransition metals in contact with passivators probably also applies to steel in concentrated refrigerating brines (NaCl or $CaCl_2$) to which chromates are added as inhibitors (approximately 1.5–3.0 g $Na_2Cr_2O_7$/liter adjusted with NaOH to form CrO_4^{2-}). In the presence of so large a Cl^- concentration, passivity of the kind discussed under Definition 1 (Section 6.1) does not take place. The reduction in corrosion rate is not as pronounced as when chlorides are absent [14] (see Table 17.1), and any reduction that occurs apparently results from formation of a surface diffusion barrier of chromate reduction products and iron oxides. Chromates are not adequate inhibitors for the hot concentrated brine solutions that, in the past, were sometimes mistakenly proposed as antifreeze solutions for engine cooling systems.

17.2.2 Applications of Passivators

Chromates are applied mostly as inhibitors for recirculating cooling waters (e.g., of internal combustion engines, rectifiers, and cooling towers). The concentration of Na_2CrO_4 used for this purpose is about 0.04–0.2%, with the higher concentrations being employed at higher temperatures or in fresh waters of chloride concentration above 10 ppm. The pH is adjusted, if necessary, to 7.5–9.5 by addition of NaOH. Periodic analysis (colorimetric) is required to ensure that the concentration remains above the critical level ($10^{-3} M$ or 0.016% Na_2CrO_4 at room temperature). Combinations of chromates with polyphosphates or other inhibitors may permit the concentration of chromates to fall below the critical level. This reduction in chromate concentration results in some sacrifice of inhibiting efficiency, but there may be adequate protection against pitting for the treatment of very large volumes of water employing cooling towers [15].

Chromates (Cr^{6+}) are toxic and carcinogenic and cause a rash on prolonged contact with the skin, and so they must be used with caution and with full regard to disposal requirements.

Corrosion rates of mild steel as a function of chromate and chloride concentration at various temperatures are shown in Table 17.1 [14].

Nitrites are used as inhibitors for antifreeze cooling waters (see Section 18.2.1) because, unlike chromates, they have little tendency to react with alcohols or ethylene glycol. They are not so well-suited to cooling tower waters because they are gradually decomposed by bacteria [16]. They are used to inhibit cutting oil–water emulsions employed in the machining of metals (0.1–0.2%). In pipelines transporting gasoline and other petroleum products, where water is a very minor phase, sufficient nitrite or chromate solution may be continuously injected to give a 2% concentration in the water phase [17]. In this connection, gasoline is corrosive to steel because, on reaching lower temperatures underground, it releases dissolved water which, in contact with large quantities of oxygen dissolved in the gasoline (solubility of O_2 in gasoline is six times that in water), corrodes steel, forming voluminous rust products that may clog the line. Sodium nitrite enters the water phase and effectively inhibits rusting. Chromates used for the same purpose have the disadvantage that they tend to react with some constituents of the gasoline.

The corrosion rates of steel in contact with water–gasoline mixtures containing increasing amounts of $NaNO_2$ are listed in Table 17.2 [18]. The minimum amount of $NaNO_2$ for effective inhibition is 0.06% or $7 \times 10^{-3} M$ which, because of impurities present in the water, is higher than the critical concentration in distilled water. Nitrites are inhibitors only above about pH 6.0. In more acid environments, they decompose, forming volatile nitric oxide and nitrogen

TABLE 17.2. Corrosion Rates of Mild Steel in Sodium Nitrite Solutions Containing Gasoline [18]

Rotating Bottle Tests Using Pipeline Water, pH 9, and Gasoline; 14-Day Exposure, Room Temperature	
% $NaNO_2$	Corrosion Rate (mm/y)
0.0	0.11
0.02	0.08
0.04	0.02
0.06	0.0
0.10	0.0

TABLE 17.3. Critical Concentrations of NaCl or Na_2SO_4 above which Pitting of Armco Iron in Chromate or Nitrite Solutions Occurs [11][a]

5-Day Tests, 25°C, Stagnant Solutions			
		Critical Concentrations	
		NaCl	Na_2SO_4
Na_2CrO_4	200 ppm	12 ppm	55 ppm
	500	30	120
$NaNO_2$	50 ppm	210	20
	100	460	55
	500	>2000	450

[a]See also Refs. 4 and 5.

peroxide. In common with other passivators, they tend to induce pitting at concentrations near the critical in presence of Cl^- or SO_4^{2-} ions. In this regard, nitrates are less sensitive to Cl^- than to SO_4^{2-}, contrary to the situation for chromates [3–5] (Table 17.3).

17.3 PICKLING INHIBITORS

Most pickling inhibitors function by forming an adsorbed layer on the metal surface, probably no more than a monolayer in thickness, which essentially blocks discharge of H^+ and dissolution of metal ions. For example, both iodide and quinoline are reported to inhibit corrosion of iron in hydrochloric acid by this mechanism [19]. Some inhibitors block the cathodic reaction (raise hydrogen overpotential) more than the anodic reaction, or vice versa; but adsorption appears to be general over all the surface rather than at specific anodic or cathodic sites, and both reactions tend to be retarded. Hence, on addition of an inhibitor to an acid, the corrosion potential of steel is not greatly altered (<0.1 V), although the corrosion rate may be appreciably reduced (Fig. 17.3).

Compounds serving as pickling inhibitors require, by and large, a favorable polar group or groups by which the molecule can attach itself to the metal surface. These include the organic N, amine, S, and OH groups. The size, orientation, shape, and electric charge of the molecule play a part in the effectiveness of inhibition. For example, the corrosion of iron in $1N$ hydrochloric acid was found to be inhibited by derivatives of thioglycolic acid and 3-mercaptopropionic acid to an extent that varied systematically with chain length [20]. Whether a compound adsorbs on a given metal and the relative strength of the adsorbed bond often depend on factors such as surface charge of the metal [21]. For inhibitors that adsorb better at increasingly active potential as measured from the point of zero surface charge (potential of minimum ionic adsorption), cathodic polar-

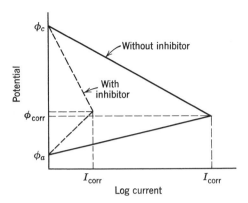

Figure 17.3. Polarization diagram for steel corroding in pickling acid with and without inhibitor.

ization in presence of the inhibitor provides better protection than either the equivalent cathodic protection or use of an inhibitor alone. Antropov [22] demonstrated this for iron and zinc in sulfuric acid containing various organic inhibitors.

The anion of the pickling acid may also take part in the adsorbed film or double-layer structure, accounting for differing efficiencies of inhibition for the same compound in HCl compared to H_2SO_4. For example, the corrosion rate of steel at 20°C inhibited with 20 g/liter quinoline is 26 gmd in $2N$ H_2SO_4, but only 4.8 gmd in $2N$ HCl, whereas the corrosion rates in the absence of inhibitor are 36 and 24 gmd, respectively [23]. In addition, specific electronic interaction of polar groups with the metal (chemisorption) may account for a given compound being a good inhibitor for iron, but not for zinc, or vice versa. The latter factor in certain cases may be more important than the steric factor (diffusion-barrier properties) of a closely packed, oriented layer of high-molecular-weight molecules. This is shown by the outstanding inhibition provided by the simple molecule carbon monoxide, CO, dissolved in HCl to which 18–8 stainless steel is exposed [24]; in $6.3N$ HCl at 25°C, the inhibition efficiency is 99.8%, where the inhibition efficiency is defined as

$$\text{Percent inhibition efficiency} = (\text{rate}_{\text{no inhib}} - \text{rate}_{\text{with inhib}}) \times 100/\text{rate}_{\text{no inhib}} \quad (17.1)$$

Very effective inhibition of corrosion of iron is also provided by a small amount of iodide in dilute H_2SO_4 [25]. Both CO and iodide chemisorb on the metal surface, interfering mostly with the anodic reaction [26]. Kaesche [27] showed that $10^{-3}\,M$ KI is a much more effective inhibitor for iron in $0.5M$ Na_2SO_4 solution at pH 1 (89% efficient) compared to pH 2.5 (17% efficient), indicating that adsorption of iodide, particularly in this pH range, is pH-dependent.

The interaction of factors just mentioned, plus perhaps others, enter into accounting for the fact that some compounds (e.g., *o*-tolylthiourea [28] in 5%

H_2SO_4) are better inhibitors at elevated temperatures than at room temperature, presumably because adsorption increases or the film structure becomes more favorable at higher temperatures. Others (e.g., quinoline derivatives) are more efficient in the lower-temperature range [28].

Typical effective organic pickling inhibitors for steel are quinolinethiodide, o- and p-tolylthiourea, propyl sulfide, diamyl amine, formaldehyde, and p-thiocresol.

Inhibitors containing sulfur, although effective otherwise, sometimes induce hydrogen embrittlement of steel in the event that the compound itself or hydrolysis products (e.g., H_2S) are formed that favor entrance of H atoms into the metal (see Section 5.5). In principle, inhibitors containing arsenic or phosphorus can behave similarly.

17.3.1 Applications of Pickling Inhibitors

Pickling inhibitors used in practice are seldom pure compounds. They are usually mixtures, which may be a byproduct, for example, of some industrial chemical process for which the active constituent is unknown. They are added to a pickling acid in small concentration, usually on the order of 0.01–0.1%. The typical effect of concentration of an inhibitor on the reaction between steel and 5% H_2SO_4 is illustrated [28] in Fig. 17.4, showing that, above a relatively low concentration, presumably that necessary to form an adsorbed monolayer, an additional amount of inhibitor has little effect on further reducing the rate.

Inhibitors are commonly used in the acid pickling of hot-rolled steel products in order to remove mill scale. The advantages of using an inhibitor for this purpose are (1) saving of steel, (2) saving of acid, and (3) reduction of acid fumes

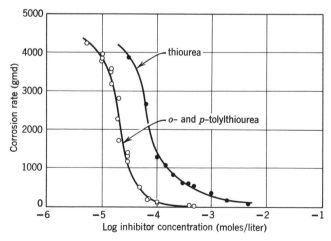

Figure 17.4. Effect of inhibitor concentration on corrosion of 0.1% C steel in 5% H_2SO_4, 70°C [28].

caused by hydrogen evolution. Inhibited dilute sulfuric or hydrochloric acid is also used to clean out steel water pipes clogged with rust, to clean boiler tubes encrusted with $CaCO_3$ or iron oxide scales, and to activate oil wells underground, the inhibitor protecting steel oil-well tubing. For example, boiler scale can be removed by using 0.1% hexamethylene tetramine in 10% HCl at a maximum temperature of 70°C (160°F) [29].

17.4 SLUSHING COMPOUNDS

Slushing compounds are used to protect steel surfaces temporarily from rusting during shipment or storage. They consist of oils, greases, or waxes containing small amounts of organic additives. The latter are polar compounds that adsorb on the metal surface in the form of a closely packed oriented layer. In this respect, the mechanism of inhibition by organic additives is similar to that of inhibition by pickling inhibitors, except that for use as slushing compounds, additives must suitably adsorb in the near-neutral range of pH, whereas pickling inhibitors adsorb best in the low pH range, calling for somewhat different properties.

Whether an oil or a wax is chosen as the vehicle depends on (1) the relative length of time protection is desired, the wax usually providing longer life and (2) the ease of removal before the protected machine part is put into service, with an oil being easier to wipe off or dissolve in solvents. The thickness of an applied coat varies from 0.1 mm to more than 2.5 mm (5 to more than 100 mils).

Suitable organic additives for use in slushing compounds include organic amines, zinc naphthenate, various oxidation products of petroleum, alkali and alkaline-earth metal salts of sulfonated oils, and various other compounds [30]. A substance that has been used successfully is lanolin, obtained from wool scouring; its active constituents are various high-molecular-weight fatty alcohols and acids. Sometimes, lead soaps are added to slushing compounds, with these soaps reacting to form relatively insoluble $PbCl_2$ with any NaCl from perspiration transferred to steel surfaces through handling.

17.5 VAPOR-PHASE INHIBITORS

Substances of low but significant vapor pressure, the vapor of which has corrosion-inhibiting properties, are called vapor-phase inhibitors. They are used to protect critical machine parts (e.g., ball bearings or other manufactured steel articles) temporarily against rusting by moisture during shipping or storage. They have the advantage over slushing compounds of easy application, with the possibility of immediate use of the protected article without first removing a residual oil or grease film. They have the disadvantgage of accelerating the corrosion of some nonferrous metals, discoloring some plastics, and requiring relatively effective sealing of a package against loss of the inhibiting vapor. The latter requirement is relatively easily achieved, however, by using wrapping paper impregnated on the inside surface with the inhibitor and incorporating a vapor-barrier coating on the outside.

The mechanism of inhibition has not been studied in detail, but it appears to be one of adsorbed film formation on the metal surface that provides protection against water or oxygen, or both. In the case of volatile nitrites, the inhibitor may also supply a certain amount of NO_2^- that passivates the surface.

Detailed data have been presented for dicyclohexylammonium nitrite [31], which is one of the most effective of the vapor-phase inhibitors. This substance is white, crystalline, almost odorless, and relatively nontoxic. It has a vapor pressure of 0.0001 mmHg at 21°C (70°F), which is about one-tenth the vapor pressure of mercury itself.* One gram saturates about 550 m³ (20,000 ft³) of air, rendering the air relatively noncorrosive to steel. The compound decomposes slowly; nevertheless, in properly packaged paper containers at room temperature, it effectively inhibits corrosion of steel over a period of years. However, it should be used with caution in contact with nonferrous metals. In particular, corrosion of zinc, magnesium, and cadmium is accelerated.

Cyclohexylamine carbonate has the somewhat higher vapor pressure of 0.4 mmHg at 25°C and its vapor also effectively inhibits steel [32]. The higher vapor pressure provides more rapid inhibition of steel surfaces either during packaging or on opening and again on closing a package, during which time the concentration of vapor may fall below that required for protection. The vapor is stated to reduce corrosion of aluminum, solder, and zinc, but it has no inhibiting effect on cadmium, and it increases corrosion of copper, brass, and magnesium.

Ethanolamine carbonate and various other compounds have also been described as vapor-phase inhibitors [23, 32]. A combination of urea and sodium nitrite has found practical application, including use in impregnated paper. The mixture probably reacts in the presence of moisture to form ammonium nitrite, which is volatile, although unstable, and conveys inhibiting nitrite ions to the metal surface.

17.5.1 Inhibitor to Reduce Tarnishing of Copper

Benzotriazole (BTA), $C_6H_5N_3$, is a corrosion inhibitor widely used for copper and its alloys. Typical protective films formed on copper are reported to consist of various multilayers. Film thickness is normally controlled by the pH of the solution, ranging from less than 0.01 μm in near-neutral and mildly basic solutions to 0.5 μm in acidic solutions [33]. For a wide variety of conditions [34], this inhibitor is reported to form a Cu^+ surface complex following its adsorption on Cu_2O that is present on copper surfaces.

---

*I. Rosenfeld et al. (*Symposium sur les Inhibiteurs de Corrosion*, University of Ferrara, Italy, 1961, p. 344) report the lower value of 0.00001 mmHg obtained for dicyclohexylammonium nitrite carefully purified by multiple recrystallization from alcohol.

REFERENCES

1. H. Uhlig and P. King, *J. Electrochem. Soc.* **106**, 1 (1959).
2. W. Robertson, *J. Electrochem. Soc.* **98**, 94 (1951).
3. S. Matsuda and H. Uhlig, *J. Electrochem. Soc.* **111**, 156 (1964).
4. A. Mercer and I. Jenkins, *Br. Corros. J.* **3**, 130 (1968).
5. A. Mercer, I. Jenkins, and J. Rhoades-Brown, *Br. Corros. J.* **3**, 136 (1968).
6. M. Pryor and M. Cohen, *J. Electrochem. Soc.* **100**, 203 (1953).
7. D. Brasher and A. Mercer, *Br. Corros. J.* **3**, 120 (1968).
8. F. Wormwell, *Chem. Ind. (London)* **1953**, 556.
9. H. Uhlig, D. Triadis, and M. Stern, *J. Electrochem. Soc.* **102**, 59 (1955).
10. G. Hatch, *Mater. Prot.* **8**, 31 (1969).
11. S. Matsuda, D. Sc. thesis, Department of Metallurgy, M.I.T., Cambridge, MA, 1960.
12. H. Gatos, in *Symposium sur les Inhibiteurs de Corrosion*, University of Ferrara, Italy, 1961, p. 257.
13. G. Butler and D. Owen, *Corros. Sci.* **9**, 603 (1969).
14. B. Roetheli and G. Cox, *Ind. Eng. Chem.* **23**, 1084 (1931).
15. H. Kahler and P. Gaughan, *Ind. Eng. Chem.* **44**, 1770 (1952).
16. T. P. Hoar, *Corrosion* **14**, 103t (1958).
17. J. Bregman, Inhibition in the petroleum industry, in *Proceedings of the Third European Symposium on Corrosion Inhibitors*, University of Ferrara, Italy, 1970, p. 365.
18. A. Wachter and S. Smith, *Ind. Eng. Chem.* **35**, 358 (1943).
19. K. Rajagopalan and G. Venkatachari, *Corrosion* **36**, 320 (1980).
20. M. Carroll, K. Travis, and J. Noggle, *Corrosion* **31**, 123 (1975).
21. L. Antropov, *Kinetics of Electrode Processes and Null Points of Metals*, Council of Scientific & Industrial Research, New Delhi, 1960, pp. 48–82.
22. L. Antropov, Inhibitors of metallic corrosion and the phi-scale of potentials, *First International Congress on Metallic Corrosion*, Butterworths, London, 1961, p. 147.
23. I. Putilova, S. Balezin, and V. Barannik, *Metallic Corrosion Inhibitors*, G. Ryback, translator, Pergamon Press, New York, 1960, p. 63.
24. H. Uhlig, *Ind. Eng. Chem.* **32**, 1490 (1940).
25. K. Hager and M. Rosenthal, *Ordnance* **35**, 479 (1951).
26. K. Heusler and G. Cartledge, *J. Electrochem. Soc.* **108**, 732 (1961).
27. H. Kaesche, in *Symposium sur les Inhibiteurs de Corrosion*, University of Ferrara, Italy, 1961, p. 137.
28. T. Hoar and R. Holliday, *J. Appl. Chem.* **3**, 502 (1953).
29. W. Cerna, *Proceedings of the Seventh Annual Water Conference*, Pittsburgh, PA, 1947.
30. H. Baker and W. Zisman, *Ind. Eng. Chem.* **40**, 2338 (1948).
31. A. Wachter, T. Skei, and N. Stillman, *Corrosion* **7**, 284 (1951).
32. E. Stroud and W. Vernon, *J. Appl. Chem.* **2**, 178 (1952).
33. I. Ogle and G. Poling, *Can. Metall. Q.* **14**, 37 (1975).
34. D. Chadwick and T. Hashemi, *Corros. Sci.* **18**, 39 (1978).

GENERAL REFERENCES

B. P. Boffardi, Corrosion inhibitors in the water treatment industry, in *ASM Handbook*, Vol. 13A, *Corrosion: Fundamentals, Testing, and Protection*, ASM International, Materials Park, OH, 2003, pp. 891–906.

R. L. Martin, Corrosion inhibitors for oil and gas production, in *ASM Handbook*, Vol. 13A, *Corrosion: Fundamentals, Testing, and Protection*, ASM International, Materials Park, OH, 2003, pp. 878–886.

S. Papavinasam, Corrosion inhibitors, in *Uhlig's Corrosion Handbook*, 2nd edition, R. W. Revie, editor, Wiley, New York, 2000, pp. 1089–1105.

S. Papavinasam, Evaluation and selection of corrosion inhibitors, in *Uhlig's Corrosion Handbook*, 2nd edition, R. W. Revie, editor, Wiley, New York, 2000, pp. 1169–1178.

J. D. Poindexter, Corrosion inhibitors for crude oil refineries, in *ASM Handbook*, Vol. 13A, *Corrosion: Fundamentals, Testing, and Protection*, ASM International, Materials Park, OH, 2003, pp. 887–890.

V. S. Sastri, *Corrosion Inhibitors: Principles and Applications*, Wiley, Chichester, 1998.

K. L. Vasanth, Vapor phase corrosion inhibitors, in *ASM Handbook*, Vol. 13A, *Corrosion: Fundamentals, Testing, and Protection*, ASM International, Materials Park, OH, 2003, pp. 871–877.

18

TREATMENT OF WATER AND STEAM SYSTEMS

18.1 DEAERATION AND DEACTIVATION

In accord with principles described in Sections 7.1 and 7.2, corrosion of iron is negligible at ordinary temperatures in water that is free of dissolved oxygen. An effective practical means, consequently, for reducing corrosion of iron or steel in contact with fresh water or seawater is to reduce the dissolved oxygen content. In this way, corrosion of copper, brass, zinc, and lead is also minimized.

Removal of dissolved oxygen from water is accomplished either by chemically reacting oxygen beforehand, called *deactivation*, or by distilling it off in suitable equipment, called *deaeration*. Deactivation can be carried out in practice by slowly flowing hot water over a large surface of steel laths or sheet contained in a closed tank called a deactivator. The water remains in contact long enough to corrode the steel, and, by this process, most of the dissolved oxygen is consumed. Subsequent filtration removes suspended rust. Water so treated is much less corrosive to a metal pipe distribution system. Deactivators of this kind have been employed in some buildings, but, since use of deactivation requires regular attention in addition to periodic renewal of the steel sheet, this approach to

Corrosion and Corrosion Control, by R. Winston Revie and Herbert H. Uhlig
Copyright © 2008 John Wiley & Sons, Inc.

oxygen removal is usually too cumbersome compared with the use of chemical inhibitors or deaeration.

Deactivation of industrial waters (not potable waters, because the chemicals used are toxic) is possible by using *oxygen scavengers* (strong reducing agents), such as sodium sulfite (Na_2SO_3) and hydrazine (N_2H_2), which function as corrosion inhibitors by removing the oxygen. Since hydrazine has been identified as a possible carcinogen, there have been considerable efforts to replace it.

Sodium sulfite scavenges oxygen in accord with the reaction

$$Na_2SO_2 + \frac{1}{2}O_2 \rightarrow Na_2SO_4 \qquad (18.1)$$

for which Na_2SO_3 reacts with oxygen in the weight ratio of 8:1 (8 kg Na_2SO_3 to 1 kg O_2). The reaction is relatively fast at elevated temperatures, but slow at ordinary temperatures. It can be accelerated by adding catalysts, such as Cu^{2+} or Co^{2+}, salts [1, 2].

The rapid decrease with time of dissolved oxygen in the San Joaquim, California, river water after treatment with 80 ppm Na_2SO_3 (0.67 lb/1000 gal) plus copper or cobalt salts is shown in Fig. 18.1. Water so treated using $CoCl_2$ as catalyst was found by Pye [1] to be noncorrosive to a steel heat-exchanger system that, without treatment, had previously suffered serious corrosion and loss of heat transfer. Tests showed a reduction in corrosion rate from 0.2 mm/y (0.008 ipy, pitting factor = 7.4) before treatment to 0.004 mm/y (0.00016 ipy) afterward.

Hydrazine (N_2H_4), supplied as a concentrated aqueous solution, also reacts with dissolved oxygen, according to

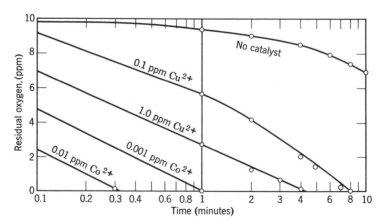

Figure 18.1. Effect of cobalt and copper salts on reaction rate of Na_2SO_3 with dissolved oxygen at room temperature [1]. [Reprinted from *Journal AWWA* **39** (11) (November 1947), by permission. Copyright 1947, American Water Works Association.]

$$N_2H_4 + O_2 \rightarrow N_2 + 2H_2O \tag{18.2}$$

in the weight ratio of $1:1$. This reaction is slow at ordinary temperatures, but can be accelerated by using catalysts (e.g., activated charcoal, metal oxides, alkaline solutions [3] of Cu^{2+} and Mn^{2+}) and by raising the temperature. At elevated temperatures, decomposition occurs slowly at 175 °C (345 °F) and more rapidly at 300 °C (570 °F), producing ammonia [4]:

$$3N_2H_4 \rightarrow N_2 + 4NH_3 \tag{18.3}$$

The normal reaction products—nitrogen, water, and a small amount of NH_3—are all volatile, and, unlike sulfite addition, no dissolved solids accumulate in the treated water.

Two additional oxygen scavengers are carbohydrazide, $(NH_2NH)_2CO$, and diethylhydroxylamine (DEHA), $(C_2H_5)_2NOH$; the reactions with oxygen are

$$(NH_2NH)_2CO + 2O_2 \rightarrow 2N_2 + 3H_2O + CO_2 \tag{18.4}$$

and

$$4(C_2H_5)_2NOH + 9O_2 \rightarrow 8CH_3COOH + 2N_2 + 6H_2O \tag{18.5}$$

Both these oxygen scavengers have the additional advantage that they form protective films on both iron and copper [5].

Deaeration is accomplished by spraying water or flowing it over a large surface, countercurrent to steam. Oxygen distills off and also some dissolved carbon dioxide (Fig. 18.2). Water is heated in the process and is suitable, therefore, as feedwater for boilers. Steam deaerators of this kind are standard equipment for all high-pressure stationary boilers. On the other hand, if the water is to be used cold, dissolved gases are distilled off by lowering the pressure, employing a mechanical pump or steam ejector instead of a countercurrent flow of steam. This is called vacuum deaeration. Equipment of this kind has been designed to deaerate several million gallons of water per day.

In principle, it is more difficult and more expensive to remove the last traces of dissolved oxygen by distillation compared to the first 90–95%, and it is more difficult at low temperatures than at high temperatures. To achieve a low enough oxygen level in cold water, it is often necessary to use multiple-stage vacuum treatment. Fortunately, acceptable levels of dissolved oxygen for corrosion control in cold water are higher than in hot water or in steam. Allowable levels established through experience are given in Table 18.1 [6].

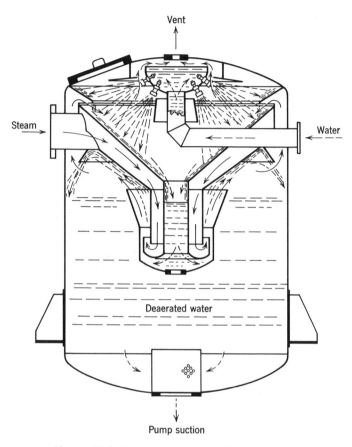

Figure 18.2. Sketch of one type of steam deaerator.

TABLE 18.1. Approximate Allowable Oxygen Concentration in Deaerated Water for Corrosion Control in Steel Systems

	Maximum Oxygen Concentration	
	ppm	mL/liter
Cold water	0.3	0.2
Hot water	0.1	0.07
Low-pressure boilers (<1.7 MPa, <250 psi)	0.03	0.02
High-pressure boilers	0.005	0.0035

18.2 HOT- AND COLD-WATER TREATMENT

1. *Hot-Water Heating Systems.* These are closed steel systems in which the initial corrosion of the system soon consumes dissolved oxygen; corrosion is negligible once the dissolved oxygen is consumed. A continuing minor reaction of steel with water produces hydrogen, with a characteristic odor due to traces of hydrocarbon gases originating from the reaction of carbides in steel with water. The hydrogen reaction can be minimized by treating the water with NaOH or Na_3PO_4 to a pH of 8.5.

2. *Municipal Water Supplies.* Usually, hard waters of positive saturation index are relatively noncorrosive and do not require treatment of any kind for corrosion control. Soft waters, on the other hand, cause rapid accumulation of rust in ferrous piping, are readily contaminated with toxic quantities of lead salts on passing through lead piping, and cause blue staining of bathroom fixtures by copper salts originating from slight corrosion of copper or brass piping.

Chemical treatment of potable waters is limited to small concentrations of inexpensive, nontoxic chemicals, such as polyphosphates, orthophosphates, and silicates. Zinc polyphosphate has been used for more than 60 years in controlling corrosion of steel in municipal water systems. Zinc orthophosphate is effective in controlling corrosion of steel, copper, and lead.

Raising the saturation index is a potentially useful means for reducing the corrosion rate in either flowing or stagnant portions of the distribution system, and it is also effective for reducing corrosion of hot-water systems. This treatment requires addition of lime $[Ca(OH)_2]$ or both lime and soda ash (Na_2CO_3) to the water in amounts that raise the saturation index to about +0.5 (see Section 7.2.6.1). For the treatment to be successful, the water must be low in colloidal matter and in dissolved solids other than calcium salts. Corrosion of copper, lead, and brass is also reduced by this treatment. In hot-water systems, the possibility of excess deposition of $CaCO_3$, which causes scaling, must be taken into account in arriving at the proper proportions of added chemicals.

Sodium silicate treatment in the amount of about 4–15 ppm SiO_2 is sometimes used by individual owners of buildings in soft-water areas. This treatment reduces "red water" caused by suspended rust resulting from corrosion of ferrous piping, and it also eliminates blue staining by water that has passed through copper or brass piping.

The conditions under which protection exists or is optimum are not entirely understood, but it is clear that dissolved calcium and magnesium salts have an effect and that some protection may result alone from the alkaline properties of sodium silicate. In the presence of silicate, passivity of iron may be observed at pH 10 with an accompanying reduction of the corrosion rate to 0.1–0.7 gmd [7]. Sodium hydroxide induces similar passivity and corresponding low corrosion rates at the somewhat higher pH range of 10–11. Under other conditions (e.g., pH 8), a protective diffusion-barrier film is formed, apparently containing SiO_2

and consisting perhaps of an insoluble iron silicate. Laboratory tests in distilled water at 25 °C showed a reduction in the corrosion rate of iron on the order of 85–90% when sodium silicate was added to bring the pH to 8 [7]. However, no inhibition was obtained in the laboratory at the same SiO_2 content using Cambridge tap water (pH 8.3, 44 ppm Ca, 10 ppm Mg, 16 ppm Cl^-). If larger quantitites of sodium silicate were added to raise the pH of the water to 10 or 11, the range in which passivity of iron occurs, a marked decrease in the corrosion rate was observed.

Domestic or industrial hot-water heaters of galvanized steel through which hot aerated water passes continuously are not protected reliably in all types of water by nontoxic chemical additions such as silicates or polyphosphates. Adjustment of the saturation index to a more positive value, as discussed earlier, is sometimes helpful. Often, cathodic protection or use of nonferrous metals, such as copper or 70% Ni–Cu (Monel), is the best or only practical measure.

18.2.1 Cooling Waters

Once-through cooling waters (usually obtained from rivers, lakes, or wells) usually cannot be treated chemically, both because of the large quantities of inhibitors required and because of the problem of water pollution. Sometimes, additions of about 2–5 ppm sodium or calcium polyphosphate are made to help reduce corrosion of steel equipment. In such small concentrations, polyphosphates are not toxic, but water disposal may continue to be a problem because of the need to avoid accumulation of phosphates in rivers and lakes. Adjustment of the saturation index to a more positive value is sometimes a practical possibility. Otherwise, a protective coating or metals more corrosion resistant than steel must be used.

Recirculating cooling waters, as for engine-cooling systems, may be treated with sodium chromate (Na_2CrO_4) in the amount of 0.04–0.2% (or the equivalent amount of $Na_2Cr_2O_7 \cdot 2H_2O$ plus alkali to pH 8). Chromates inhibit corrosion of steel, copper, brass, aluminum, and soldered components of such systems. Since chromate is consumed slowly, additions must be made periodically in order to maintain the concentration above the critical. For diesel or other heavy-duty engines, 2000 ppm sodium chromate (0.2%) can be used to reduce damage by cavitation-erosion as well as by aqueous corrosion (see Section 7.2.5.1). Because of the toxicity of hexavalent chromium (Cr^{+6}), chromates must be handled, used, and disposed of appropriately and with regard to regulatory requirements.

Chromates cannot be used in the presence of antifreeze solutions, because of their tendency to react with organic substances. Many proprietary inhibitor mixtures are on the market that are usually dissolved beforehand in methanol or in ethylene glycol in order to simplify the packaging problem, but this also limits the available number of suitable inhibitors. In the United States, borax ($Na_2B_4O_7 \cdot 10H_2O$) is a common ingredient. Sulfonated oils, which produce an oily protective coating, and mercaptobenzothiazole, which specifically inhibits corrosion of copper and at the same time removes the accelerating influence of dissolved Cu^{2+} on corrosion of other portions of the system, are sometimes added to borax. One specification calls for a final concentration in the antifreeze solution of 1.7% borax, 0.1% mercaptobenzothiazole, and 0.06% Na_2HPO_4, with the latter being added specifically to protect aluminum. Ethanolamine phosphate is also used as an inhibitor for engine-cooling systems containing ethylene glycol.

For industrial water cooled by recirculation through a spray or tray-type tower, chromates are the most reliable from the standpoint of efficient inhibition. However, the critical concentration is high; and as the sulfate and chloride concentrations increase through evaporation of the water, chromates tend to cause pitting or may cause increased galvanic effects at dissimilar metal couples. Windage losses (loss of spray by wind) must be carefully avoided because chromates are toxic. Toxicity also makes it difficult to dispose of chromate solutions whenever it becomes necessary to reduce the concentration of accumulated chlorides and sulfates.

Because of pollution problems that can be caused by chromates, numerous inhibitor systems have been developed as alternatives—for example, organic phosphonic acids, which are effective in alkaline waters and are biodegradable [8]. Nontoxic inhibitor formulations containing mixtures of azoles and water-soluble phosphates (e.g., disodium phosphate and sodium tripolyphosphate) have been developed [9]. Sodium molybdate, which is less toxic than sodium chromate, has also been reported to be a useful component of inhibitor formulations for use in recirculating aqueous systems [9, 10]. Other nontoxic, chromate-free inhibitor formulations are based on mixtures of sorbitol, benzotriazole or tolytriazole, and water-soluble phosphates [9].

Sodium polyphosphate is often used in a concentration of about 10–100 ppm, sometimes with added zinc salts to improve inhibition. The pH value is adjusted to 5–6 in order to minimize pitting and tubercle formation, as well as scale deposition. Polyphosphates decompose slowly into orthophosphates, which, in the presence of calcium or magnesium ions, precipitate insoluble calcium or magnesium orthophosphate, causing scale formation on the warmer parts of the system. Unlike chromates, they favor algae growth, which necessitates the addition of algaecides to the water. Corrosion inhibition with polyphosphates is less effective than that by chromates, but polyphosphates in low concentration are not toxic, and the required optimum amount of inhibitor is less than that for chromates.

18.3 BOILER-WATER TREATMENT

18.3.1 Boiler Corrosion

Steam boilers are constructed according to various designs, but they consist essentially of a low-carbon steel or low-alloy steel container for water that is heated by hot gases. The steam may afterward pass to a superheater of higher-alloy steel at higher temperature than the boiler itself. For maximum heat transfer, a series of boiler tubes is usually used, with the hot gases passing either around the outer surface or, less frequently, through the inner surface of the tubes. The steam, after doing work or completing some other kind of service, eventually reaches a condenser that is usually constructed of copper-base alloy tubes. Steam is cooled on one side of the tubes by water passing along the opposite side of a quality ranging from fresh to polluted, brackish, or seawater. The condensed steam then returns to the boiler and the cycle repeats.

The history and causes of corrosion in power-station boilers have been reviewed by Mann [11]; case histories of corrosion and its prevention in industrial boilers have been presented by Frey [12].

Some boilers are equipped with an embrittlement detector by means of which the chemical treatment of a water can be evaluated continuously in terms of its potential ability to induce stress-corrosion cracking (Fig. 18.3) [13]. A specimen of plastically deformed boiler steel is stressed by setting a screw; adjustment of this screw regulates a slight leak of hot boiler water in the region where the specimen is subject to maximum tensile stress and where boiler water evaporates. A boiler water is considered to have no embrittling tendency if specimens do not crack within successive 30-, 60-, and 90-day tests. Observation of the detector is

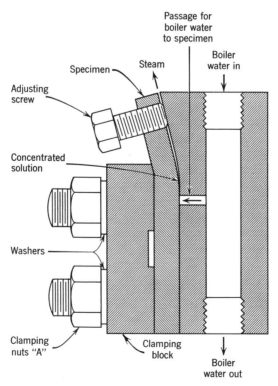

Figure 18.3. Embrittlement detector which, when attached to an operating boiler, detects tendency of a boiler water to induce stress-corrosion cracking.

a worthwhile safety measure because the tendency toward cracking is more pronounced in the plastically deformed test piece than in any portion of the welded
boiler; hence, water treatment can be corrected, if necessary, before the boiler
is damaged.

Corrosion of boiler and superheater tubes is sometimes a problem on the
hot combustion gas side, especially if vanadium-containing oils are used as fuel.
This matter is discussed in Sections 11.8 and 11.9. On the steam side, since
modern boiler practice ensures removal of dissolved oxygen from the feedwater,
a reaction occurs between H_2O and Fe, resulting in a protective film of magnetite
(Fe_3O_4) as follows:

$$3Fe + 4H_2O \rightarrow Fe_3O_4 + 4H_2 \tag{18.6}$$

The mechanism of this reaction, so far as it is understood, indicates that Fe_3O_4
is formed only below about 570 °C (1060 °F) and that above this temperature,
FeO forms instead. The latter then decomposes on cooling to a mixture of magnetite and iron in accord with

$$4FeO \rightarrow Fe_3O_4 + Fe \tag{18.7}$$

Measurements of hydrogen accumulation in boilers as a function of time, as well
as laboratory corrosion rate determinations, indicate that growth of the oxide
obeys the parabolic equation [14]. Hence, the rate is diffusion-controlled, in line
with the mechanism depending on ion and electron migration through solid reaction products as described in Chapter 11.

At lower temperatures (e.g., room temperature to about 100 °C) and probably at higher temperatures before a relatively thick surface film develops, experiments show that $Fe(OH)_2$ is the initial reaction product and not Fe_3O_4 [15]. The
mechanism of corrosion in this temperature range follows that described for
anode and cathode interaction on the plane of the metal surface in contact with
an electrolyte. Ferrous hydroxide eventually decomposes, at a rate depending on
temperature, into magnetite and hydrogen in accord with a reaction first described
by Schikorr (Schikorr reaction)[16]

$$3Fe(OH)_2 \rightarrow Fe_3O_4 + H_2 + 2H_2O \tag{18.8}$$

Any factors that disturb the protective magnetite layer on steel, either chemically or mechanically, induce a higher rate of reaction, usually in a localized
region, causing pitting or sometimes grooving of the boiler tubes. In this regard,
the specific damaging chemical factor of excess OH^- concentration is discussed
later; mechanical damage, on the other hand, may take place each time the
boiler is cooled down. Differential contraction of the oxide and metal causes
some degree of spalling of the oxide, thereby exposing fresh metal. Accordingly,
it is observed that rate of hydrogen evolution is momentarily high after a boiler
is started up again, with hydrogen production then falling to normal values

presumably after a reasonable thickness of oxide has again built up at damaged areas.

Conditions of boiler operation that lead to metal oxide or inorganic deposits (from the boiler itself or from condenser leakage) on the water side of boiler tubes cause local overheating accompanied by additional precipitation of solutes from the water. Pitting usually results, or so-called plug-type oxidation occurs, which accentuates local temperature rise, leading eventually to stress rupture of the tube. Furthermore, hydrogen resulting from the H_2O–Fe corrosion reaction may enter the steel causing decarburization, followed by microfissuring along grain boundaries and eventual blowout of the affected tube. The latter type of failure may take place without any major loss of tube wall thickness. In the absence of deposits within boiler tubes, these types of damage are not observed [17].

18.3.2 Boiler-Water Treatment for Corrosion Control

Feedwaters of boilers are chemically treated both to reduce corrosion of the boiler and auxiliary equipment and to reduce formation of inorganic deposits on the boiler tubes (scaling), which interfere with heat transfer. If steam is used for power production, concentrations of silica and silicates in feedwaters must also be reduced in order to minimize volatilization of SiO_2 with steam, causing formation of damaging deposits on turbine blades. Control of scale formation usually requires removing all calcium and magnesium salts by various means, including use of ion-exchange resins or adding substances to the water that favor precipitation of sludges rather than adherent continuous scales on the metal surface. Details are discussed in standard references on boiler-water treatment.

For corrosion control, the basic treatment consists of removal of dissolved gases, addition of alkali, and addition of inhibitors, as described in the following paragraphs.

1. *Removal of Dissolved Oxygen and Carbon Dioxide.* For high-pressure boilers, any remaining dissolved oxygen in the feedwater combines quantitatively with the metals of the boiler system, usually causing pitting of the boiler tubes and general attack elsewhere. Removal of oxygen is accomplished by steam deaeration of the water, followed by addition of a scavenger, such as sodium sulfite or hydrazine (see Section 18.1). Final oxygen concentration is usually held below values ($<0.005\,ppm\ O_2$) analyzable by chemical methods (e.g., the Winkler method).

Deaeration is accompanied by some reduction of carbon dioxide content, particularly if the water is acidified before the deaeration process to liberate carbonic acid from the dissolved carbonates. Carbonic acid is corrosive to steel in the absence of dissolved oxygen and more so in its presence [18], but addition of alkali to boiler water limits any corrosion caused by carbon dioxide to the boiler itself by converting dissolved carbon dioxide to carbonates. At prevailing boiler temperatures, however, carbonates dissociate as follows:

$$Na_2CO_3 + H_2O \rightarrow CO_2 + 2NaOH \qquad (18.9)$$

bringing hot carbonic acid into contact with the condenser and return-line systems. Steel return-line systems suffer serious corrosion, therefore, if the carbon dioxide content of boiler water is high. Soluble $FeCO_3$ is formed, and it returns with the condensate to the boiler, where it decomposes into $Fe(OH)_2 + CO_2$, with the carbon dioxide again being available for further corrosion. The copper-alloy condenser system also suffers corrosion should dissolved oxygen be present together with carbon dioxide, but attack of copper-base alloys is negligible in the absence of oxygen. Since carbon dioxide is not used up in the corrosion process, it accumulates increasingly in the boiler with each addition of feedwater unless an occasional blowdown (intentional release of some boiler water) is arranged.

2. *Addition of Alkali.* Alkali addition to boiler waters is standard practice for most high-pressure boilers in the United States and abroad. Feedwater for a high-pressure boiler is treated to a minimum pH (measured at room temperature) of 8.5 to minimize corrosion of steel, with the optimal pH range being 9.2–9.5. For copper alloys, the preferred pH range is 8.5–9.2. Since both steel and copper alloys are common in boiler systems, a compromise range of 8.8–9.2 is recommended [19].

It is apparent from Figure 18.4 that excess alkali can be damaging to a boiler in that the corrosion rate increases rapidly as pH is increased above 13. The

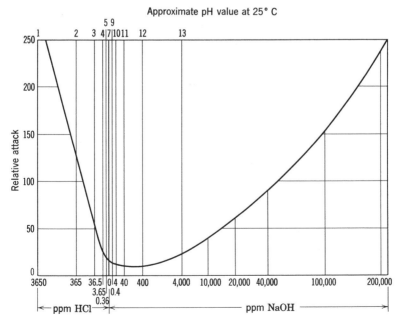

Figure 18.4. Corrosion of iron by water at 310 °C (590 °F) at various values of pH measured at 25 °C [E. Partridge and R. Hall, *Trans. Am. Soc. Mech. Eng.* **61**, 597 (1939)].

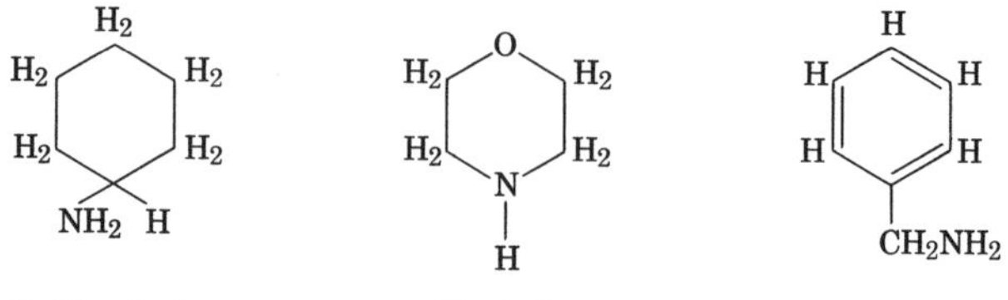

Figure 18.5. Structural formulas for three neutralizing-type amines.

danger is not so much that the initial pH of the boiler water may be too high as that accidental concentration of an alkaline boiler water at a crevice, such as is formed between riveted plates, at a weld, beneath a cracked oxide scale, or at a hot spot on a scaled tube surface, may reach OH$^-$ concentration levels above the safe range. For this reason, it is held advisable to add buffer ions, such as PO_4^{3-} (Na_3PO_4), which limit the increase in pH that a water can achieve no matter how concentrated it becomes. Such ions are also useful in avoiding similar high OH$^-$ concentrations that lead to stress-corrosion cracking of any portion of the boiler under high residual or applied stress. Goldstein and Burton [17] reported that 5–10 ppm phosphate at pH 9.5–10.0 was more effective in reducing corrosion of high-pressure boiler tubes under a variety of operating conditions than either NaOH or NH_3 treatment.

3. *Addition of Inhibitors.* It is possible to add inhibitors for controlling two kinds of corrosion in boiler systems, namely, stress-corrosion cracking and return-line corrosion. The first can be minimized by addition of phosphates, as mentioned previously.

Corrosion caused by dissolved carbon dioxide in steam condensate is minimized by adding a volatile amine to the boiler water. There are two categories of volatile amines used for this purpose: (1) neutralizing amines and (2) filming amines. The first group includes cyclohexylamine, benzylamine, and morpholine (Fig. 18.5). When any of these is added to boiler water in sufficient amount, it neutralizes carbonic acid and raises the pH of steam condensate to an alkaline value, thereby making the condensate less corrosive. Operating by a different principle, volatile octadecylamine, hexadecylamine, and dioctadecylamine are typical of the filming-type inhibitors, which protect against corrosion by building up a protective organic film on the condenser surface. The filming amines more nearly fit the definition of an inhibitor, whereas the other amines are actually, for the most part, neutralizers.

18.3.3 Mechanisms

Considerable progress has been made in elucidating the mechanisms and causes of boiler corrosion. Advances in this area have been assisted by the development of experimental techniques for studies at high temperature and high pressure [20]. Potential–pH, or Pourbaix, diagrams have been developed for some systems at elevated temperature, so that equilibria under specific potential and pH conditions may be more readily predicted [21].

In high-temperature aqueous environments, iron and iron alloys form a characteristic double oxide layer [22], with the layers in deoxygenated solutions consisting of magnetite, Fe_3O_4. The outer layer consists of loosely packed crystals >1 μm in diameter, whereas the inner protective layer consists of densely packed crystallites 0.05–0.2 μm in diameter that adhere tightly to the underlying metal. In solutions of too high or too low pH, however, the protective layer of magnetite dissolves or is undermined, accounting for higher corrosion rates. The effects of dissolved O_2 are more complex.

Removal of dissolved oxygen is generally agreed upon as a necessary step in all boiler-water treatment. Oxygen is damaging primarily because it initiates pitting, possibly through action of differential aeration cells. The limiting safe concentration of dissolved O_2 is reported to depend inversely on the Cl^- concentration in the boiler water. In other words, dissolved neutral chlorides are not a corrosion problem in the absence of dissolved O_2, and the presence of dissolved O_2 is not a problem in sufficiently pure water [11].

Traces of O_2, even if not damaging to the steel boiler directly, are nevertheless effective in causing attack of the condenser system, especially if CO_2 or NH_3 is also present in the condensate. Such corrosion is sufficient to return small amounts of copper salts to the boiler followed by deposition of metallic copper. Although the condensers may not be damaged appreciably by such corrosion, the question remains whether pitting of the boiler is initiated by copper contamination of boiler waters.

Whether a particular boiler is damaged by a given water treatment is not evidence in itself that the treatment, in general, is good or bad. Statistical analysis of many boilers, or a fundamental investigation of corrosion mechanisms, is required in order to obtain a final answer. There are many interacting factors—including feedwater composition, condenser leakage, boiler design, and boiler operation, varying from one boiler to another—that determine whether oxygen and copper contamination in specific instances are damaging.

Alkali Additions. The optimum alkalinity of a boiler water depends, in part, on the extent to which contaminants slowly leak into the boiler from the cooling water of the condenser system (usually at the interface of condenser tubes and tubesheet). Leakage will vary with design and age of the condenser system; ensuing effects on the boiler depend on the composition of a specific cooling water. If, for example, seawater is used for cooling, magnesium chloride, which is a natural constituent of seawater, hydrolyzes to HCl and causes acid attack of the boiler. Periodic sodium hydroxide additions to the boiler water neutralize the acid and prevent such attack [23]. In this regard, NH_4OH is less effective than $NaOH + Na_3PO_4$ as a neutralizing agent in the quantities of each usually used in boiler-water treatment.

Excess alkali, on the other hand, is damaging because it may slowly dissolve the magnetite film in accord with

$$Fe_3O_4 + 4NaOH \rightarrow 2NaFeO_2 + Na_2FeO_2 + 2H_2O \qquad (18.10)$$

forming sodium hypoferrite or sodium ferroate (Na_2FeO_2) and sodium ferrite ($NaFeO_2$), both of which are soluble in hot concentrated NaOH. In addition, concentrated alkali reacts directly and more rapidly with iron to form hydrogen and sodium ferroate

$$Fe + 2NaOH \rightarrow Na_2FeO_2 + H_2 \qquad (18.11)$$

Reactions of this kind account in part for pitting and grooving of boiler tubes and are those that account for the excessive corrosion rate of iron at high values of pH, as Fig. 18.4 shows. As mentioned earlier, corrosive concentrations of NaOH are usually built up by evaporation of alkaline boiler water at various crevices where liquid flow is impeded and heat transfer is restricted. In the absence of conditions that favor concentration of alkaline solutes, damage by corrosion is expected to be minimal. Furthermore, phosphate additions avoid the high values of pH in concentrated boiler water that are necessary for reactions (18.10) and (18.11).

Oxygen Scavengers. The use of sodium sulfite as a scavenger in high-pressure boilers is limited by the decomposition of sulfites at high temperatures into sulfides or SO_2. It is said to be satisfactory below 12.4 MPa (1800 psi) steam pressure. One possible path for decomposition is the following:

$$4Na_2SO_3 + 2H_2O \rightarrow 3Na_2SO_4 + 2NaOH + H_2S \qquad (18.12)$$

The use of hydrazine is not subject to similar objections, but its slow reaction with oxygen and partial decomposition into NH_3 should be taken into account in calculating proper dosages. Also, NH_3 is a possible cause of stress-corrosion cracking of copper-base alloys in the condenser system, should oxygen accidentally contaminate the condensate.

REFERENCES

1. D. Pye, *J. Am. Water Works Assoc.* **39**, 1121 (1947).
2. S. Pirt, D. Callow, and W. Gillett, *Chem. Ind. (London)* **1957**, 730.
3. H. Gaunt and E. Wetton, *J. Appl. Chem.* **16**, 171 (1966).
4. M. Baker and V. Marcy, *Trans. Am. Soc. Mech. Eng.* **78**, 299 (1956).
5. B. P. Boffardi, Corrosion inhibitors in the water treatment industry, in *ASM Handbook*, Vol. 13A, *Corrosion: Fundamentals, Testing, and Protection*, ASM International, Materials Park, OH, 2003, p. 892.

6. See *ASME Consensus on Operating Practices for the Control of Feedwater and Boiler Water Quality in Modern Boilers*, United Engineering Center, New York, 1979.

7. S. Matsuda, Sc.D. thesis, Department of Metallurgy, M.I.T., Cambridge, MA, 1960.

8. T. P. Hoar, in *Third European Symposium on Corrosion Inhibitors*, Università degli Studi di Ferrara, Italy, 1971, p. 619.

9. *Industrial Water Treatment Chemicals and Processes: Developments since 1978*, M. Collie, editor, Chemical Technology Review No. 217, Noyes Data Corp., Park Ridge, NJ, 1983, pp. 36–42.

10. D. Robitaille and J. Bilek, *Chem. Eng.* **83**(27), 77 (1976).

11. G. Mann, *Brit. Corros. J.* **12**, 6 (1977).

12. D. Frey, *Mater. Perf.* **20**(2), 49 (1981).

13. ASTM D807-05, *Standard Practice for Assessing the Tendency of Industrial Boiler Waters to Cause Embrittlement (USBM Embrittlement Detector Method)*, ASTM International, West Conshohocken, PA.

14. E. Ulrich, in *Passivierende Filme und Deckschichten, Anlaufschichten*, H. Fischer, K. Hauffe, and W. Wiederholt, editors, Springer, Berlin, 1956, p. 308.

15. V. Linnenbom, *J. Electrochem. Soc.* **105**, 322 (1958).

16. G. Schikorr, *Z. Elektrochem.* **35**, 65 (1929); *Z. Anorg. Allgm. Chem.* **212**, 33 (1933).

17. P. Goldstein and C. Burton, *Trans. Am. Soc. Mech. Engrs., Series A* **91**, 75 (1969).

18. G. Skaperdas and H. Uhlig, *Ind. Eng. Chem.* **34**, 748 (1942).

19. Ref. 5, p. 893.

20. J. Dobson, *Advances in Corrosion Science and Technology*, Vol. **7**, M. Fontana and R. Staehle, editors, Plenum Press, New York 1980, p. 177.

21. H. Townsend, in *Proceedings of the Fourth International Congress on Metallic Corrosion*, National Association of Corrosion Engineers, Houston, TX, 1972, p. 477.

22. G. Mann, in *High Temperature High Pressure Electrochemistry in Aqueous Solutions*, D. de G. Jones and R. Staehle, editors, National Association of Corrosion Engineers, Houston, TX, 1976, p. 34.

23. H. Masterson, J. Castle, and G. Mann, *Chem. Ind. (London)* **1969**, 1261.

GENERAL REFERENCES

B. P. Boffardi, Corrosion inhibitors in the water treatment industry, in *ASM Handbook*, Vol. 13A, *Corrosion: Fundamentals, Testing, and Protection*, ASM International, Materials Park, OH, 2003, pp. 891–906.

Corrosion in fossil fuel power plants, in *ASM Handbook*, Vol. 13, *Corrosion*, ASM International, Materials Park, OH, 1987, pp. 985–1010.

D. Frey, Case histories of corrosion in industrial boilers, *Mater. Perf.* **20**(2), 49 (1981).

R. Garnsey, The chemistry of steam-generator corrosion, *Combustion* **52**(2), 36 (1980).

High Temperature High Pressure Electrochemistry in Aqueous Solutions, D. de G. Jones and R. Staehle, editors, National Association of Corrosion Engineers, Houston, TX, 1976.

G. R. Holcomb, Corrosion in supercritical water—Ultrasupercritical environments for power production, in *ASM Handbook*, Vol. 13C, *Corrosion: Environments and Industries*, ASM International, Materials Park, OH, 2006, pp. 236–245.

R. M. Latanision and D. B. Mitton, Corrosion in supercritical water—Waste destruction environments, in *ASM Handbook*, Vol. 13C, *Corrosion: Environments and Industries*, ASM International, Materials Park, OH, 2006, pp. 229–235.

G. Navitsky and F. Gabrielli, Boiler water treatment, feedwater treatment, and chemical cleaning of drum-type utility steam generators, *Combustion* **52**(2), 19 (1980).

19

ALLOYING FOR CORROSION RESISTANCE; STAINLESS STEELS

19.1 INTRODUCTION

The development of alloys for controlling corrosion in specific aggressive environments is certainly one of the great metallurgical developments of the twentieth century. The basis upon which alloys resist corrosion and the causes of corrosion susceptibility of alloys are explored in this and subsequent chapters. In general, the corrosion behavior of alloys depends on the interaction of:

1. The alloy of specific chemical composition and metallurgical structure.
2. The film on the alloy surface.
3. The environment, whether it is sufficiently aggressive to break down the protectiveness of the surface film, thereby initiating localized corrosion.
4. The alloy/environment combination, controlling whether the film self-repairs after breakdown and, if not, the type and rate of corrosion that propagates after initiation has occurred.

The beneficial effect of alloyed chromium or aluminum on the oxidation resistance of iron has already been described (Section 11.11), in addition to the

Corrosion and Corrosion Control, by R. Winston Revie and Herbert H. Uhlig
Copyright © 2008 John Wiley & Sons, Inc.

beneficial effect of small alloying additions of copper, chromium, or nickel on atmospheric resistance (Section 9.5).

Alloys of gold with copper or silver retain the corrosion resistance of gold above a critical alloy concentration called the *reaction limit* by Tammann [1]. Below the reaction limit, the alloy corrodes, for example, in strong acids, leaving a residue of pure gold either as a porous solid or as a powder. This behavior of noble-metal alloys is known as *parting* and is probably similar in mechanism to dezincification of copper–zinc alloys (see Section 20.2.2).

Pickering and Wagner [2] proposed that the predominant mechanism of parting in a gold–copper alloy containing, for example, 10 at.% Au occurs by injection of divacancies at the corroding alloy surface that readily diffuse into the alloy at room temperature and tend to fill with copper atoms. The latter diffuse in the opposite direction toward the surface, where they enter into aqueous solution. A mechanism involving dissolution of the alloy and subsequent redeposition of gold is not likely because dissolved gold in any amount is not detected. Instead, X-ray examination of the residual depleted porous layer shows that interdiffusion of gold and copper takes place to form solid–solution alloys varying in composition from that of the original alloy to pure gold. In other words, the evidence supports solid-state diffusion of copper from within the alloy to the surface, where it alone undergoes dissolution. Above the critical gold composition for parting, it is possible that conditions for divacancy formation are no longer favorable, or perhaps the vacancies tend to fill with a higher proportion of gold compared to copper atoms; hence, copper no longer corrodes preferentially.

Various other noble-metal alloys (e.g., Pt–Ni, Pt–Cu, and Pt–Ag) also exhibit reaction limits in HNO_3. Corresponding alloy compositions vary with the corrosive medium, but reaction limits in general, whatever the environment, tend to fall between 25 and 50 at.% of the noble-metal component (Table 19.1).

At higher temperatures, the same reaction limits apply, with the exception that some attack may initiate above the reaction limit composition after long exposure times. For example, at 100°C, exposure of the gold–silver alloys containing more than 50 at.% Au for one or more weeks to nitric acid resulted in measurable attack [3].

Rapid quenching of certain alloys consisting of, for example, iron, cobalt, or nickel base and containing one or more metalloid additions, such as phosphorus,

TABLE 19.1. Reaction Limits [1]

Corrosive Medium	Cu–Au Alloys		Ag–Au Alloys	
	At.% Au	Wt.% Au	At.% Au	Wt.% Au
$(NH_4)_2S_2$	24.5–25.5	50.2–51.5	32	46.5
H_2CrO_4	50	75.5	49.2	63.9
HNO_3 (specific gravity 1.3)	50	75.5	48.0–49.0	62.8–63.7
H_2SO_4	49–50	74.5–75.5	50	64.7

silicon, carbon, or boron, produces amorphous materials that are commonly referred to as *glassy alloys*. Cooling from the liquid at rates of about 10^6 K degrees per second "freezes" the atoms in nearly the same positions that they occupied in the liquid [4]. More recently, alloys have been designed with glass-formation cooling rates as low as 1 K/s [5]. As a result of their amorphous state, these alloys do not have the long-range order that is characteristic of crystalline materials. Instead, these materials are chemically and structurally homogeneous, without defects, multiple phases, and grain boundaries, so that the corrosion resistance may be much greater than that of crystalline alloys of the same composition. Some of the glassy-alloy compositions are especially resistant to acid media, including HCl and oxidizing solutions (e.g., $FeCl_3$). Chromium, if present, may be above or below 12%.

Alloying is an especially effective means for improving corrosion resistance if passivity results from the combination of a metal, otherwise active, with a normally passive metal. The alloying element may reduce either $i_{critical}$ [e.g., chromium alloyed with iron (Fig. 6.11, Section 6.8)], or $i_{passive}$ [e.g., nickel alloyed with copper (Fig. 6.14, Section 6.8.1)]. For >12% Cr–Fe alloys, $i_{critical}$ is reduced to such small values that any small corrosion current exceeds $i_{critical}$ and accounts for initiation of passivity in aerated aqueous solutions. Similarly, if passivity results from a noble metal added in small amount to an active metal or alloy, the corrosion rate may be reduced by orders of magnitude. The noble metal stimulates the cathodic reaction and, in this way, increases the anodic current density to the critical value for passivation. Practical examples are palladium or platinum alloyed with stainless steels or with titanium (see Sections 6.4 and 25.3), which resist sulfuric acid at concentrations and temperatures otherwise extremely corrosive. Carbon in steel can act similarly by creating cathodic sites of cementite (Fe_3C) on which reduction of HNO_3 (or HNO_2 in HNO_3) proceeds rapidly, allowing passivity to establish itself in less concentrated acid than is the case for pure iron. In principle, second phases of any kind (e.g., intermetallic compounds) can induce passivity in multicomponent alloys by the same mechanism.

In homogeneous single-phase alloys, passivity usually occurs at and above a composition that is specific to each alloy and that depends on the environment, as explained in Section 6.8. For Ni–Cu alloys, the critical composition is 30–40% Ni; for Cr–Co, Cr–Ni, and Cr–Fe alloys, it is 8%, 14%, and 12% Cr, respectively. The stainless steels are ferrous alloys that contain at least 10.5% Cr. Stainless steels are passive in many aqueous media, similar to the passivity of pure chromium itself, and they are the most important of the passive alloys.

19.2 STAINLESS STEELS

Stainless steels depend on the integrity of a thin passive film to maintain low corrosion rates in aqueous solutions. Any breakdown of this film is likely to result in localized corrosion, at pits and crevices. Stress-corrosion cracking may also occur. A wide range of stainless steels have been developed for applications in many environments. In this section, the properties and behavior of stainless steels are reviewed, and the principles for successful use of these important alloys are described.

19.2.1 Brief History*

In 1820, J. Stodart and M. Faraday in England published a report [6] on the corrosion resistance of various iron alloys they had prepared; apparently, it was this report in which the chromium–iron alloys were first mentioned. The maximum chromium content, however, was below that required for passivity, and they narrowly missed discovering the stainless steels. In France in 1821, Berthier [7], whose attention had been drawn to the work of Stodart and Faraday, found that iron alloyed with considerable chromium was more resistant to acids than was unalloyed iron. His alloys were obtained by direct reduction of the mixed oxides, producing what today is called ferrochromium (40–80% Cr). The alloys were brittle and were high in carbon, and they had no value as structural materials. Berthier prepared some steels using ferrochromium as a component, but the chromium content again was too low to overlap the useful passive properties characteristic of the stainless steels.

Although a variety of chromium–iron alloys was produced in subsequent years by several investigators who took advantage of the high-strength and high-hardness properties imparted by chromium, the inherent corrosion resistance of the alloys was not observed, largely because the accompanying high carbon content impaired corrosion resistance. In 1904, Guillet [8] of France produced low-carbon chromium alloys overlapping the passive composition range. He studied the metallurgical structure and mechanical properties of the Cr–Fe alloys and also of the Cr–Fe–Ni alloys that are now called austenitic stainless steels. But recognition of the outstanding property of passivity in such alloys, initiating at a minimum of 12% Cr, was apparently first described by Monnartz [9] of Germany, who began his research in 1908 and published a detailed account of the chemical properties of Cr–Fe alloys in 1911. This research included a description of the beneficial effect of oxidizing compared to reducing environments on corrosion resistance, the necessity of maintaining low carbon contents, and the effects of small quantities of alloying elements (e.g., Ti, V, Mo, and W).

The commercial utility of the Cr–Fe hardenable stainless steels as possible cutlery materials was recognized in about 1913 by H. Brearley of Sheffield, England. Looking for a better gun-barrel lining, he noticed that the 12% Cr–Fe alloys did not etch with the usual nitric acid etching reagents and that they did not rust over long periods of exposure to the atmosphere. The austenitic Cr–Fe–Ni stainless steels, on the other hand, were first exploited in Germany in 1912–1914 based on research of E. Maurer and B. Strauss of the Krupp Steel Works.

---

*Based largely on the account by Carl Zapffe, in *Stainless Steels*, American Society for Metals, Cleveland, OH, 1949, pp. 5–25.

The 18% Cr, 8% Ni (18–8) austenitic stainless steel is the most popular of all the stainless steels now produced.

19.2.2 Classes and Types

There are five main classes of stainless steels, designated in accord with their crystallographic structure. Each class consists of several alloys of somewhat differing composition having related physical, magnetic, and corrosion properties. These were given type numbers by the American Iron and Steel Institute (AISI). Although AISI stopped issuing designations for new stainless steels several decades ago, the AISI numbers remain in use today. The Unified Numbering System (UNS) was introduced in the 1970s to include all alloys, including stainless steels. A listing of many stainless steels that are produced commercially is given in Table 19.2.

The five main classes of stainless steels are martensitic, ferritic, austenitic, precipitation-hardenable, and duplex.

1. *Martensitic.* The name of the first class derives from the analogous martensite phase in carbon steels. Martensite is produced by a shear-type phase transformation on cooling a steel rapidly (quenching) from the austenite region (face-centered cubic structure) of the phase diagram. It is the characteristically hard component of quenched carbon steels, as well as of the martensitiic stainless steels. In stainless steels, the structure is body-centered cubic, and the alloys are magnetic. Typical applications include cutlery, steam turbine blades, and tools.

2. *Ferritic.* Ferritic steels are named after the analogous ferrite phase, or relatively pure-iron component of carbon steels cooled slowly from the austenite region. The ferrite, or alpha, phase for pure iron is the stable phase existing below 910 °C. For low-carbon Cr–Fe alloys, the high-temperature austenite (or gamma) phase exists only up to 12% Cr; above this Cr content, the alloys are ferritic at all temperatures up to the melting point. They can be hardened moderately by cold working, but not by heat treatment. Ferritic stainless steels are body-centered cubic in structure and are magnetic. Uses include trim on autos and applications in synthetic nitric acid plants.

3. *Austenitic.* Austenitic stainless steels are named after the austenite, or γ, phase, which, for pure iron, exists as a stable structure between 910 °C and 1400 °C. This phase is face-centered cubic and nonmagnetic, and it is readily deformed. It is the major or only phase of austenitic stainless steels at room temperature, existing as a stable or metastable structure depending on composition. Alloyed nickel is largely responsible for the retention of austenite on quenching the commercial Cr–Fe–Ni alloys from high temperatures, with increasing nickel content accompanying increased stability of the retained austenite. Alloyed Mn, Co, C, and N also contribute

TABLE 19.2. Types and Compositions of Stainless Steels

AISI Type or Common Name	UNS No.	Composition (%)								Remarks
		Cr	Ni	C	Mn (max)	P (max)	S (max)	Si	Other Elements	
Class: Martensitic—Body-Centered Tetragonal, Magnetic, Heat-Treatable										
403	S40300	11.5–13.0	—	0.15 max	1.0	0.04	0.03	0.5		Turbine quality
410	S41000	11.5–13.5	—	0.15 max	1.0	0.04	0.03	1.0		
414	S41400	11.5–13.5	1.25–2.5	0.15 max	1.0	0.04	0.03	1.0		
416	S41600	12.0–14.0	—	0.15 max	1.25	0.06	0.15 min	1.0	0.6 Mo max	Easy machining, nonseizing
416Se	S41623	12.0–14.0	—	0.15 max	1.25	0.06	0.06	1.0	0.15 Se min	Easy machining, nonseizing
420	S42000	12.0–14.0	—	Over 0.15	1.0	0.04	0.03	1.0		
431	S43100	15.0–17.0	1.25–2.5	0.20 max	1.0	0.04	0.03	1.0		
440A	S44002	16.0–18.0	—	0.6–0.75	1.0	0.04	0.03	1.0	0.75 Mo max	
440B	S44003	16.0–18.0	—	0.75–0.95	1.0	0.04	0.03	1.0	0.75 Mo max	
440C	S44004	16.0–18.0	—	0.95–1.2	1.0	0.04	0.03	1.0	0.75 Mo max	Highest attainable hardness
Class: Ferritic—Body-Centered Cubic; Magnetic, not Heat-Treatable										
405	S40500	11.5–14.5	—	0.08 max	1.0	0.04	0.03	1.0	0.1–0.3 Al.	
430	S43000	16.0–18.0	—	0.12 max	1.0	0.04	0.03	1.0		
430F	S43020	16.0–18.0	—	0.12 max	1.25	0.06	0.15 min	1.0	0.6 Mo max	Easy machining, nonseizing
430FSe	S43023	16.0–18.0	—	0.12 max	1.25	0.06	0.06	1.0	0.15 Se min	Easy machining, nonseizing

446	S44600	23.0–27.0	—	0.20 max	1.5	0.04	0.03	1.0	0.25 N max	Resistant to high-temperature oxidation

Class: Austenitic—Face-Centered Cubic, Nonmagnetic, not Heat-Treatable

201	S20100	16.0–18.0	3.5–5.5	0.15 max	5.5–7.5	0.06	0.03	1.0	0.25 N max	
202	S20200	17.0–19.0	4.0–6.0	0.15 max	7.5–10	0.06	0.03	1.0	0.25 N max	
301	S30100	16.0–18.0	6.0–8.0	0.15 max	2.0	0.045	0.03	1.0		
302	S30200	17.0–19.0	8.0–10.0	0.15 max	2.0	0.045	0.03	1.0		
302B	S30215	17.0–19.0	8.0–10.0	0.15 max	2.0	0.045	0.03	2.0–3.0		Resistant to high-temperature oxidation
303	S30300	17.0–19.0	8.0–10.0	0.15 max	2.0	0.20	0.15 min	1.0	0.6 Mo max	Easy machining, nonseizing
303 Se	S30323	17.0–19.0	8.0–10.0	0.15 max	2.0	0.20	0.06	1.0	0.15 Se min	Easy machining, nonseizing
304	S30400	18.0–20.0	8.0–10.5	0.08 max	2.0	0.045	0.03	1.0		
304L	S30403	18.0–20.0	8.0–12.0	0.03 max	2.0	0.045	0.03	1.0		Extra-low carbon
305	S30500	17.0–19.0	10.5–13.0	0.12 max	2.0	0.045	0.03	1.0		Lower rate of work hardening than 302 or 304
308	S30800	19.0–21.0	10.0–12.0	0.08 max	2.0	0.045	0.03	1.0		
309	S30900	22.0–24.0	12.0–15.0	0.20 max	2.0	0.045	0.03	1.0		
309 S	S30908	22.0–24.0	12.0–15.0	0.08 max	2.0	0.045	0.03	1.0		
310	S31000	24.0–26.0	19.0–22.0	0.25 max	2.0	0.045	0.03	1.5		
310 S	S31008	24.0–26.0	19.0–22.0	0.08 max	2.0	0.045	0.03	1.5		
314	S31400	23.0–26.0	19.0–22.0	0.25 max	2.0	0.045	0.03	1.5–3.0		
316	S31600	16.0–18.0	10.0–14.0	0.08 max	2.0	0.045	0.03	1.0	2.0–3.0 Mo	
316L	S31603	16.0–18.0	10.0–14.0	0.03 max	2.0	0.045	0.03	1.0	2.0–3.0 Mo	Extra-low carbon
317	S31700	18.0–20.0	11.0–15.0	0.08 max	2.0	0.045	0.03	1.0	3.0–4.0 Mo	
321	S32100	17.0–19.0	9.0–12.0	0.08 max	2.0	0.045	0.03	1.0	Ti: 5 × C min	Stabilized grade

TABLE 19.2. *Continued*

AISI Type or Common Name	UNS No.	Composition (%)								Remarks
		Cr	Ni	C	Mn (max)	P (max)	S (max)	Si	Other Elements	
347	S34700	17.0–19.0	9.0–13.0	0.08 max	2.0	0.045	0.03	1.0	Cb + Ta: 10 × C min	Stabilized grade
348	S34800	17.0–19.0	9.0–13.0	0.08 max	2.0	0.045	0.03	1.0	Cb + Ta: 10 × C min 0.20 Co max	Stabilized grade 0.1 Ta max when radiation conditions require low Ta
Class: Precipitation-Hardenable										
15–5 PH	S15500	14.00–15.5	3.50–5.50	0.07	1.00	0.04	0.03	1.00	2.5–4.5 Cu 0.15–0.45 Nb	
17–7 PH	S17700	16.0–18.0	6.50–7.75	0.09	1.00	0.04	0.04	1.00	0.75–1.50 Al	
Class: Duplex—Austenitic/Ferritic										
2205	S32205	21.0–23.0	4.5–6.5	0.03 max	2.0	0.03	0.02	1.0	2.5–3.5 Mo 0.014–0.2 N	
2304	S32304	21.5–24.5	3.0–5.5	0.03 max	2.5	0.04	0.04	1.0	0.05–0.6 Mo 0.05–0.6 Cu 0.05–0.2 N	
Zeron 100	S32760	24.0–26.0	6.0–8.0	0.03 max	1.0	0.03	0.01	1.0	3.0–4.0 Mo 0.50–1.00 Cu 0.50–1.00 W 0.20–0.30 N	

to the retention and stability of the austenite phase. Austenitic stainless steels can be hardened by cold working, but not by heat treatment. On cold working, but not otherwise, the metastable alloys (e.g., 201, 202, 301, 302, 302B, 303, 303Se, 304, 304L, 316, 316L, 321, 347, 348; see Table 19.2) transform in part to the ferrite phase (hence, the descriptor, "metastable") having a bond-centered-cubic structure that is magnetic. This transformation also accounts for a marked rate of work hardening. Alloys 305, 308, 309, and 309S, on the other hand, work harden at a relatively low rate and become only slightly magnetic, if at all. Alloys containing higher chromium and nickel (e.g., 310, 310S, 314) are essentially stable austenitic alloys and do not transform to ferrite or become magnetic when deformed. Uses of austenitic stainless steels include general-purpose applications, architectural and automobile trim, and various structural units for the food and chemical industries.

4. *Precipitation Hardenable.* The *precipitation-hardening stainless steels* achieve high strength and hardness through low-temperature heat treatment after quenching from high temperatures. These Cr–Fe alloys contain less nickel (or none) than is necessary to stabilize the austenite phase, and, in addition, they contain alloying elements, such as aluminum or copper, that produce high hardness through formation and precipitation of intermetallic compounds along slip planes or grain boundaries. They are applied whenever the improved corrosion resistance imparted by alloyed nickel is desirable, or, more important, when hardening of the alloy is best done after machining operations, using low-temperature heat treatment [e.g., 480°C (900°F)] rather than a high-temperature quench as is required in the case of the martensitic stainless steels.

5. *Duplex.* The duplex stainless steels contain both austenite and ferrite, typically with the austenite:ferrite ratio ~ 60:40. The mixed austenite–ferrite microstructure, consisting of a continuous ferrite matrix with austenite islands (Fig. 19.1 [10]), is achieved by including in the composition a balance of elements that stabilize austenite (e.g., nickel and nitrogen) and those that stabilize ferrite (e.g., chromium and molybdenum). The 22% chromium duplex stainless steel was developed in the late 1970s and was later modified and designated as Alloy 2205 (UNS designation S32205). Because of good mechanical properties, high resistance to chloride stress-corrosion cracking, good erosion and wear resistance, and low thermal expansion, duplex stainless steels are used in many applications, including pressure vessels, storage tanks (e.g., for phosphoric acid), and heat exchangers [10]. For seawater service, duplex stainless steels of higher molybdenum content (e.g., Zeron 100) have been developed [11].

In general, the highest resistance to uniform corrosion is obtained with the nickel-bearing austenitic types, and, in general, the highest nickel-composition alloys in this class are more resistant than the lowest nickel compositions. For

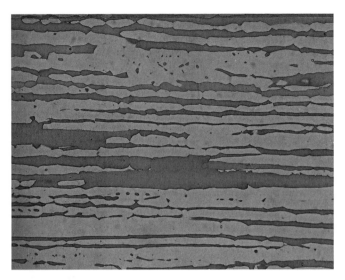

Figure 19.1. Typical microstructure of a duplex stainless steel. The grains are elongated in the rolling direction (~300×) [10]. (Reprinted with permission of John Wiley & Sons, Inc.)

optimum corrosion resistance, austenitic alloys must be quenched (rapidly cooled by water or by an air blast) from about 1050–1100°C (1920–2000°F). The molybdenum-containing austenitic alloys (316, 316L, 317) have improved corrosion resistance to chloride-containing environments, dilute nonoxidizing acids, and crevice corrosion.

The ferritic stainless steels have optimum corrosion resistance when cooled slowly from above 925°C (1700°F) or when annealed at 650–815°C (1200–1500°F).* Recently developed ferritic stainless steels of controlled purity contain molybdenum additions and low carbon and nitrogen. Typical analyses include 26% Cr, 1% Mo, <0.01% C, <0.02% N; 18% Cr, 2% Mo, <0.025% C, <0.025% N; 28% Cr, 4% Mo, <0.01% C, <0.02% N. Alloyed titanium or columbium is sometimes included in the first two alloys in order to raise the tolerable carbon and nitrogen contents. All these alloys can be quenched from above 925 °C without loss of corrosion resistance. Their corrosion properties are generally better than those of the conventional ferritic, and some austenitic, stainless steels listed in Table 19.2—for example, in NaCl solutions, HNO_3, and various organic acids, provided they retain passivity. Should they lose passivity, locally or generally, their corrosion rates tend to be higher than those of the active nickel-containing austenitic stainless steels of equivalent chromium and molybdenum contents [12].

The martensitic stainless steels, on the other hand, have optimum corrosion resistance as quenched from the austenite region. In this state, they are very hard and brittle. Ductility is improved by annealing (650–750°C for 403, 410, 416,

---

*For additional details on heat treating stainless steels, see *ASM Handbook*, Vol. 4, *Heat Treating*, ASM International, Materials Park, OH, 1991, pp. 769–792.

416Se; 650–730°C for 414; 620–700°C for 431; 680–750°C for 440 A, B, C, and 420), but at some sacrifice of corrosion resistance. Resistance to pitting and rusting in 3% NaCl at room temperature reaches a minimum after tempering a 0.2–0.3% C, 13% Cr stainless steel at 500°C, as well as 650°C for a similar steel containing 0.06% C [13]. In general, the tempering range 450–650°C (840–1200°F) should be avoided if possible. Decrease of corrosion resistance probably results, in part, from the transformation of martensite containing interstitial carbon into a network of chromium carbides attended by depletion of chromium in the adjoining metallic phase.

Cold working any of the stainless steels usually has only minor effect on uniform corrosion resistance if temperatures are avoided that are high enough to permit appreciable diffusion during or after deformation. Phase changes brought about by cold working the metastable austenitic alloys are not accompanied by major changes in corrosion resistance.* Furthermore, quenched austenitic stainless steel (face-centered cubic) of 18% Cr, 8% Ni composition has approximately the same corrosion resistance as quenched ferritic stainless steel (body-centered cubic), of the same chromium and nickel composition but with lower carbon and nitrogen content [14]. However, if a similar alloy containing a mixture of austenite and ferrite is briefly heated at, for example, 600°C, composition differences are established between the two phases setting up galvanic cells that accelerate corrosion. Composition gradients, whatever their cause, are more important in establishing uniform corrosion behavior than are structural variations of an otherwise homogeneous alloy. This statement is true for metals and alloys in general.

19.2.3 Intergranular Corrosion

Improper heat treatment of ferritic or austenitic stainless steels causes the grain boundaries, which separate the individual crystals, to become especially susceptible to corrosion. Corrosion of this kind leads to catastrophic reduction in mechanical strength. The specific temperatures and times that induce susceptibility to intergranular corrosion are called *sensitizing* heat treatments. They are very much different for the ferritic and the austenitic stainless steels. In this respect, the transition in sensitizing temperatures for steels containing 18% Cr occurs at about 2.5–3% Ni [15]; stainless steels containing less than this amount of nickel are sensitized in the temperature range typical of nickel-free ferritic steels, whereas those containing more nickel respond to the temperature range typical of the austenitic stainless steels.

19.2.3.1 Austenitic Stainless Steels. For austenitic alloys, the sensitizing temperature range is approximately 425–875°C (800–1610°F). The degree of

---

*After transformation, the alloys, otherwise resistant, become susceptible to hydrogen cracking; also, ferritic and martensitic stainless steels on cold working tend to become more susceptible to hydrogen cracking.

damage to commercial alloys caused by heating in this range depends on time, with a few minutes at the higher temperature range of 750°C (1380°F) being equivalent to several hours at a lower (or still higher) temperature range (Fig. 19.2) [16, 17]. Slow cooling through the sensitizing temperature range or prolonged welding operations induce susceptibility, but rapid cooling avoids damage. Hence, austenitic stainless steels must be quenched from high temperatures. Spot welding, in which the metal is rapidly heated by a momentary electric current followed by a naturally rapid cooling, does not cause sensitization. Arc welding, on the other hand, can cause damage, with the effect being greater the longer the heating time, especially when heavy-gage material is involved. Sensitizing temperatures are reached some millimeters away from the weld metal itself, with the latter being at the melting point or above. Hence, on exposure to a corrosive environment, failure of an austenitic stainless steel weld—called *weld decay*—occurs in zones slightly away from the weld rather than at the weld itself (see Fig. 19.3).

The extent of sensitization for a given temperature and time depends very much on carbon content. An 18–8 stainless steel containing 0.1% C or more may be severely sensitized after heating for 5 min at 600°C, whereas a similar alloy containing 0.06% C is affected less, and for 0.03% C the alloy heat treated similarly may suffer no appreciable damage on exposure to a moderately corrosive environment. The higher the nickel content of the alloy, the shorter the time for sensitization to occur at a given temperature, whereas alloying additions of molybdenum increase the time [16].

The physical properties of stainless steels after sensitization do not change greatly. Because precipitation of carbides accompanies sensitization, the alloys become slightly stronger and slightly less ductile. Damage occurs only on exposure to a corrosive environment, with the alloy corroding along grain boundaries at a rate depending on severity of the environment and the extent of sensitization. In seawater, a sensitized stainless-steel sheet may fail within weeks or

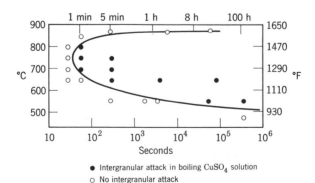

Figure 19.2. Effect of time and temperature on sensitization of 18.2% Cr, 11.0% Ni, 0.05% C, 0.05% N stainless steel [16]. (Reprinted with permission of ASM International®. All rights reserved. www.asminternational.org.)

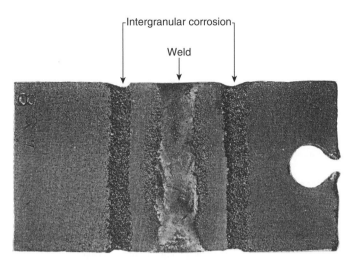

Figure 19.3. Example of weld decay (2×). After sensitization, specimen was exposed to 25% HNO₃.

months; but in a boiling solution containing $CuSO_4{\cdot}5H_2O$ (13 g/liter) and H_2SO_4 (47 mL concentrated acid/liter), which is used as an accelerated test medium, failure occurs within hours.

19.2.3.2 Theory and Remedies. Since intergranular corrosion of austenitic stainless steels is associated with carbon content, a low-carbon alloy (<0.02% C) is relatively immune to this type of corrosion [18]. Nitrogen, normally present in commercial alloys to the extent of a few hundredths percent, is less effective than carbon in causing damage (Fig. 19.4) [19]. At high temperatures [e.g., 1050°C (1920 °F)], carbon is almost completely dispersed throughout the alloy, but within (or somewhat above) the sensitizing temperature range it rapidly diffuses to the grain boundaries where it combines preferentially with chromium to form chromium carbides (e.g., $M_{23}C_6$, in which M represents the presence of some small amount of iron along with chromium). This reaction depletes the adjoining alloy of chromium to the extent that the grain-boundary material may contain less than the 12% Cr necessary for passivity. The affected volume of alloy normally extends some small distance into the grains on either side of the boundary itself, causing apparent grain-boundary broadening of the etched surface. The chromium-depleted alloy sets up passive–active cells of appreciable potential difference, with the grains constituting large cathodic areas relative to small anodic areas at the grain boundaries. Electrochemical action results in rapid attack along the grain boundaries and deep penetration of the corrosive medium into the interior of the metal.

If the alloy is rapidly cooled through the sensitizing zone, carbon does not have time either to reach the grain boundaries or to react with chromium if some carbon is already concentrated at the grain boundaries. On the other hand, if the

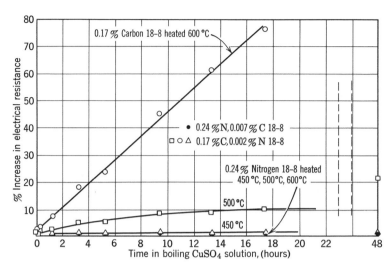

Figure 19.4. Intergranular corrosion measured by change in electrical resistance of 18–8 stainless steels containing nitrogen or carbon immersed in 10% CuSO$_4$ + 10% H$_2$SO$_4$. All specimens sensitized at stated temperatures for 217 h [19]. (Reproduced with permission. Copyright 1945, The Electrochemical Society.)

alloy remains within the sensitizing temperature zone an especially long time (usually several thousand hours), chromium again diffuses into the depleted zones, reestablishing passivity of the grain boundaries and eliminating susceptibility to preferred attack. Although nitrogen forms chromium nitrides, it is less effective than carbon in causing damage, perhaps in part because nitrides precipitate more generally throughout the grains or they form islands along grain boundaries, interrupting a continuous path along which the corrosive agent can proceed [19]. Carbides, on the other hand, form continuous paths of chromium-depleted alloy.

Other experimental data can be cited that supports the chromium-depletion mechanism of intergranular corrosion of austenitic stainless steels. For example, carbides isolated from grain boundaries of sensitized stainless steels have shown an expected high chromium content. Along the same lines, corrosion products of grain-boundary alloy obtained by choosing corrosive conditions that avoid attack of carbides show a lower chromium content than corresponds to the alloy. In this respect, Schafmeister [20], using cold concentrated sulfurous acid acting on sensitized 18% Cr, 8.8% Ni, 0.22% C stainless steel for 10 days, found only 8.7% Cr in the alloy that had corroded from grain-boundary regions. Accompanying analyses for Ni and Fe of 8.4% and 83%, respectively, showed no appreciable depletion in nickel and an increase in iron content.

Microprobe scans of sensitized stainless steels have indicated chromium depletion and nickel enrichment at grain boundaries [21]. Radioactive C^{14} introduced into an austenitic 18% Cr, 12.8% Ni, 0.12% C stainless steel has demonstrated enrichment of carbon at grain boundaries observable immediately after quenching from 1000 °C to 1350 °C. This indicates that time for sensitization may depend, in part, on the reaction rate of chromium with carbon already at grain boundaries and not alone on the time required for additional carbon to diffuse to grain boundaries.

There are at least three effective means for avoiding susceptibility to intergranular corrosion:

1. Heat Treatment at 1050–1100°C (1920–2000°F) Followed by Quenching. At high temperature, precipitated carbides are dissolved, and rapid cooling prevents their re-formation. This treatment is recommended, for example, after welding operations. It is not always a possible treatment, however, because of the size of the welded structure or because of a tendency of the alloy to warp at high temperatures.

2. Reduction of Carbon Content. Carbon content can be reduced in the commercial production of stainless steels, but at extra cost. Alloys of low carbon content (e.g., <0.03% C) are designated by letter L—for example, Types 304L and 316L. These alloys can be welded or otherwise heated in the sensitizing temperature range with much less resultant susceptibility to intergranular corrosion. However, they are not immune.

3. Addition of Titanium or Niobium (Columbium). By alloying austenitic alloys with a small amount of an element (e.g., Ti or Nb) having higher affinity for carbon than does chromium, carbon is restrained from diffusing to the grain boundaries, and any which is already at the boundary reacts with titanium or columbium instead of with chromium. Alloys of this kind are called *stabilized grades* (e.g., Types 321, 347, 348). Except as noted below, they can be welded or otherwise heated within the sensitizing zone without marked susceptibility to intergranular corrosion. Optimum resistance to intergranular corrosion at 675°C (1250°F) is obtained by a prior *carbide stabilizing heat treatment* for several hours at about 900°C (1650°F) [17, 22]. This heat treatment converts available carbon to stable carbides of titanium or niobium in a temperature range of lower carbon solubility compared to that of the usual higher quench temperatures.

In welding operations, the weld rod usually contains niobium rather than titanium; the latter tends to oxidize at elevated temperatures with the danger of its residual concentration becoming too low to stabilize the weld alloy against corrosion. Niobium, on the other hand, is lost by oxidation to a lesser extent.

If, during welding the stabilized grades of stainless steel, the region of the base metal adjacent to the weld is heated to temperatures at which the carbides of titanium and niobium dissolve [above about 1230°C (2250°F)] and the cooling rate is sufficiently rapid to prevent the formation of the stabilizing carbides, then these steels can be sensitized if they are subsequently heated in the temperature range in which chromium carbides precipitate. In such a situation, a narrow region immediately adjacent to the weld is susceptible to intergranular corrosion, which, if it occurs, is called *knife-line attack*. This type of intergranular corrosion can be prevented by selecting the welding process parameters to avoid temperatures at which carbides of titanium and niobium dissolve and by using a *carbide-stabilizing heat treatment* as described above.

19.2.3.3 Intergranular Corrosion of Nonsensitized Alloys.

Sigma phase is a hard, brittle, intermetallic compound, rich in chromium and molybdenum, with a tetragonal structure that can precipitate at grain boundaries in Type 316 stainless steel when heated in the range 540–1000 °C (1000–1800°F). Sigma phase impairs corrosion resistance in highly oxidizing nitric acid environments. The mechanism of corrosion may be based on the high molybdenum content of sigma phase, which results in preferential direct attack on the sigma phase particles in preference to the molybdenum-depleted zones surrounding them [23].

Sigma phase may also form in Types 310, 430, and 446 stainless steels, but in these alloys the phase forms at a rate so low that it is important only when the alloys are used in the temperature range in which sigma phase is formed [23].

In strongly oxidizing media (e.g., boiling 5N HNO_3 + Cr^{6+}), austenitic stainless steels, including stabilized grades, quenched from 1050°C, undergo slow intergranular attack [24]. Stress is not necessary. The attack occurs only in the transpassive region; hence, oxidizing ions, such as Cr^{6+} (0.1–0.5 N $K_2Cr_2O_7$), Mn^{7+}, or Ce^{4+}, are necessary additions to the boiling nitric acid. The rate of attack increases with the amount of alloyed nickel [25], being more than 10 times higher for a 78% Ni, 17% Cr, bal. Fe alloy compared to a similar 10% Ni alloy (70-h test). This trend is opposite to the beneficial effect of nickel on the resistance of stainless steels to stress-corrosion cracking.

Grain-boundary attack of 19% Cr, 9% Ni stainless steel is most rapid after quenching from 1100°C to 1200°C; it is less pronounced on quenching from either 900°C or 1400°C [26]. High-purity alloys are immune. Alloyed carbon, nitrogen, oxygen, or manganese added in small concentrations have no effect, whereas silicon and phosphorus (>100 ppm) are damaging. Silicon causes increased intergranular attack in the intermediate range of 0.1–2% for the 14% Cr, 14% Ni stainless steel; in larger or smaller amounts the alloy is not susceptible [27]. The necessity of strongly oxidizing conditions and presence of phosphorus for intergranular attack was confirmed for a quenched low carbon 20% Cr, 20% Ni, 0.1% P stainless steel [28]. When polarized anodically in 1 N H_2SO_4 or in 27% HNO_3 at 40°C, intergranular attack occurred only at very noble potentials in the transpassive region. With only 0.002% P in the alloy, such attack was not observed at any potential.

Similar intergranular attack of 15% Cr, 6% Fe, bal. Ni (Inconel 600) in high-temperature water (350°C) or steam (600–650°C) [25] or of stabilized 18–8 stainless steel in potassium hydroxide solutions (pH 11) at 280°C [29] has been reported. This matter is of prime interest in view of the common use of Inconel 600 and stainless steels in the construction of nuclear reactors for power production. Contaminants in the water [such as traces of dissolved O_2, NaOH, or dissolved lead (from heat-transfer tube–tube sheet leakage)], and the presence of crevices at which superheating occurs accompanied by concentration of solute accelerate S.C.C. [30]. Laboratory tests indicate that such failures also occur in longer times in relatively pure water [31] (see also Section 23.3.3). Increased nickel content favors intergranular corrosion [32] of 18–20% Cr stainless steels at 200–300°C in water containing Cl⁻ and dissolved oxygen, as it does in boiling HNO_3–Cr^{6+}. Silicon (>0.3%) and phosphorus (>0.023%) in 14% Cr, 14% Ni stainless steel exposed to 0.01% $FeCl_3$ at 340°C, analogous to results in HNO_3 + Cr^{6+}, cause intergranular corrosion. Unlike the latter medium, the high-temperature 0.01% $FeCl_3$ solution (21-day exposure) caused intergranular corrosion [33] of the nonsensitized stainless steel when it contained >0.05% C.

The overall results confirm that intergranular attack is the result of specific impurities in the alloy segregating at grain-boundary regions. The extent of their damaging effect depends on the chemical environment to which the alloy is exposed. Although applied stress has been stated to increase the observed attack in various media, stress is not always necessary; hence, the observed damage in such instances is better described as intergranular corrosion rather than as stress-corrosion cracking.

19.2.3.4 Ferritic Stainless Steels.

The sensitizing range for ferritic stainless steels lies above 925°C (>1700°F), and immunity is restored by heating for a short time (approximately 10–60 min) at 650–815°C (1200–1500°F). These temperatures are the opposite of those applying to austenitic stainless steels. The same accelerating media (i.e., boiling $CuSO_4$–H_2SO_4 or 65% HNO_3) produce intergranular corrosion in either class, and the extent and rapidity of damage are similar. In welded sections, however, damage to ferritic steels occurs to metal immediately adjacent to the weld and to the weld metal itself, whereas in austenitic steels the damage by weld decay localizes some small distance away from the weld.

Chromium content of ferritic steels, whether high or low (16–28% Cr), has no appreciable influence on susceptibility to intergranular corrosion [34]. Similar to the situation for austenitic steels, lowering the carbon content is helpful, but the critical carbon content is very much lower. A type 430 ferritic stainless steel containing only 0.009% C was still susceptible [34]. Only on decarburizing low-carbon steels containing 16% or 24% Cr in H_2 at 1300°C for 100 h was immunity obtained in the $CuSO_4$–H_2SO_4 test solution [35]. Similarly, a low-carbon (~0.002%) 25% Cr–Fe alloy was reported to be immune [36]. Addition of titanium in the amount of eight times carbon content or more provided immunity to the $CuSO_4$ test solution, but not to boiling 65% HNO_3 [34].

Niobium additions, it was stated, behaved similarly, and only heat treatment, as described previously, was effective in conferring immunity to nitric acid. It was reported [37], however, that columbium additions (8× C + N content), but not titanium additions, minimize the observed intergranular corrosion of welds exposed to boiling 65% HNO_3. This behavior may be explained by the observed marked reactivity of titanium carbides, but not niobium carbides, with HNO_3 along grain boundaries where such carbides are concentrated [38].

The Mo–Cr stainless steels of controlled purity, described earlier, although in some instances containing more than 0.01% C, are immune to intergranular corrosion. This is accounted for by their molybdenum content, which slows down diffusion of carbon and nitrogen, and by their titanium or niobium content which, if present, reacts preferentially with carbon and nitrogen.

Intergranular corrosion of ferritic stainless steels is best explained on the basis of the chromium-depletion theory that is generally accepted for the austenitic stainless steels [39]. Differences in sensitization temperatures and times compared to the austenitic steels are explained by the much higher diffusion rates of carbon, nitrogen, and chromium in the ferritic body-centered-cubic lattice compared to the austenitic face-centered-cubic lattice. Accordingly, chromium carbides and nitrides, in solution at higher temperatures, precipitate rapidly (within seconds) below 950°C along grain boundaries, depleting the adjoining alloy of chromium to resultant compositions below stainless-steel requirements and which corrode at correspondingly higher rates than the grains. But accompanying high diffusion rates of chromium account for restored immunity to intergranular corrosion on heating the alloy for several minutes in the range 650–815°C (rather than weeks or months required to restore sensitized austenitic stainless steels), thereby restoring the grain-boundary alloy to normal stainless-steel composition.

19.2.4 Pitting and Crevice Corrosion

In environments containing appreciable concentrations of Cl^- or Br^-, in which stainless steels otherwise remain essentially passive, all the stainless steels tend to corrode at specific areas and to form deep pits. Ions such as thiosulfate $(S_2O_3^{2-})$ may also induce pitting. In the absence of passivity, such as in deaerated alkali-metal chlorides, nonoxidizing metal chlorides (e.g., $SnCl_2$ or $NiCl_2$), or oxidizing metal chlorides at low pH, pitting does not occur. This holds even though, in acid environments, general corrosion may be appreciable.

Stainless steels exposed to seawater develop deep pits within a matter of months, with the pits usually initiating at crevices or other areas of stagnant electrolyte (*crevice corrosion*). Susceptibility to pitting and crevice corrosion is greater in the martensitic and ferritic steels than in the austenitic steels; it decreases in the latter alloys as the nickel content increases. The austenitic 18–8 alloys containing molybdenum (types 316, 316L, 317) are still more resistant to seawater; however, crevice corrosion and pitting of these alloys eventually develop within a period of 1–2.5 years.

Stainless steels exposed at room temperature to chloride solutions containing active depolarizing ions, such as Fe^{3+}, Cu^{2+}, or Hg^{2+}, develop visible pits within hours. These solutions have sometimes been used as accelerated test media to assess pitting susceptibility.

Many extraneous anions, some more effective than others, act as pitting inhibitors when added to chloride solutions. For example, as mentioned in Section 6.6, addition of 3% $NaNO_3$ to a 10% $FeCl_3$ solution completely inhibited pitting of 18–8 stainless steel, as well as avoiding general attack, over a period of at least 25 years. Hence, so long as passivity does not break down at crevices because of dissolved oxygen depletion, or for other reasons, localized corrosion does not initiate in inhibited solutions no matter how long the time. Similarly, in neutral chlorides (e.g., NaCl solutions), addition of alkalies inhibits pitting. In aerated 4% NaCl solution at 90°C, at which temperature the pitting rate of 18–8 is highest, addition of 8 g NaOH per liter was found to eliminate pitting [40]. In refrigerating brines, addition of 1% Na_2CO_3 was effective for at least 5 years [41].

Pits develop more readily in a stainless steel that is metallurgically inhomogeneous. Similarly, the pitting tendency of an austenitic steel increases when the alloy is heated briefly in the carbide precipitation (sensitization) range. Pitting resulting from crevice corrosion is also favored whenever a stainless steel is covered by an organic or inorganic film or by marine fouling organisms, which partially shield the surface from access to oxygen. The movement of flowing seawater tends to keep the entire surface in contact with aerated water and uniformly passive, reducing any tendency for localized corrosion.

19.2.4.1 Theory of Pitting. Pitting corrosion is usually considered to consist of two stages:

1. Initiation, in which the integrity of the passive film is breeched at localized areas and pits form.
2. Propagation of the pit, at a rate that is often found to increase with time, because of the increasing acidity inside the pit.

The initiation of pits on an otherwise fully passive 18–8 surface, as discussed earlier (Section 6.5.2), requires that the corrosion potential exceed the critical potential [0.21 V(S.H.E.) in 3% NaCl]. The oxygen potential in air at pH 7 (0.8 V) or the ferric–ferrous potential ($\phi° = 0.77$ V) is sufficiently noble to induce pitting. However, the stannic–stannous ($\phi° = 0.15$ V) and the chromic–chromous potentials ($\phi° = -0.41$ V) are too active; hence, pitting of 18–8 stainless steels is not observed in deaerated stannic or chromic chloride solutions. In sufficient concentration, the nitrate ion shifts the critical potential to a value that is noble to the ferric–ferrous oxidation–reduction potential; hence, pitting in 3% $NaNO_3$ + 10% $FeCl_3$ is not observed. Other anions shift the critical potential similarly, with their effectiveness in this regard decreasing in the order $OH^- > NO_3^- > SO_4^{2-} > ClO_4^-$. Increasing amounts of alloyed chromium, nickel, molybdenum, and rhenium in stainless steels also shift the critical potential in the noble direction, accounting for increased resistance to pitting.

In addition to the critical pitting potential (CPP) (noble to which pits initiate), critical pitting temperatures (CPT) have been measured. Stable pitting occurs at temperatures above the CPT [42]. Below this temperature, stable pitting does not take place at any potential. The CPT has been explained as the temperature below which the rapid dissolution required for pitting cannot take place.

Once a pit initiates, a passive–active cell is set up of 0.5–0.6 V potential difference. The resultant high current density accompanies a high corrosion rate of the anode (pit) and, at the same time, polarizes the alloy surface immediately surrounding the pit to values below the critical potential. Through flow of current, chloride ions transfer into the pit forming concentrated solutions of Fe^{2+}, Ni^{2+}, and Cr^{3+} chlorides, which, by hydrolysis, account for an acid solution (Fig. 19.5). The measured pH of the confined anodic corrosion products of 18–8 stainless steel in 5% NaCl at 200 A/m^2 (0.02 A/cm^2) is 1.5 [43]. The high Cl$^-$ concentration and low pH ensure that the pit surface remains active. At the same time, the high specific gravity of such corrosion products causes leakage out of the pit in the direction of gravity, inducing breakdown of passivity wherever the products come into contact with the alloy surface. This accounts for a shape of pits elongated in the direction of gravity, as is often observed in practice. An 18–8 stainless-steel sheet exposed to seawater for one year was observed to develop an elongated narrow pit reaching 60 mm (2.5 in.) from its point of origin (Fig. 19.6). The mechanism of growth was demonstrated in the laboratory [43] by continuously flowing a fine stream of concentrated FeCl$_2$ solution over an 18–8 surface slightly inclined to the vertical and totally immersed in FeCl$_3$, resulting in formation of a deep groove under the FeCl$_2$ stream within a few hours (Fig. 19.6). A similar groove does not form on an iron surface, because a passive–active cell is not established.

A pit stops growing only if the surface within the pit is again passivated, bringing the pit and the adjacent alloy to the same potential. Extraneous anions, such as SO_4^{2-}, have no effect; on the other hand, dissolved oxygen or passivator ions (e.g., NO_3^-) reinitiate passivity on entering a pit. Successful repassivation depends on factors such as pit geometry and stirring rate.

The factors accounting for rates of corrosion at crevices follow the same principles as those described for pit growth. The higher the electrolyte conductivity and the larger the cathode area outside the crevice, the higher the rate of attack at the anode. The initiation of crevice corrosion, however, does not depend

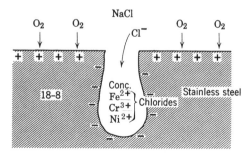

Figure 19.5. Passive–active cell responsible for pit growth in stainless steel exposed to chloride solution.

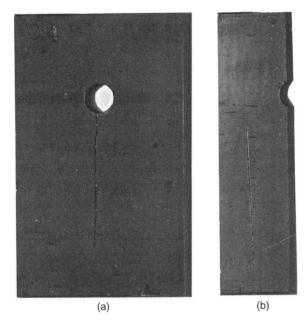

Figure 19.6. (a) Elongated pit in 18–8 stainless steel specimen 75 × 125 mm (3 × 5 in.) exposed to Boston Harbor seawater for 1 year; pit began at crevice formed between bakelite rod and interior surface of hole. (b) Artificial elongated pit formed by flowing 50% $FeCl_2$ in a fine stream over 18–8 stainless steel immersed in 10% $FeCl_3$, 4 hr.

on exceeding the critical pitting potential. Instead, it depends only on factors influencing the breakdown of passivity within the crevice. This breakdown may occur, for example, by oxygen depletion in the crevice caused by slow uniform alloy corrosion, followed by setting up a differential aeration cell that results in accumulation of acid anodic corrosion products in the crevice. These changes in electrolyte composition eventually destroy passivity, thereby establishing a still larger potential difference between active metal in the crevice and passive metal outside, analogous to cells operating in pitting corrosion. This mechanism for initiation of crevice corrosion indicates that chlorides are not essential to its occurrence, accounting for the observation that crevice corrosion occurs in solutions of sulfates, nitrates, acetates, and so on, as well as of chlorides.

Cathodic protection effectively avoids crevice corrosion, provided that the alloy surrounding the crevice is polarized to the open-circuit potential of the active (nonpassive) alloy surface within the crevice. This contrasts with the more lenient requirement of polarizing below the critical potential in order to avoid pit initiation.

Alloy additions that are effective in helping retain passivity in the presence of both low dissolved oxygen concentration and acid corrosion products help reduce or avoid crevice attack. Additions of molybdenum to 18–8 stainless steel (type 316) and palladium additions to titanium (see Fig. 25.2, Section 25.2) are in this category.

19.2.4.2 Reducing or Avoiding Pitting Corrosion. There are four principal ways of reducing or avoiding pitting corrosion:

1. Cathodically protect at a potential below the critical pitting potential. An impressed current can be used; or in good conducting media (e.g., seawater), stainless steel can be coupled to an approximately equal or greater area of zinc, iron, or aluminum [44]. Austenitic stainless steels used to weld mild-steel plates, or 18–8 propellers on steel ships, do not pit.
2. Add extraneous anions (e.g., OH⁻ or NO_3^-) to chloride environments.
3. Reduce oxygen concentration of chloride environments (e.g., NaCl solutions).
4. For type 304, operate at the lowest temperature possible.*

19.2.4.3 Reducing or Avoiding Crevice Corrosion. There are three ways to reduce or avoid crevice corrosion:

1. Avoid crevices between metals or between metals and nonmetals. Periodically remove contaminating surface films using alkaline cleaners with stainless-steel wool or the equivalent.
2. Avoid stagnant solutions. Circulate, stir, and aerate electrolytes in contact with stainless steels. Ensure uniform composition of electrolyte at all regions of the metal surface.
3. Cathodically protect, ideally to the potential of corroding active metal at the crevice. Approaching such a potential reduces the corrosion rate, but not to zero. In seawater, use of sacrificial iron and similar less-noble-metal anodes have proved useful [45].

19.2.5 Stress-Corrosion Cracking and Hydrogen Cracking

In the presence of an applied or residual *tensile* stress, stainless steels may crack transgranularly when exposed to certain environments (Fig. 19.7). Compressive stresses, to the contrary, are not damaging. The higher the tensile stress, the shorter is the time to failure. Although at low stress levels, time to cracking may be long, there is, in general, no practical minimum stress below which cracking will not occur, given sufficient time in a critical environment.

Damaging environments that cause cracking may differ for austenitic, compared to martensitic or ferritic, stainless steels. For austenitic steels, the two major damaging ions are hydroxyl and chloride (OH⁻ and Cl⁻). A boiling, relatively concentrated chloride that hydrolyzes to a slightly acid pH, such as $FeCl_2$ or $MgCl_2$, can cause cracking of thick sections of stressed 18–8 within hours. A

---

*In aerated 4% NaCl, maximum weight loss by pitting occurs at 90°C. [H. Uhlig and M. Morrill, *Ind. Eng. Chem.* **33**, 875 (1941)]. To avoid S.C.C. and to minimize pitting, operating temperatures in NaCl solutions should be kept below 60–80°C.

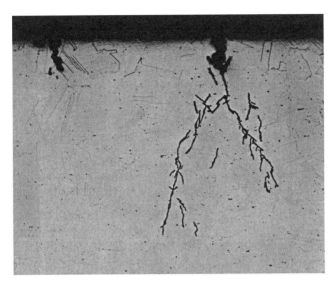

Figure 19.7. Stress-corrosion cracking of 18–8, type 304 stainless steel exposed to calcium silicate insulation containing 0.02–0.5% chlorides, 100°C (250×). Note that the cracks for this environment start at a pit. An undulating path accounts for disconnected cracks as viewed in one plane [47]. (Copyright ASTM International. Reprinted with permission.)

solution of concentrated $MgCl_2$ boiling at about 154°C, for example, is used as an accelerated test medium. The presence of dissolved oxygen in such solutions is not necessary for cracking to occur, but its availability hastens damage, as does the presence of oxidizing ions, such as Fe^{3+}. Pitting is not a preliminary requirement for initiation of cracks. In NaCl and similar neutral solutions, on the other hand, cracking of austenitic stainless steels is observed only if dissolved oxygen or other oxidizing agent is present [46], and the amount of chloride needed to cause damage can be extremely small (Fig. 19.8). Cracking of 18–8 stainless tubes in heat exchangers has been observed in practice after contact with cooling waters containing 25 ppm Cl^- or less, and cracking has also been induced by small amounts of chlorides contained in magnesia insulation wrapped around stainless-steel tubes [47]. In these instances, if small pits form initially at which chlorides concentrate (Figs. 19.5 and 19.7) (see Section 19.2.4.1), the cracking tendency is accentuated by concentrated $FeCl_2$ and analogous metal chlorides within the pits. Hence, oxygen may induce stress-corrosion cracking in sodium chloride solutions because pitting occurs when it is present. Another contributing factor is a shift of the corrosion potential in the presence of oxygen, but not in its absence, to values that are noble to the critical potential for stress-corrosion cracking. For this situation, stress-corrosion cracking can occur regardless of whether corrosion pits develop.

Cracking by alkaline solutions requires relatively high concentrations of OH^-; hence, cracking of 18–8 usually occurs not in alkaline boiler water, but

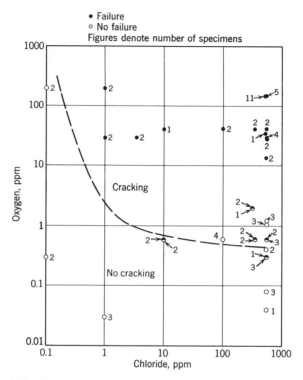

Figure 19.8. Relation between chloride and oxygen content of boiler water on stress-corrosion cracking of austenitic 18–8-type stainless steels exposed to steam phase with intermittent wetting; pH 10.6, 50 ppm PO_4^{3-}, 242–260°C (467–500°F), 1–30 days' exposure [W. Williams and J. Eckel, *J. Am. Soc. Nav. Eng.* **68**, 93 (1956)].

rather within the splash zone above the water line where dissolved alkalies concentrate by evaporation. Failures can occur in such solutions in the absence of dissolved oxygen [48]. There is no evidence that transgranular stress-corrosion cracking occurs in pure water or pure steam.

Cracking of stressed, nickel-free ferritic steels, in general, does not occur in the chloride media described previously, these steels being sufficiently resistant to warrant their practical use in preference to austenitic stainless steels in chloride-containing solutions. They (e.g., type 430) are also resistant to 55% $Ca(NO_3)_2$ solution boiling at 117°C and to 25% NaOH solution boiling at 111°C [49]. When alloyed with more than 1.5% Ni, the 18% Cr–Fe, 0.003% C stainless steels, cold-rolled, are susceptible to transgranular stress-corrosion cracking in $MgCl_2$ boiling at 130°C. When annealed at 815°C for 1 h, only the alloy containing 2% Ni fails; both higher and lower nickel contents resist cracking up to 200 h [50]. The effect of nickel is partly explained by an observed shift of potentials such that, at and above 1.5% Ni, the corrosion potential of the cold-rolled material becomes noble to the critical potential; hence cracking occurs. The reverse order of potentials applies to alloys of lower nickel content [50].

In slightly or moderately acidic solutions, the martensitic steels, when heat-treated to high hardness values, are very sensitive to cracking, particularly in the presence of sulfides, arsenic compounds, or oxidation products of phosphorus or selenium. The specific anions of the acid make little difference so long as hydrogen is evolved, contrary to the situation for austenitic steels, for which only specific anions are damaging. Also, cathodic polarization, rather than protecting against cracking, accelerates failure. All these facts suggest that the martensitic steels under these conditions fail not by stress-corrosion cracking but by hydrogen cracking (see Section 8.4). The more ductile ferritic stainless steels undergo hydrogen blistering instead when they are cathodically polarized in seawater, especially at high current densities. Austenitic stainless steels are immune to both hydrogen blistering and cracking.

Galvanic coupling of active metals to martensitic stainless steels may also lead to failure because of hydrogen liberated on the stainless (cathodic) surface. Such failures have been demonstrated in laboratory tests [51]. As mentioned in Section 8.4, practical instances have been noted of self-tapping martensitic stainless-steel screws cracking spontaneously soon after being attached to an aluminum roof in a seacoast atmosphere. Similarly, hardened martensitic stainless-steel propellers coupled to the steel hull of a ship have failed by cracking soon after being placed in service. Severely cold-worked 18–8 austenitic stainless steels may also fail under conditions that would damage the martensitic types [52]. Here again, sulfides accelerate damage, and since the alloy on cold working undergoes a phase transformation to ferrite, the observed effect is probably another example of hydrogen cracking.

Martensitic stainless steels and also the precipitation-hardening types, heat-treated to approximately >200,000 psi (>1,400 MPa) yield strength, have been reported to crack spontaneously in the atmosphere, in salt spray, and immersed in aqueous media, even when not coupled to other metals [53–55]. Martensitic stainless-steel blades of an air compressor [56] have failed along the leading edges where residual stresses were high and where condensation of moisture occurred. Stressed to about 75% of the yield strength, and exposed to a marine atmosphere, life of ultra-high-strength 12% Cr martensitic stainless steels was in the order of 10 days or less [57]. It is generally accepted that ultra-high strength steels crack by a hydrogen embrittlement mechanism [55].

19.2.5.1 Metallurgical Factors.

Austenitic stainless steels containing more than about 45% Ni are immune to stress-corrosion cracking in boiling $MgCl_2$ solution and probably in other chloride solutions as well (Fig. 19.9) [58]. Edeleanu and Snowden [48] noted that high-nickel stainless steels were more resistant to cracking in alkalies. Increasing the amount of nickel in austenitic stainless steels shifts the critical potential for stress-corrosion cracking in $MgCl_2$ solution in the noble direction more rapidly than it shifts the corresponding corrosion

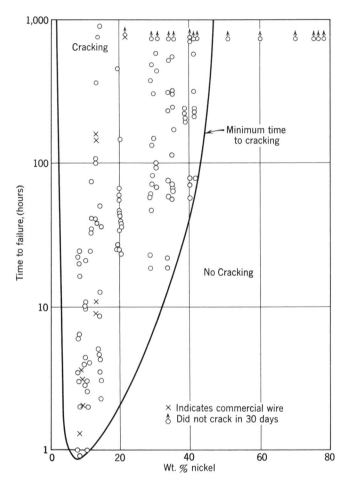

Figure 19.9. Stress-corrosion cracking of 15–26% Cr–Fe–Ni alloy wires in boiling 42% MgCl₂. Low-nickel or nickel-free alloys are ferritic and do not crack. Austenitic alloys do not crack above 45% Ni [58].

potential; hence, the alloys become more resistant [59]. At and above approximately 45% Ni, the alloys resist stress-corrosion cracking regardless of applied potential, indicating that not environmental, but metallurgical factors, such as unfavorable dislocation arrays or decreasing interstitial nitrogen solubility, become more important.

For austenitic steels that are resistant to transformation on cold working (e.g., type 310), nitrogen is the element largely responsible for stress-cracking susceptibility, whereas additions of carbon decrease susceptibility (Fig. 19.10) [60]. The effect is related to alloy imperfection structure rather than to any shift of either critical or corrosion potential [59]. Stabilizing additions effective in preventing intergranular corrosion, such as titanium or columbium, have no

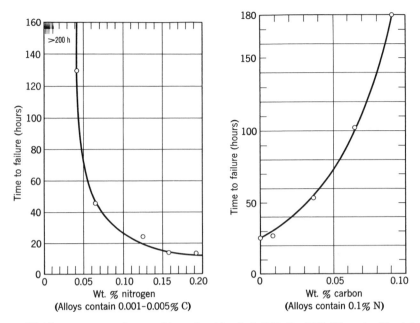

Figure 19.10. Stress-corrosion cracking of cold-rolled 19% Cr, 20% Ni austenitic stainless steels in boiling MgCl₂ (154°C) as affected by carbon or nitrogen content [60].

beneficial effect on stress-corrosion cracking, nor do alloying additions such as 2–3% Mo in type 316 stainless steel. Chloride S.C.C. occurs primarily above about 90°C (190°F), but, in acidified chloride solutions, S.C.C. can occur at lower temperatures [61, 62].

19.2.6 Cracking of Sensitized Austenitic Alloys in Polythionic Acids

During shutdown of certain 18–8 stainless-steel oil-refinery equipment operating in the carbide precipitation temperature range, it was found that rapid intergranular cracking occurred in the presence of a tensile stress (but not otherwise). The cause was traced to polythionic acids ($H_2S_xO_6$ where $x = 3,4,5$) formed by reaction of residual metallic sulfide films on the equipment surface with moist air at room temperature [63–66]. Such acids can be produced in the laboratory by bubbling H_2S through water saturated with SO_2. It is important to the mechanism of failure that these acids are readily reduced cathodically by operating corrosion cells.

One function of polythionic acids is to act as cathodic depolarizers, thereby stimulating dissolution of chromium-depleted grain-boundary material. Another possible function is that their cathodic reduction products (H_2S or analogous compounds) stimulate absorption to interstitial hydrogen by chromium-depleted alloy. The latter alloy along grain boundaries, ferritic in structure, is subject to hydrogen cracking when stressed, unlike the grains austenitic in structure that are resistant. The function of >2 ppm sulfur as Na_2S or as cathodic reduction products of sulfites (SO_3^{2-}) or of thiosulfates ($S_2O_3^{2-}$) to induce hydrogen cracking of a 0.77% C high-strength steel and also of ferritic and martensitic stainless steels was demonstrated in seawater [67]. It is expected that polythionic acids would act in an analogous manner.

The mechanism of failure has been described as S.C.C. by electrochemical dissolution along an active path exposed through application of stress [68]. Alternatively, the mechanism may be one of progressive hydrogen cracking along the grain boundaries of sensitized alloy. An acid environment is necessary to failure because reaction with alloy supplies the required hydrogen; it also favors formation of H_2S (rather than HS^- or S^{2-}), which is the primary catalyst poison stimulating absorption of atomic hydrogen by the alloy. Aqueous SO_2 solutions are also found to cause intergranular cracking of sensitized 18–8, similar to that by polythionic acids, because SO_3^{2-} is readily reduced at cathodic sites forming H_2S or similarly acting reduction products, whereas sulfuric acid is much less effective in causing cracking because SO_4^{2-} is not similarly reduced.

Cracking in polythionic acids is most pronounced in the potential range 0.04–0.34 V (S.H.E.) [68]. This potential range lies above the values associated with hydrogen ion discharge and would seem to rule out hydrogen cracking. However, the relevant potential is not the one measured at the alloy surface, but the one prevailing at the crack tip, which can be appreciably more active. In practice, use of stabilized stainless steels avoids intergranular cracking of the kind described.

To Reduce or Eliminate Cracking

Austenitic Stainless Steels

a. Cathodically protect. The critical potential of 18–8 stainless steel in $MgCl_2$ at 130°C is –0.128 V (S.H.E.). Coupling of stressed 18–8 to a small area of nickel ($\phi_{corr} = -0.18$ V) prevents cracking in this medium, or (as a porous nickel coating) in water containing 50 ppm Cl^- at 300°C [69].

b. In acid environments, eliminate Cl^-. In neutral or slightly alkaline chloride environments, eliminate dissolved oxygen and other oxidizing ions. Add extraneous ions (e.g., NO_3^-, I^-, acetates).

c. Avoid high concentrations of OH^-. High concentrations of alkalies occurring initially or incidentally through concentration at crevices or in vapor zones are damaging. Add buffering ions (e.g., PO_4^{3-}).

d. Use alloys containing >45% Ni or reduce the nitrogen content (and other detrimental impurities, if present) to the lowest possible value (<0.04% N in the case of 20% Cr, 20% Ni, 0.001% C stainless steel) (Fig. 19.10).

e. Substitute ferritic alloys (e.g., type 430 or controlled purity Cr–Mo steels). However, ferritic alloys may become embrittled by hydrogen or they may blister when galvanically coupled to more active metals in certain environments.

f. Operate below 60–80°C (140–175°F).

Martensitic, Precipitation-Hardening, or Ferritic Stainless Steels

a. Avoid excess current if cathodic protection is applied.

b. Avoid galvanic coupling to more active metals.

c. Temper martensitic or precipitation-hardening steels to the lowest possible hardness values. Exposed to the atmosphere, hardness should be below about Rockwell C 40. Maximum susceptibility of types 410 and 420 stainless steels to cracking in salt spray or to hydrogen cracking occurs after tempering for 2 h at 425–550°C (800–1000°F); minimum susceptibility to hydrogen cracking occurs after tempering for 2 h at 260°C (500°F) [70].

19.2.7 Galvanic Coupling and General Corrosion Resistance

Since stainless steels are passive and exhibit a noble potential, they can be coupled successfully to metals that are either passive or inherently noble. This includes metals and alloys like silver; silver solder; copper; nickel; 70% Ni–Cu alloy; 76% Ni, 16% Cr, 7% Fe alloy; and, usually, aluminum in environments in which it remains passive.

As discussed in Section 19.2.4, stainless steels are best employed under fully aerated or oxidizing conditions, which favor the passive state. Whether used in handling chemicals or exposed to the atmosphere, the alloy surface should always be kept clean and free of surface contamination. Otherwise, crevice corrosion may cause pitting and localized rusting. Austenitic stainless steels cooled too slowly through the sensitizing temperature zones tend to rust in the atmosphere.

In brief, stainless steels are resistant to:

1. Nitric acid, over a wide range of concentrations and temperatures.
2. Very dilute sulfuric acid at room temperature when aerated; higher concentrations (e.g., 10%) and at boiling temperatures if Fe^{3+}, Cu^{2+}, or nitric acid is added as an inhibitor [71]; or at lower temperatures if small amounts of Cu, Pt, or Pd are alloyed (see Section 6.4). Also resistant to cold or hot sulfuric acid if anodically protected.
3. Many organic acids, including almost all food acids and acetic acid (but not boiling glacial acetic acid).
4. Sulfurous acid (in the absence of SO_4^{2-} or Cl^-).
5. Alkalies, except under stress in hot concentrated caustic solutions.

6. The atmosphere. Types 302 and 304 have been used successfully as architectural trim of storefronts and buildings (e.g., the Chrysler and Empire State buildings in New York City). These and type 430 are used for auto trim.

Stainless steels are *not* resistant to:

1. Dilute or concentrated HCl, HBr, and HF; also not resistant to salts that hydrolyze to these acids.
2. Oxidizing chlorides (e.g., $FeCl_3$, $HgCl_2$, $CuCl_2$, and NaOCl).
3. Seawater, except for brief periods or when cathodically protected.
4. Photographic solutions, especially fixing solutions containing thiosulfates (pitting occurs).
5. Some organic acids, including oxalic, formic, and lactic acids.
6. Stressed austenitic alloys (e.g., type 304) in waters containing Cl^- plus O_2 at temperatures above 60–80°C.

REFERENCES

1. G. Tammann, *Lehrbuch der Metallkunde*, Leipzig, 1932, pp. 428–433.
2. H. Pickering and C. Wagner, *J. Electrochem. Soc.* **114**, 698 (1967); H. Pickering, *J. Electrochem. Soc.* **115**, 143 (1968).
3. W. Katz, in *Korrosion und Korrosionsschutz*, F. Tödt, editors, de Gruyter, Berlin, 1955, pp. 412–414.
4. J. J. Noël, Effects of metallurgical variables on aqueous corrosion, in *ASM Handbook*, Vol. 13A, *Corrosion: Fundamentals, Testing, and Protection*, ASM International, Materials Park, OH, 2003, p. 259.
5. J. R. Scully and A. Lucente, Corrosion of amorphous metals, in *ASM Handbook*, Vol. 13B, *Corrosion: Materials*, ASM International, Materials Park, OH, 2005, pp. 476–489.
6. J. Stodart and M. Faraday, *Q. J. Sci. Lit. Arts* **9**, 319 (1820); *Philos. Trans. Roy. Soc.* **112**, 253 (1822).
7. P. Berthier, *Ann. Chim. Phys.* **17**, 55 (1821); *Ann. Mines* **6**, 573 (1821).
8. L. Guillet, *Rev. Met.* **1**, 155 (1904); **2**, 350 (1905); **3**, 332 (1906).
9. P. Monnartz, *Metallurgie* **8**, 161 (1911).
10. M. L. Erbing Falkland, Duplex stainless steels, in *Uhlig's Corrosion Handbook*, 2nd edition, R. W. Revie, editor, Wiley, New York, 2000, pp. 651–666.
11. J. F. Grubb, T. DeBold, and J. D. Fritz, Corrosion of wrought stainless steels, in *ASM Handbook*, Vol. 13B, *Corrosion: Materials*, ASM International, Materials Park, OH, 2005, pp. 56–57.
12. R. Steigerwald, A. Bond, H. Dundas, and E. Lizlovs, *Corrosion* **33**, 279 (1977).
13. J. Truman, R. Perry, and G. Chapman, *J. Iron Steel Inst.* **202**, 745 (1964).

14. H. H. Uhlig, *Trans. Am. Soc. Metals* **30**, 947 (1942).

15. J. Upp, F. Beck, and M. Fontana, *Trans. Am. Soc. Metals* **50**, 759 (1958).

16. W. Binder, C. Brown, and R. Franks, *Trans. Am. Soc. Metals* **41**, 1301 (1949).

17. H. Ebling and M. Scheil, in *Advances In Technology of Stainless Steels and Related Alloys*, American Society for Testing And Materials, Philadelphia, 1965, p. 275.

18. E. Bain, R. Aborn, and J. Rutherford, *Trans. Am. Soc. Metals (Steel Treating)* **21**, 481 (1933); B. Strauss, H. Schottky, and J. Hinnüber, *Z. Anorg. Allgem. Chem.* **188**, 309 (1930). See also M. A. Streicher, Austenitic and ferritic stainless steels, in *Uhlig's Corrosion Handbook*, 2nd edition, R. W. Revie, editor, Wiley, New York, 2000, p. 613.

19. H. H. Uhlig, *Trans. Electrochem. Soc.* **87**, 193 (1945). See also Ref. 16.

20. P. Schafmeister, *Arch. Eisenhüttenw.* **10**, 405 (1936–37).

21. P. Lacombe, in Proceedings of the First International Congress on Metallic Corrosion, Butterworths, London, 1962; A. Baümel et al., Corros. Sci. 4, 89 (1964), pp. 21–35.

22. M. A. Streicher, Austenitic and ferritic stainless steels, in *Uhlig's Corrosion Handbook*, 2nd edition, R. W. Revie, editor, Wiley, New York, 2000, p. 613.

23. Ref. 22, p. 615.

24. M. Streicher, *J. Electrochem. Soc.* **106**, 161 (1959).

25. H. Coriou et al., *Corrosion* **22**, 280 (1966).

26. J. Armijo, *Corros. Sci.* **7**, 143 (1967).

27. J. Armijo, *Corrosion* **24**, 24 (1968).

28. Y. Kolotyrkin and O. Kasparova, in *Corrosion and Corrosion Protection*, R. Frankenthal and F. Mansfeld, editors, Electrochemical Society, Pennington, NJ, 1981, p. 186.

29. J. Wanklyn and D. Jones, *J. Nucl. Mater.* **2**, 154 (1959).

30. H. Copson and G. Economy, *Corrosion* **24**, 55 (1968).

31. H. Coriou et al., in *Fundamental Aspects of Stress Corrosion Cracking*, National Association of Corrosion Engineers, Houston, TX, 1969, pp. 352–359.

32. W. Hübner, M. de Pourbaix, and G. Östberg, in Proceedings of the Fourth International Congress on Metallic Corrosion, Amsterdam, September 1969, National Association of Corrosion Engineers, Houston, TX, 1972, p. 65.

33. R. Duncan, J. Armijo, and A. Pickett, *Mater. Prot.* **8**, 37 (1969).

34. R. Lula, A. Lena, and G. Kiefer, *Trans. Am. Soc. Metals* **46**, 197 (1954).

35. E. Houdremont and W. Tofaute, *Stahl Eisen* **72**, 539 (1952).

36. J. Hochmann, *Rev. Met.* **48**, 734 (1951).

37. A. Bond and E. Lizlovs, *J. Electrochem. Soc.* **116**, 1305 (1969).

38. A. Baümel, *Stahl Eisen* **84**, 798 (1964).

39. R. Frankenthal and H. Pickering, *J. Electrochem. Soc.* **120**, 23 (1973).

40. J. Matthews and H. Uhlig, *Corrosion* **7**, 419 (1951).

41. P. Schafmeister and W. Tofaute, *Tech. Mitt. Krupp* **3**, 223 (1935).

42. P.-E. Arnvig and A. D. Bisgård, Paper 437, CORROSION/96, NACE, Houston, Texas, 1996; P. Ernst and R. C. Newman, *Corrosion Sci.* **49**(9), 3705 (2007).

43. H. H. Uhlig, *Trans. Am. Inst. Min. Metall. Eng.* **140**, 411 (1940).

44. F. L. LaQue, in *Corrosion Handbook*, H. H. Uhlig, editor, Wiley, New York, 1948, p. 416.

45. T. Lee and A. Tuthill, *Mater. Perf.* **22**(1), 48 (1983).

46. P. Elliott, in *ASM Handbook*, Vol. 13A, *Corrosion: Fundamentals, Testing, and Protection*, ASM International, Materials Park, OH, 2003, p. 915; W. Williams and J. Eckel, *J. Am. Soc. Nav. Eng.* **68**, 93 (February 1956).

47. A. Dana, Jr., *Am. Soc. Testing Mater. Bull.* **225**, 46 (1957).

48. C. Edeleanu and P. Snowden, *J. Iron Steel Inst.* **186**, 406 (1957).

49. A. Bond, J. Marshall, and H. Dundas, in *Stress Corrosion Testing*, Special Technology Publication No. 425, American Society for Testing and Masterials, Philiadelphia, 1967, p. 116.

50. R. Newberg and H. Uhlig, *J. Electrochem. Soc.* **119**, 981 (1972).

51. H. H. Uhlig, *Metal Progr.* **57**, 486 (1950).

52. F. Bloom, *Corrosion* **11**, 351t (1955).

53. E. Phelps and R. Mears, *First International COngres On Metallic Corrosion*, Butterworths, London, 1962, p. 319.

54. J. Truman, R. Berry, and G. Chapman, *J. Iron Steel Inst.* **202**, 745 (1964).

55. B. Craig, in *ASM Handbook*, Vol. 13A, *Corrosion: Fundamentals, Testing, and Protection*, ASM International, Materials Park, OH, 2003, p. 373.

56. W. Badger, *Soc. Automot. Eng. Trans.* **62**, 307 (1954).

57. E. Phelps, in *Fundamental Aspects of Stress Corrosion Cracking*, National Association of Corrosion Engineers, Houston, TX, 1969, p. 398.

58. H. Copson, in *Physical Metallurgy of Stress Corrosion Fracture*, T. Rhodin, editor, Interscience, New York, 1959, p. 247.

59. H. Lee and H. Uhlig, *J. Electrochem. Soc.* **117**, 18 (1970).

60. H. Uhlig and J. Sava, quoted in Fracture, Vol. 3, H. Liebowitz, editor, Academic Press, New York, 1971, p. 656.

61. B. Phull, Evaluating stress-corrosion cracking, in *ASM Handbook*, Vol. 13A, *Corrosion: Fundamentals, Testing, and Protection*, ASM International, Materials Park, OH, 2003, p. 604.

62. J. D. Harston and J. C. Scully, *Corrosion* **25**, 493 (1969).

63. C. Samans, *Corrosion* **20**, 256t (1964).

64. G. Cragnolino and D. Macdonald, *Corrosion* **38**, 406 (1982).

65. H. Nishida, K. Nakamura, and H. Takahashi, *Mater. Perf.* **23**(4), 38 (1984).

66. R. D. Kane, Corrosion in petroleum refining and petrochemical operations, in *ASM Handbook*, Vol. 13C, *Corrosion: Environments and Industries*, ASM International, Materials Park, OH, 2006, pp. 991–992.

67. H. Uhlig, *Mater. Perf.* **16**(1), 22 (1977).

68. I. Matsushima, in Proceedings of the Sixth International Congress on Metallic Corrosion, Sydney, Australia, 1975, Australasian Corrosion Association, Melbourne, Australia, p. 514.

69. P. Neumann and J. Griess, *Corrosion* **19**, 345t (1963).

70. P. Lillys and A. Nehrenberg, *Trans. Am. Soc. Metals* **48**, 327 (1956).

71. J. Cobb and H. Uhlig, *J. Electrochem. Soc.* **99**, 13 (1952).

GENERAL REFERENCES

B. D. Craig, *Fundamental Aspects of Corrosion Films in Corrosion Science*, Plenum Press, New York, 1991.

Glassy Alloys

J. R. Scully and A. Lucente, Corrosion of amorphous metals, in *ASM Handbook*, Vol. 13B, *Corrosion: Materials*, ASM International, Materials Park, OH, 2005.

Stainless Steels

M. Blair, Corrosion of cast stainless steels, *ASM Handbook*, Vol. 13B, *Corrosion: Materials*, ASM International, Materials Park, OH, 2005, pp. 78–87.

M. L. Erbing Falkland, Duplex stainless steels, in *Uhlig's Corrosion Handbook*, 2nd edition, R. W. Revie, editor, Wiley, New York, 2000, pp. 651–666.

J. F. Grubb, Martensitic stainless steels, in *Uhlig's Corrosion Handbook*, 2nd edition, R. W. Revie, editor, Wiley, New York, 2000, pp. 667–676.

J. F. Grubb, T. DeBold, and J. D. Fritz, Corrosion of wrought stainless steels, in *ASM Handbook*, Vol. 13B, *Corrosion: Materials*, ASM International, Materials Park, OH, 2005, pp. 54–77.

H. S. Khatak and B. Raj, editors, *Corrosion of Austenitic Stainless Steels: Mechanism, Mitigation and Monitoring*, Narosa Publishing House, New Delhi, 2002.

A. J. Sedriks, *Corrosion of Stainless Steels*, 2nd edition, Wiley, New York, 1996.

M. A. Streicher, Austenitic and ferritic stainless steels, in *Uhlig's Corrosion Handbook*, 2nd edition, R. W. Revie, editor, Wiley, New York, 2000, pp. 601–650.

20

COPPER AND COPPER ALLOYS

20.1 COPPER

$$Cu^{2+} + 2e^- \rightarrow Cu \qquad \phi^\circ = 0.337\,V$$

Copper dissolves anodically in most aqueous environments forming the divalent ion Cu^{2+}. Equilibrium relations at the metal surface indicate that the reaction $Cu + Cu^{2+} \rightleftarrows 2Cu^+$ is displaced far to the left (see Problem 1, p. 381). On the other hand, if complexes are formed, as, for example, between Cu^+ and Cl^- in a chloride solution, the continuous depletion of Cu^+ by conversion to $CuCl_2^-$ favors the univalent ion as the major dissolution product. Comparatively, when copper is heated in air at elevated temperatures, a Cu_2O film develops that is covered by a thin film of CuO, which is formed as the film thickness increases [1].

In Fig. 20.1, the theoretical potential–pH domains of corrosion, immunity, and passivation of copper at 25°C are shown, considering that passivation occurs by formation of the oxides Cu_2O and CuO [2].

Copper and its alloys, although toxic to fouling organisms, such as seaweed and shellfish, are subject to microbiologically influenced corrosion, undergoing severe attack by sulfide if conditions allow the growth of sulfate reducing bacteria [3, 4].

Corrosion and Corrosion Control, by R. Winston Revie and Herbert H. Uhlig
Copyright © 2008 John Wiley & Sons, Inc.

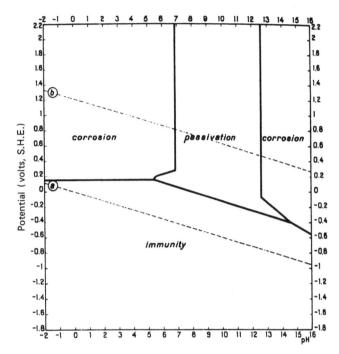

Figure 20.1. Theoretical potential–pH domains of corrosion, immunity, and passivation for copper, assuming that passivation results from the formation of the oxides, Cu_2O and CuO [2]. (M. Pourbaix, *Atlas of Electrochemical Equilibria in Aqueous Solutions*, 2nd English edition, p. 389. Copyright NACE International 1974 and CEBELCOR.)

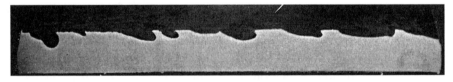

Figure 20.2. Longitudinal cross section of undercut pits associated with impingement attack of condenser alloy by salt water (7×) (flow of water from left to right).

Copper is a metal widely used because of good corrosion resistance combined with mechanical workability, excellent electrical and thermal conductivity, and ease of soldering or brazing. It is noble to hydrogen in the Emf Series, and it is thermodynamically inert with no tendency to corrode in water and in non-oxidizing acids free of dissolved oxygen. In oxidizing acids and in aerated solutions of ions that form copper complexes (e.g., CN^-, NH_4^+), corrosion can be severe. Copper is also characterized by sensitivity to corrosion by high-velocity water and by aqueous solutions, called *impingement attack* (Fig. 20.2). The rate increases with dissolved oxygen content, whereas in air-free high-velocity water up to at least 7.5 m/s (25 ft/s), impingement attack is negligible. In groundwaters, the corrosion rate increases with increasing salinity (concentration of dissolved salts) [5]. The rate of corrosion increases with dissolved oxygen concentration and temperature. In an open system, the rate of corrosion may show a maximum as temperature is increased because the solubility of oxygen decreases with temperature [5].

20.1.1 Corrosion in Natural Waters

Copper is resistant to seawater, with the corrosion rate in temperate climates being about 0.5–1.0 gmd (0.001–0.002 ipy) in quiet water and somewhat higher in moving water. In tropical climates, the rate increases by $1^1/_2$–2 times. It is one of the very few metals that remains free of fouling organisms, with normal corrosion being sufficient to release copper ions in concentrations that poison marine life.

In seawater and in fresh waters, corrosion resistance depends on the presence of a surface oxide film through which oxygen must diffuse in order for corrosion to continue. Upon exposure to distilled water at room temperature, the oxide film on copper is found to be a mixture of Cu_2O and CuO [6, 7]. Visible light markedly retards the rate of oxide formation [6]. The film is easily disturbed by high-velocity water or is dissolved by either carbonic acid or the organic acids that are found in some fresh waters and soils, leading to a high corrosion rate. For example, a hot water in Michigan, zeolite softened with resultant high concentration of $NaHCO_3$, perforated copper water pipe within 6–30 months [8]. The same water unsoftened, on the other hand, was not nearly as corrosive because a protective film of $CaCO_3$ containing some silicate was deposited on the metal surface. Copper piping used to transport well water in homes in some areas of New Jersey [9] developed pinhole leaks (pits) within 10 years.

Even if the corrosion rate is not excessive and the copper is durable, a water containing carbonic or other acids nevertheless may corrode copper and copper-base alloys sufficiently to cause blue staining of bathroom fixtures and cause an increase in the corrosion rate of iron, galvanized steel, or aluminum surfaces with which such water comes into contact. Accelerated corrosion in this instance is caused by a replacement reaction in which copper metal is deposited on the base metal, forming numerous small galvanic cells. Treatment of acid waters or waters of negative saturation index with lime or with sodium silicate reduces the corrosion rate sufficiently to overcome both staining and accelerated corrosion of other metals, except aluminum. Aluminum is an exception because of its sensitivity to extremely small amounts of Cu^{2+} in solution, with the usual water treatment being inadequate to reduce Cu^{2+} pickup to nondamaging levels. The U.S. Environmental Protection Agency Lead and Copper Rule, implemented in 1991, limits total copper in potable water to a maximum of 1.3 mg L^{-1} [10].

It is possible to avoid the damaging effects of Cu^{2+} contamination of water by using copper piping coated on the inside surface with tin (tinned copper). For this reason, copper tanks for holding distilled water or for water storage on ships should be tinned [11]. The tin coating must be pore free in order to avoid accelerated attack of copper at exposed areas, since tin (or copper–tin intermetallic compounds) is cathodic to copper.

Copper piping is usually satisfactory for transporting seawater and hard waters, whether hot or cold. Hard waters form a protective film of calcium compounds on the copper surface that protects against corrosion. In soft waters, particularly those containing appreciable amounts of free carbon dioxide, and in carbonated waters in general, the corrosion rate of copper is significant. The initial corrosion rate in distilled water may be 0.051–0.16 mm/year (12–7 mdd), whereas the rate in an aggressive supply water, as high as 0.26 mm/year (62 mdd), may result in sufficient copper to stain plumbing fixtures [11].

In addition to the effects just described, a pitting type of corrosion may be induced in waters having relatively good conductivity if dirt or rust from other parts of the system accumulate on the copper surface. Differential aeration cells are formed, supplemented in some cases by turbulent flow, which initiates impingement attack. The resulting corrosion is sometimes called *deposit attack*. Periodic cleaning of the piping or tubing usually prevents corrosion by such deposits.

Even when the copper surface is free of deposits, pitting nevertheless may occur in waters of certain compositions or as a result of pipe fabrication methods. In this respect, the types of pits that can develop on copper have been divided into categories [12, 13]. The first category, type 1 pitting, usually takes place in hard or moderately hard waters and is usually restricted to the cold portions of the system. Pits of this type contain a higher proportion of cuprous chloride than do pits formed in soft water, and mounds consist largely of cupric carbonate. The surrounding surface scale is largely $CaCO_3$ stained green, beneath which is a highly cathodic film of carbon [14] that was the residue of the lubricant used during pipe manufacture. Type 1 pitting has been largely eliminated by improvements in the tube manufacturing processes [11].

In the second category are *soft waters* containing small amounts of manganese salts. With these waters, pitting, designated as Type 2 pitting, usually occurs in the hottest part of the system, above 60°C, in waters with low bicarbonate/sulfate ratio, < 1. In this case, corrosion products within the pit are Cu_2O plus a small amount of chloride, whereas mounds around the pits may contain basic copper sulfate, and the pipe surface is covered by a black scale rich in manganese oxide.

To protect copper hot-water tanks from pitting, sacrificial anodes of aluminum are sometimes used, particularly in certain areas of Great Britain [15].

In summary, copper is resistant to:

1. Seawater.
2. Fresh waters, hot or cold. Copper is especially suited to convey soft, aerated waters that are low in carbonic and other acids.
3. Deaerated, hot or cold, dilute H_2SO_4, H_3PO_4, acetic acid, and other non-oxidizing acids.

4. Atmospheric exposure.

5. Halogens under specific conditions—for example, fluorine below about 400°C, dry HF below 600°C, dry Cl_2 below 150°C, dry Br_2 at 25°C, iodine below 375°C [16].

Copper is *not* resistant to:

1. Oxidizing acids, for example, HNO_3, hot concentrated H_2SO_4, and aerated nonoxidizing acids (including carbonic acid).

2. NH_4OH (plus O_2). A complex ion, $Cu(NH_3)_4^{2+}$, forms. Substituted NH_3 compounds (amines) are also corrosive. These compounds are those that cause stress-corrosion cracking of susceptible copper alloys.

3. High-velocity aerated waters and aqueous solutions. In corrosive waters (high in O_2 and CO_2, low in Ca^{2+} and Mg^{2+}), the velocity should be kept [8, 13] below 1.2 m/s (4 ft/s); in less corrosive waters, <65°C (<150°F), the velocity should be kept below 2.4 m/s (8 ft/s).

4. Oxidizing heavy metal salts, for example, $FeCl_3$ and $Fe_2(SO_4)_3$.

5. Hydrogen sulfide, sulfur, and some sulfur compounds.

20.2 COPPER ALLOYS

Tin bronzes are copper–tin alloys noted for their high strength. Alloys containing more than 5% Sn are especially resistant to impingement attack. The copper–silicon alloys containing 1.5–4% Si have better physical properties than copper and similar general corrosion resistance. Seawater immersion tests at Panama showed that 5% Al–Cu was the most resistant of the common copper-base alloys, losing only 20% of the corresponding weight loss of copper after 16 years [17].

20.2.1 Copper–Zinc Alloys (Brasses)

Copper–zinc alloys have better physical properties than copper alone, and they are also more resistant to impingement attack; hence, brasses are used in preference to copper for condenser tubes. Corrosion failures of brasses usually occur by dezincification, pitting, or stress-corrosion cracking. The tendency for brass to corrode in these ways, except for pitting attack, varies with zinc content, as shown in Fig. 20.3. Pitting is usually caused by differential aeration cells or high-velocity conditions. It can normally be avoided by keeping the brass surface clean at all times and by avoiding velocities and design geometry that lead to impingement attack.

Various names have become attached to brasses of different zinc content. *Muntz metal*, 40% Zn–Cu, is used primarily for condenser systems that use fresh water (e.g., Great Lakes water) as coolant. *Naval brass* is a similar composition, but containing 1% Sn. *Manganese bronze* is also similar, containing about 1%

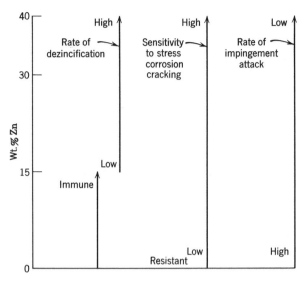

<u>Figure 20.3.</u> Trends of dezincification, stress-corrosion cracking, and impingement attack with increasing zinc content in copper–zinc alloys (brasses).

each of tin, iron, and lead; it is used for ship propellers, among other applications. Dezincification of manganese bronze propellers in seawater is avoided to some extent by the cathodic protection afforded by the steel hull.

Yellow brass, 30% Zn–Cu, is used for a variety of applications where easy machining and casting are desirable. The alloy gradually dezincifies in seawater and in soft fresh waters. This tendency is retarded by addition of 1% Sn, the corresponding alloy being called *admiralty metal* or *admiralty brass*. Addition of small amounts of arsenic, antimony, or phosphorus still further retards the rate of dezincification; the resulting alloy, called *inhibited admiralty metal*, is used in seawater or fresh-water condensers.

Red brass, 15% Zn–Cu, is relatively immune to dezincification, but is more susceptible to impingement attack than yellow brass.

20.2.2 Dealloying/Dezincification

Dezincification, as one type of dealloying, was defined in Section 2.4. Other types of dealloying include the selective dissolution of copper in copper–gold alloys.

In brasses, dezincification takes place either in localized areas on the metal surface, called *plug type* (Fig. 20.4), or uniformly over the surface, called *layer type* (Fig. 20.5). Brass so corroded retains some strength, but has no ductility. Layer-type dezincification in a water pipe may lead to splitting open of the pipe under conditions of sudden pressure increase; and, for plug type, a plug of dezincified alloy may blow out, leaving a hole. Because dezincified areas are porous,

Figure 20.4. Plug-type dezincification in brass pipe (actual size).

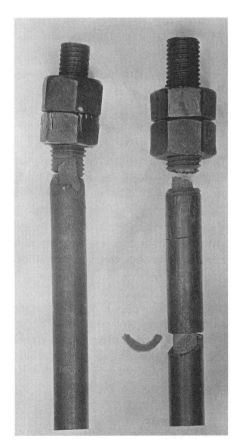

Figure 20.5. Layer-type dezincification in brass bolts (actual size).

plugs may be covered on the outside surface with corrosion products and residues of evaporated water.

Conditions of the environment that favor dezincification are (1) high temperatures, (2) stagnant solutions, especially if acid, and (3) porous inorganic scale formation. Looking at the situation metallurgically, brasses that contain 15% or less zinc are usually immune. Also, dezincification of α-brasses (up to 40% Zn) can be reduced, as mentioned earlier, by small alloying additions of tin plus a few hundredths percent arsenic, antimony, or phosphorus.

Although inhibited admiralty brass resists dezincification, conditions favoring thermogalvanic action between overheated local areas and adjoining colder areas of heat exchanger tubes can cause attack of this kind at the overheated zones. Attack is reduced by using inhibited or scaling (positive saturation index) waters [18].

The detailed mechanisms of dealloying have recently been reviewed [19]. The four main mechanisms that researchers have developed to account for the processes by which one metal in an alloy is removed by corrosion, leaving behind a porous metal, are:

1. *The ionization–redeposition mechanism*, according to which the alloy corrodes and the more noble metal (copper in copper–zinc alloys) is then redeposited to form a porous outer layer [20].
2. *The volume diffusion mechanism*, based on selective dissolution of the less noble element and volume diffusion of both elements in the solid phase [21, 22].
3. *The surface diffusion mechanism*, in which only the less noble element (zinc in copper–zinc alloys) dissolves and the remaining more noble metal is rearranged by diffusion on the surface and nucleation of islands of almost pure metal [23].
4. *The percolation model of selective dissolution*, an extension of the surface diffusion mechanism, based on preexisting interconnected paths of like elements in the binary alloy and effects of curvature on dissolution potential [24, 25].

Any one of these mechanisms may apply in specific instances of dealloying. For example, twin bands in brass, visible in the completely or incompletely dezincified layer, constituted early evidence for a volume diffusion mechanism of zinc transport from the bulk alloy to the surface [26]. In the gold–copper alloy system, copper corrodes preferentially, without dissolution of gold, leaving a porous residue of gold–copper alloy or pure gold.

20.2.3 Stress-Corrosion Cracking (Season Cracking)

Virtually all copper alloys, as well as pure copper, can be made to crack in ammonia [27]. Requirements for cracking include the presence of water, ammonia, air or oxygen, and tensile stress in the metal. Only trace amounts of ammonia are required in many cases, and cracking occurs at ambient temperature. Stress-corrosion cracking failures of copper pipe under elastomeric insulation have been attributed to wet ammoniacal environments [27, 28]. Small concentrations of hydrogen sulfide inhibit S.C.C. of brass in petroleum refinery process streams, most likely by reducing the dissolved oxygen concentration [27, 29].

When an α brass is subjected to an applied or residual tensional stress in contact with a trace of NH_3 or a substituted ammonia (amine), in the presence of oxygen (or another depolarizer) and moisture, it cracks usually along the grain boundaries (intergranular) (see Fig. 20.6). Cracking through the grains (trans-granular) may occur in specific test solutions or if the alloy is severely deformed plastically.

According to one source, both types of cracking were originally called *season cracking* because of the resemblance of stress-corrosion cracks in bar stock to those of seasoned wood. In England, the origin of the term is ascribed to the fact that years ago brass cartridge cases stored in India were observed to crack, particularly during the monsoon season.

Traces of nitrogen oxides may also cause stress-corrosion cracking, probably because such oxides are converted to ammonium salts on the brass surface by chemical reaction with the metal. In one instance of this kind, premature failure of yellow brass brackets in the humidifier chamber of an air-conditioning system was traced to this cause [30]. The air had passed through an electrostatic dust

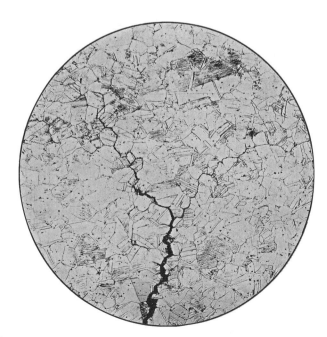

Figure 20.6. Intergranular stress-corrosion cracking of brass (75×) (specimen stored 1 year).

precipitator, the high voltage field of which generated traces of nitrogen oxides. These, in turn, formed corrosion products on the brass surface which were found by analysis to contain a high proportion of NH_4, causing intergranular cracking of the stressed brackets. Similar cracking of stressed brass could be reproduced in the laboratory over a period of days by using a spark discharge in air of 100% relative humidity.

Stress-corrosion cracking of 12% Ni, 23% Zn–Cu alloy (nickel brass) parts of Central Office Telephone Equipment in Los Angeles occurred within two years for similar reasons [31]. Pollution of Los Angeles air accounts for abnormally high concentrations of nitrogen oxides and suspended nitrates, with the latter settling down as dust on the brass parts. Similar failures have not been observed as frequently in New York City, where the air contains less nitrate, but also many more sulfate particles than Los Angeles air, indicating that sulfates may act as inhibitors.

Mattsson [32] observed that minimum cracking time of 37% Zn–Cu brass in a solution of 1 mole $NH_3 + NH_4^+$, plus 0.05 mole $CuSO_4$ per liter occurred at pH 7.3, with times to failure being longer at higher pH and considerably longer at lower pH values. Johnson and Leja [33] reported stress-corrosion cracking of brass in alkaline cupric citrate and in tartrate solutions at pH values in which complexing of Cu^{2+} is pronounced.

Figure 20.7 shows the effect of applied potential [34] on S.C.C. of 37% Zn–Cu brass in a solution similar to that proposed by Mattsson. The critical potential below which cracking does not occur is 0.095 V (S.H.E.). Since the corresponding corrosion potential is 0.26 V, S.C.C. on simple immersion of brass is spontaneous. Stress-corrosion cracking also occurs on substituting Cd^{2+} or Co^{2+} in place of Cu^{2+} additions, but at more noble potentials. It is found that, in the absence of other metal ammonium complexes, the equivalent of more than $0.003M$ Cu^{2+} must be present in the test solution for spontaneous cracking to occur. Also, addition of more than $0.005M$ NaBr or $0.04M$ NaCl to the Mattsson test solution inhibits spontaneous cracking. The latter observations are related to a shift of potentials by Cu^{2+}, Br^-, or Cl^- such that, when cracking does not occur, the corrosion potential in each case lies active to the critical potential for S.C.C.

Annealed brass, if not subject to a high applied stress, does not stress-corrosion crack. Whether residual stresses in cold-worked brass are sufficient to cause stress-corrosion cracking in an ammonia atmosphere can be checked by immersing brass in an aqueous solution of 100 g mercurous nitrate $[Hg_2(NO_3)_2]$ and 13 mL nitric acid (HNO_3, specific gravity 1.42) per liter of water. Mercury is released and penetrates the grain boundaries of the stressed alloy. If cracks do not appear with 15 min, the alloy is probably free of damaging stresses.

No alloying additions in small amounts effectively provide immunity to this type of failure in brasses. Low-zinc brasses are more resistant than high-zinc brasses.

High-zinc brasses (e.g., 45–50% Zn–Cu) having a β or $\beta + \gamma$ structure, stress-corrosion crack through the grains (i.e., transgranularly); and, unlike α brasses, only moisture is required to cause failure [35].

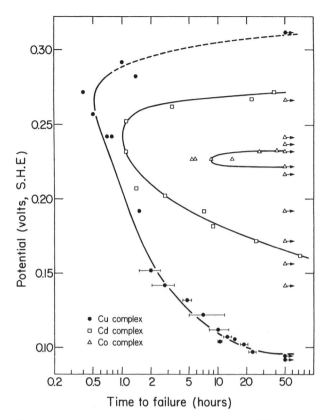

Figure 20.7. Effect of applied potential on time to failure of 37% Zn–Cu brass in 0.05*M* CuSO$_4$, CdSO$_4$, or CoSO$_4$, 1*M* (NH$_4$)$_2$SO$_4$, pH 6.5 at room temperature [34]. (Reproduced with permission. Copyright 1975, The Electrochemical Society.)

The mechanism of stress-corrosion cracking in brasses has been the subject of much study. Both high-purity alloys and single crystals of α brass crack when stressed in NH$_3$ atmospheres [36]. In support of an electrochemical mechanism, it has been noted that the grain boundaries of polycrystalline brasses are more active in potential than the grains, as measured in NH$_4$OH, but not in FeCl$_3$ solution, in which stress-corrosion cracking does not occur [37]. It has been proposed, alternatively, that a brittle oxide film forms on brasses that continuously fractures under stress exposing fresh metal underneath to further oxidation [38]. According to the generally accepted film rupture mechanism, S.C.C. initiates by film rupture as a result of plastic deformation at the crack tip. The crack propagates by localized anodic dissolution, and repassivation is inhibited by further plastic deformation at the crack tip [39–41].

There is evidence, in any event, that zinc atoms aided by plastic deformation segregate preferentially at grain boundaries. The resulting composition gradient favors galvanic action between such areas and the grains, accounting for slow intergranular attack in a variety of corrosive media without the necessity of an applied stress (intergranular corrosion). But such areas, subject to plastic deformation, may also favor adsorption of complex ammonium ions within a specific potential range, leading to rapid crack formation. Similar effects can occur along slip bands (transgranular cracking). Although segregated zinc may be essential to the observed intergranular corrosion of brasses, it is probable that the defect structure of grain boundaries or of slip bands is more important to stress-corrosion cracking. Hence, failure of copper-base alloys by cracking can occur when copper is alloyed not only with zinc, but also with a variety of other elements, such as silicon, nickel, antimony, arsenic, aluminum, phosphorus [42], or beryllium [43].

To explain the mechanism of transgranular S.C.C. in α brass, Newman and Sieradzki have proposed the film-induced cleavage model, according to which a brittle crack that initiates in a thin surface film propagates into the ductile matrix and eventually blunts and arrests, after which the process repeats itself [39, 44, 45].

Susceptibility to stress-corrosion cracking of brasses can be minimized or avoided by four main procedures:

1. *Stress-Relief Heat Treatment.* For 30% Zn–Cu brass, heating at 350°C (660°F) for 1 h may be effective, but recrystallization and some loss of strength of the alloy result. Because ammoniacal S.C.C. can occur at relatively low stress levels, stress relief may not be adequate in all cases [27].
2. *Avoiding Contact with NH_3 (or with O_2 and Other Depolarizers in the Presence of NH_3).* It is difficult to guarantee that there will be no contact with NH_3, because of the small traces of ammonia that cause cracking. Plastics containing or decomposing to traces of amines are a continuing source of damage to unannealed brass. Fertilizer washing from farm land, or air over fertilized soil, has similarly caused cracking of brass. On the other hand, brass condenser tubes do not crack in boiler-water condensate containing NH_3, because the concentration of oxygen is extremely low.
3. *Cathodic Protection.* Cathodic protection to a potential below the critical value can be provided by either impressed current, or by coating brass with a sacrificial metal (e.g., zinc).
4. *Using H2S as an Inhibitor* [27, 29]. The mechanism may, in part, involve reaction with available free oxygen.

20.2.4 Condenser Tube Alloys Including Copper–Nickel Alloys

For fresh waters, copper, Muntz metal, and admiralty metal (inhibited) are frequently used. For brackish or seawater, admiralty metal, one of the cupro-nickel alloys (10–30% Ni, bal. Cu) and aluminum brass (22% Zn, 76% Cu, 2% Al, 0.04% As) are used. For polluted waters, cupro-nickel alloys are preferred over aluminum brass because the latter is subject to pitting attack. Aluminum brass may also pit readily in unpolluted, but stagnant, seawater.

Aluminum brass resists high-velocity waters (impingement attack) better than does admiralty metal. Cupro-nickel alloys are especially resistant to high-velocity seawater when they contain small amounts of iron and sometimes manganese as well. For the 10% Ni cupro-nickel alloy, the optimum iron content is about 1.0–1.75%, with 0.75% Mn maximum; for the analogous 30% Ni composition, the amount of alloyed iron is usually less (e.g., 0.40–0.70% Fe accompanied by 1.0% Mn maximum) [46]. It is found that supplementary protective films are formed on condenser tube surfaces when iron is contained in water as a result of corrosion products upstream or when added intentionally as ferrous salts. Accordingly, the beneficial effect of iron alloyed with copper–nickel alloys is considered to result from similar availability of iron in the formation of protective films.

The susceptibility of Cu–Ni alloys to corrosion increases in aerated sulfide-polluted seawater. It appears that sulfides interfere with formation of the usual protective oxide films [47].

It is pertinent that the 30% Ni–Cu alloy is relatively resistant to stress-corrosion cracking compared to the 10 or 20% Ni–Cu alloys [42], or compared to any of the 30% Zn–Cu brasses. A detailed general account of the behavior of copper–nickel alloys, especially the 10% Ni–Cu alloy, in seawater is given by Stewart and LaQue [48].

REFERENCES

1. C. A. C. Sequeira, Corrosion of copper and copper alloys, in *Uhlig's Corrosion Handbook*, 2nd edition, R. W. Revie, editor, Wiley, New York, 2000, p. 755.

2. M. Pourbaix, Atlas of Electrochemical Equilibria in Aqueous Solutions, second English edition, National Association of Corrosion Engineers, Houston, TX, p. 389; and CEBELCOR, Brussels, 1974.

3. Ref. 1, pp. 750–753, Wiley, New York, 2000.

4. J. F. D. Stott, Evaluating microbiologically influenced corrosion, in *ASM Handbook*, Vol. 13A, *Corrosion: Fundamentals, Testing, and Protection*, ASM International, Materials Park, OH, 2003, p. 645.

5. A. Cohen, Corrosion of copper and copper alloys, in *ASM Handbook*, Vol. 13B, *Corrosion: Materials*, ASM International, Materials Park, OH, 2005, p. 137.

6. J. Krger, *J. Electrochem. Soc.* **106**, 847 (1959).

7. C. Shanley, R. Hummel, and E. Verink, *Corros. Sci.* **20**, 481 (1980).

8. M. Obrecht and L. Quill, *Heating, Piping and Air-Conditioning*, January, pp. 165–169; March, pp. 109–116; April, pp. 131–137; May, pp. 105–113; July, pp. 115–122; September, pp. 125–133 (1960); April, pp. 129–134 (1961). Also see M. Obrecht, *Corrosion* **18**, 189t (1962).

9. C. Pfeiffer, *Mental and Elemental Nutrients*, Keats Publications, New Canaan, CT, 1975, p. 330.

10. B. J. Little and J. S. Lee, *Microbiologically Influenced Corrosion*, Wiley, New York, 2007, p. 184.

11. Ref. 1, p. 747.

12. H. Campbell, *J. Appl. Chem.* **4**, 633 (1954).

13. E. Mattsson, *Br. Corros. J.* **15**, 6 (1980).

14. H. Campbell, *J. Inst. Metals* **77**, 345 (1950).

15. U. R. Evans, *An Introduction to Metallic Corrosion*, 3rd edition. Arnold, London, 1981, p. 49.

16. P. Daniel and R. Rapp, *Advances in Corrosion Science and Technology*, Vol. 5, M. Fontana and R. Staehle, editor, Plenum Press, New York, 1976, p. 114.

17. C. Southwell, C. Hummer, Jr., and A. Alexander, *Mater. Prot.* **7**, 41 (1968); *Mater. Prot.* **7**, 61 (1968).

18. V. Ereneta, *Corros. Sci.* **19**, 507 (1979).

19. S. G. Corcoran, Effects of metallurgical variables on dealloying corrosion, in *ASM Handbook*, Vol. 13A, *Corrosion: Fundamentals, Testing, and Protection*, ASM International, Materials Park, OH, 2003, pp. 287–293.

20. R. B. Abrams, *Trans. Am. Chem. Soc.* **42**, 39 (1922).

21. H. W. Pickering, *J. Electrochem. Soc.* **115**, 143 (1968).

22. H. W. Pickering and C. Wagner, *J. Electrochem. Soc.* **114**, 698 (1967).

23. A. J. Forty and G. Rowlands, *Philos. Mag. A* **43**, 171 (1981).

24. K. Sieradzki, N. Dimitrov, D. Movrin, C. McCall, N. Vasiljevic, and J. Erlebacher, *J. Electrochem. Soc.* **149**, B370 (2002).

25. K. Sieradzki, *J. Electrochem. Soc.* **140**, 2868 (1993).

26. E. Polushkin and H. Shuldener, *Min. Metall. Eng.* **161**, 214 (1945).

27. P. Elliott, Materials selection for corrosion control, in *ASM Handbook*, Vol. 13A, *Corrosion: Fundamentals, Testing, and Protection*, ASM International, Materials Park, OH, 2003, pp. 917–918.

28. R. S. Lenox and P. A. Hough, *ASHRAE J.* November (1995).

29. D. R. McIntyre and C. P. Dillon, *Guidelines for Preventing Stress Corrosion Cracking in the Chemical Process Industries*, Publication 15, Materials Technology Institute of the Chemical Process Industries, 1985.

30. H. Uhlig and J. Sansone, *Mater. Prot.* **3**, 21 (1964).

31. N. McKinney and H. Hermance, *Stress Corrosion Testing*, Special Technical Publication No. 425, p. 274, American Society for Testing and Materials, Philadelphia, PA, 1967.

32. E. Mattsson, *Electrochim. Acta* **3**, 279 (1960–1961).

33. H. Johnson and J. Leja, *Corrosion* **22**, 178 (1966).

34. H. Uhlig, K. Gupta, and W. Liang, *J. Electrochem. Soc.* **122**, 343 (1975).

35. A. Bailey, *J. Inst. Metals* **87**, 380 (1959).

36. G. Edmunds in *Symposium on Stress Corrosion Cracking of Metals*, ASTM–AIME, Philadelphia, 1945, p. 67.

37. R. Bakish and W. Robertson, *J. Electrochem. Soc.* **103**, 320 (1956).

38. E. Pugh, J. H. Craig, and A. Sedriks, in *Proceedings of Conference Fundamental Aspects of Stress Corrosion Cracking*, National Association of Corrosion Engineers, Houston, TX, 1969, p. 118; see also S. Shimodaira and M. Takano, p. 202.

39. J. A. Beavers, in *Stress-Corrosion Cracking*, R. H. Jones, editor, ASM International, Materials Park, OH, 1992, pp. 227–228.

40. F. A. Champion, *Symposium on Internal Stresses in Metals and Alloys*, Institute of Metals, London, 1948, p. 468.

41. H. L. Logan, *J. Res. Natl. Bur. Stand.* **48**(2), 99 (1952).

42. D. Thompson and A. Tracy, *Min. Metall. Eng.* **185**, 100 (1949).

43. N. Magnani, *Corrosion* **36**, 260 (1980).

44. R. H. Jones, Stress-corrosion cracking, in *ASM Handbook*, Vol. 13A, *Corrosion: Fundamentals, Testing, and Protection*, ASM International, Materials Park, OH, 2003, p. 362.

45. R. C. Newman, T. Shahrabi, and K. Sieradzki, *Scr. Metall.* **23**, 71 (1989).

46. U.S. Military Specification MIL-C-15726 E (Ships), August 20, 1965.

47. L. Eiselstein, B. Syrett, S. Wing, and R. Caligiuri, *Corros. Sci.* **23**, 223 (1983).

48. W. Stewart and F. LaQue, *Corrosion* **8**, 259 (1952).

GENERAL REFERENCES

J. A. Beavers, Stress-corrosion cracking of copper alloys, in *Stress-Corrosion Cracking*, R. H. Jones, editor, ASM International, Materials Park, OH, 1992, pp. 211–231.

A. Cohen, Corrosion of copper and copper alloys, in *ASM Handbook*, Vol. 13B, *Corrosion: Materials*, ASM International, Materials Park, OH, 2005, pp. 125–163.

C. A. C. Sequeira, Corrosion of copper and copper alloys, in *Uhlig's Corrosion Handbook*, 2nd edition, R. W. Revie, editor, Wiley, New York, 2000, pp. 729–765.

PROBLEMS

---

1. Calculate the equilibrium constant at 25°C for the reaction $Cu + Cu^{2+} \rightarrow 2Cu^+$. Since equilibrium tends to be maintained between ions and metal, which valence ion predominates when Cu is dissolved anodically? (*Data:* $Cu^{2+} + 2e^- \rightarrow Cu$, $\phi° = 0.337\,V$; $Cu^+ + e^- \rightarrow Cu$, $\phi° = 0.521\,V$.)

2. As in Problem 1, estimate, from the analogous equilibrium constant, which valence ion predominates when Cr in the active state is dissolved anodically. (*Data:* $Cr^{2+} + 2e^- \rightarrow Cr$, $\phi° = -0.91\,V$; $Cr^{3+} + 3e^- \rightarrow Cr$, $\phi° = -0.74\,V$.)

Answers to Problems

---

1. 6.1×10^{-7}; Cu^{2+}
2. Cr^{2+} predominates.

21

ALUMINUM AND ALUMINUM ALLOYS

21.1 ALUMINUM

$$Al^{3+} + 3e^- \rightarrow Al \qquad \phi° = -1.7\,V$$

Aluminum is a lightweight metal (density = $2.71\,g/cm^3$) having good corrosion resistance to the atmosphere and to many aqueous media, combined with good electrical and thermal conductivity. It is very active in the Emf Series (Section 3.8), but becomes passive on exposure to water. The theoretical potential–pH domains for immunity, corrosion, and passivation can be seen from Fig. 4.3 (Section 4.4). Although oxygen dissolved in water improves the corrosion resistance of aluminum, its presence is not necessary to achieve passivity, indicating that the Flade potential (Section 6.2) of aluminum is active with respect to the hydrogen electrode. It is usually assumed that the passive film is composed of aluminum oxide, which, for air-exposed aluminum, is estimated at about 2–10 nm (20–100 Å) in thickness. The observed corrosion behavior of aluminum is sensitive to small amounts of impurities in the metal; all these impurities, with the exception of magnesium, tend to be cathodic to aluminum. In general, the high-purity metal is much more corrosion-resistant than commercially pure

Corrosion and Corrosion Control, by R. Winston Revie and Herbert H. Uhlig
Copyright © 2008 John Wiley & Sons, Inc.

aluminum, which, in turn, is usually more resistant than aluminum alloys.* Some of the aluminum alloys produced commercially in the United States are listed in Table 21.1. For a complete list, the reference given in the table should be consulted.

21.1.1 Clad Alloys

Pure aluminum is soft and weak; it is alloyed, therefore, largely to obtain increased strength. In order to make use of the good corrosion resistance of pure aluminum, a high-strength alloy may be sandwiched between pure aluminum or one of the more corrosion-resistant alloys (e.g., 1% Mn–Al) active in the Galvanic Series with respect to the inner alloy. This combination is called alclad or clad. The clad structure, metallurgically bonded at the two interfaces, provides cathodic protection to the inner alloy by sacrificial action of the outer layers, similar to that provided by a zinc coating on steel. In addition to cathodically protecting against pitting corrosion, the less noble coatings also protect against intergranular corrosion and stress-corrosion cracking, especially if the high-strength inner alloy should become sensitized through fabrication procedures or by accidental exposure to high temperatures.

21.1.2 Corrosion in Water and Steam

Aluminum tends to pit in waters containing Cl^-, particularly at crevices or at stagnant areas where passivity breaks down through the action of differential aeration cells. The mechanism is analogous to that described for the stainless steels (Section 19.2.4.1) including an observed critical potential below which pitting corrosion does not initiate [3, 4]. Traces of Cu^{2+} (as little as 0.1 ppm) or of Fe^{3+} in water react with aluminum, depositing metallic copper or iron at local sites. The copper or iron, being efficient cathodes, shift the corrosion potential in the noble direction to the critical potential, thereby both initiating pitting and, by galvanic action, stimulating pit growth. For this reason, aluminum is not a satisfactory material for piping to handle potable or industrial waters, which all contain traces of heavy metal ions. On the other hand, for distilled water or for a water from which heavy metal ions have been removed, aluminum is satisfactory. Aluminum piping of high-purity metal or of type 1100 has been used satisfactorily for distilled water lines over many years.

Extraneous anions in dilute chloride solutions act as pitting inhibitors by shifting the critical pitting potential to more noble values [4]. The following ions are effective in decreasing order: nitrate > chromate > acetate > benzoate > sulfate.

---

*One exception is known with reference to intergranular disintegration of high-purity aluminum in steam or pure water at temperatures above 125°C (260°F). The presence of iron as an impurity in lower-grade metal avoids this type of attack or raises the temperature at which it occurs (>200°C for type 1100) [1, 2].

TABLE 21.1. Composition Limits for Some Wrought Aluminum Alloys[a]

Alloy		Maximum Amounts Specified (%)[b]							
AA No.[c]	UNS No.	Si	Fe	Cu	Mn	Mg	Cr	Zn	Ti
1100	A91100	0.95 (Si + Fe)		0.05–0.20	0.05			0.10	
2017	A92017	0.2–0.8	0.7	3.5–4.5	0.40–1.0	0.40–0.8	0.10	0.25	0.15
2024	A92024	0.50	0.50	3.8–4.9	0.30–0.9	1.2–1.8	0.10	0.25	0.15
3003	A93003	0.6	0.7	0.05–0.2	1.0–1.5			0.10	
3004	A93004	0.30	0.7	0.25	1.0–1.5	0.8–1.3		0.25	
5050	A95050	0.40	0.7	0.20	0.10	1.1–1.8	0.10	0.25	
5052	A95052	0.25	0.40	0.10	0.10	2.2–2.8	0.15–0.35	0.10	
5056	A95056	0.30	0.40	0.10	0.05–0.20	4.5–5.6	0.05–0.20	0.10	
6053	A96053	d	0.35	0.10		1.1–1.4	0.15–0.35	0.10	
6061	A96061	0.40–0.8	0.7	0.15–0.40	0.15	0.8–1.2	0.04–0.35	0.25	0.15
6063	A96063	0.20–0.6	0.35	0.10	0.10	0.45–0.9	0.10	0.10	0.10
7072	A97072	0.7 (Si + Fe)		0.10	0.10	0.10		0.8–1.3	
7075	A97075	0.40	0.50	1.2–2.0	0.30	2.1–2.9	0.18–0.28	5.1–6.1	0.20

[a]Aluminum Standards and Data 2003, The Aluminum Association, Inc., Washington, D.C., March 2003, pp. 6-5-6-6.
[b]Composition in percent by weight maximum except when composition ranges are given.
[c]Code for major alloying element: 1XXX, >99.00% Al; 2XXX, Cu; 3XXX, Mn; 4XXX, Si; 5XXX, Mg; 6XXX, Si + Mg; 7XXX, Zn.
[d]45–65% of the Mg content.

In uncontaminated seawater in the tropical environment of Panama, commercially pure aluminum (1100) or 0.6% Si, 0.8% Mg, 0.2% Cu–Al alloy (6061-T) corroded at decreasing rates with time. After 16 years of exposure, the total weight losses were 67 and 63 g/m^2, and the corresponding observed deepest pits were 0.84 and 2.0 mm (0.033 and 0.079 in.), respectively, for small-size test panels [5]. At the same location in fresh water, probably contaminated with heavy metals, the corresponding weight losses for 16 years were higher—347 and 103 g/m^2— and the deepest pits measured 2.8 mm (0.11 in.) each.

Various commercial aluminum alloys immersed in seawater at Key West, Florida, for 368 days pitted or not according to their corrosion potentials [6]. Alloys with corrosion potentials ranging from −0.4 to −0.6 V (S.H.E.) (most of which contained some alloyed copper) showed a mean depth of pitting equal to 0.15–0.99 mm (0.006–0.039 in.), whereas those alloys of more active potential (−0.7 to −1.0 V) were essentially not pitted. These potential ranges can be compared with the critical pitting potential of pure aluminum measured in 3% NaCl, equal to −0.45 V (see Section 6.6). Coupling of susceptible alloy panels to a smaller-area panel of an active aluminum alloy (see Section 13.3.1), which polarized the couple to approximately −0.85 V (S.H.E.), succeeded in largely preventing pitting over the same period of time. These field tests support the concept of critical potentials below which, in the absence of crevices, neither aluminum nor its alloys undergo pitting corrosion.

As the temperature increases, the corrosion rate of type 1100 in deaerated distilled water increases, but the value at 70°C is still very low, being about 0.007 gmd [1]. This rate is reduced to 0.004 gmd in the presence of 10^{-3}N H$_2$O$_2$ and still lower by 2 ppm K$_2$Cr$_2$O$_7$. At 150°C the rate in distilled water is about 0.04 gmd, and at 200°C it is 0.25 gmd, with the attack in all these instances being uniform and without marked pitting. At higher temperatures (e.g., 275°C), but usually only after several days of exposure, the rate becomes more rapid by a factor of 10–20 times, and corrosion proceeds mostly along grain boundaries, causing rapid failure. At 315°C, complete disintegration to Al$_2$O$_3$ may occur within a few hours. It is important to note that this catastrophic corrosion does not occur at the cited high temperatures if the aluminum is coupled to a sufficient area of stainless steel or of zirconium, both of which are cathodic to aluminum. The same beneficial effect is obtained by small alloying additions of nickel and iron. For example, the corrosion rate at 350°C for type 1100 aluminum alloyed with 1% Ni is about 1.8 gmd [1].

The beneficial effect of cathodic areas can be explained [7] as one of anodic passivation or anodic protection of aluminum in the same manner as alloying additions or coupling of platinum or palladium to stainless steels or to titanium passivate these metals in acids (Section 6.4).

At 500–540°C, the oxidation rate of pure aluminum in steam is appreciably lower than at 300–450°C [8], presumably because a more protective oxide film forms within the higher temperature range.

21.1.3 Effect of pH

Aluminum corrodes more rapidly both in acids and in alkalies compared to distilled water, with the rates in acids depending on the nature of the anion. Figure 21.1 provides data at 70–95°C [1], showing that the minimum rate, when using sulfuric acid for adjustment of pH in the acid region, occurs between pH 4.5 and 7. At room temperature, the minimum rate occurs in the pH range approximating 4–8.5. Corrosion rates of aluminum in the alkaline region greatly increase with pH, unlike iron and steel, which remain corrosion-resistant. The reason for this difference is that Al^{3+} is readily complexed by OH^-, forming AlO_2^- in accord with

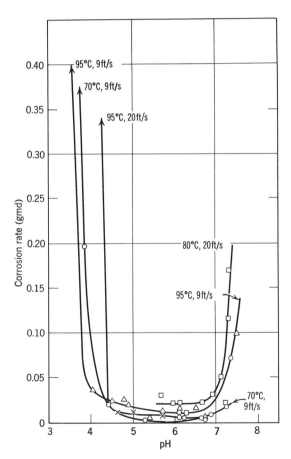

Figure 21.1. Effect of pH on corrosion of commercially pure aluminum (1100), aerated solutions; pH, measured at room temperature, was adjusted with H_2SO_4 or NaOH. Most solutions contained $1–10 \times 10^{-5}$ N H_2O_2, 68 ppm $CaSO_4$, 30 ppm $MgSO_4$, 1–2 ppm NaCl (J. Draley and W. Ruther).

$$Al + NaOH + H_2O \rightarrow NaAlO_2 + \frac{3}{2}H_2 \qquad (21.1)$$

This reaction proceeds rapidly at room temperature, whereas for iron a similar reaction forming $NaFeO_2$ and Na_2FeO_2 requires concentrated alkali and high temperatures.

21.1.4 Corrosion Characteristics

Aluminum is characterized by (1) sensitivity to corrosion by alkalies and (2) pronounced attack by traces of copper ions in aqueous media. In addition, aluminum is subject to rapid attack by (3) mercury metal and mercury ions and (4) anhydrous chlorinated solvents (e.g., CCl_4, ethylene dichloride, and propylene dichloride) [9, 10].

The rate of attack can be appreciable in either dilute or concentrated alkalies. For this reason, when aluminum is cathodically protected, overprotection must be avoided in order to ensure against damage to the metal by accumulation of alkalies at the cathode surface. Lime, $Ca(OH)_2$, and some of the strongly alkaline organic amines (but not NH_4OH) are corrosive. Fresh Portland cement contains lime and is also corrosive; hence, aluminum surfaces in contact with wet concrete may evolve hydrogen visibly. The corrosion rate is reduced when the cement sets, but continues if the concrete is kept moist or contains deliquescent salts (e.g., $CaCl_2$).

A drop of mercury in contact with an aluminum surface rapidly breaks down passivity accompanied by amalgamation (i.e., formation of an aluminum amalgam). In the presence of moisture, the amalgamated metal quickly converts to aluminum oxide, causing perforation of piping or sheet. Mercury ions present in solution in only trace amounts similarly accelerate corrosion, producing intolerably high rates of attack.

Aluminum tanks in aircraft fuel systems are subject to corrosion resulting from the growth of microbiological constituents in kerosene fuel [11–16]. Until recently, the most important of these constituents was the fungus *Cladosporium resinae* [11]. Fungi produce highly corrosive organic acids that corrode aluminum fuel tanks. Temperature and the availability of water are reported to be the principal factors determining whether and where microbial growth occurs. Growth initiates in water deposits at the interface with the fuel, with the resulting fungus adhering to the tank lining. The growth rate is controlled by temperature and is greatest at about 30–35°C. Subsequent corrosion has been explained on the basis of water-soluble organic acids produced by the metabolism of various microorganisms and also on the basis of oxygen depletion beneath the growing microbial layer (differential aeration cell).

The changes in jet fuel and jet fuel additives that have taken place in recent years have also brought about a shift in the microbial community in aircraft fuel. The bacteria isolated from aircraft jet fuel tanks have been found to be closely related to *Bacillus*, and they do have the potential to cause microbiologically influenced corrosion (MIC) [11, 17, 18].

Although MIC may be controlled by adding a biocide to the fuel, this solution may create new problems; for example, biocides may selectively enrich a population of bacteria capable of biocide resistance [12].

21.1.4.1 Reaction with Carbon Tetrachloride. Severe accidents have been caused by the rapid reaction of aluminum with anhydrous chlorinated solvents, such as in the degreasing of castings, in ball milling of aluminum flake with carbon tetrachloride (CCl_4), or even in the case of aluminum used as a container at room temperature for mixed chlorinated solvents. The reaction with CCl_4, for example, has been shown to be

$$2Al + 6CCl_4 \rightarrow 3C_2Cl_6 + 2AlCl_3 \tag{21.2}$$

forming hexachloroethane, C_2Cl_6, with evolution of heat. The corrosion rate for 99.99% Al in boiling anhydrous CCl_4 is very high—for example, 3750 gmd (20 ipy) [19]. Should the temperature reach the melting point of aluminum, the reaction may proceed explosively. An induction time of about 55 min, during which corrosion is negligible, precedes the rapid rate (Fig. 21.2). This induction time is lengthened either by the presence of water in the CCl_4 (480 min) or by some alloying additions—for example, Mn or Mg (30 h for type 5052). On the other hand, addition of $AlCl_3$ or $FeCl_3$ to CCl_4 decreases the induction time to zero without an appreciable effect on the corrosion rate. The corrosion rate of aluminum in water-saturated CCl_4, following a prolonged induction time, is about twice that in the anhydrous solvent (Fig. 21.3).

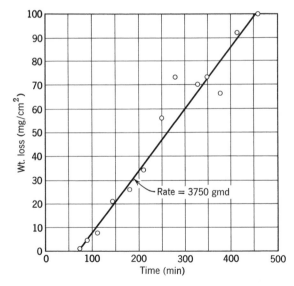

Figure 21.2. Weight loss as a function of time for 99.99% Al in boiling, distilled CCl_4 containing 0.0011% H_2O [19]. (Reproduced with permission. Copyright 1952, The Electrochemical Society.)

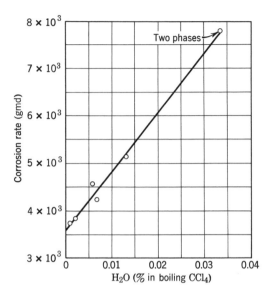

Figure 21.3. Effect of water on corrosion rate of 99.99% Al in boiling CCl_4 [19]. (Reproduced with permission. Copyright 1952, The Electrochemical Society.)

In practice, volatile inhibitors are often added to chlorinated solvents in order to suppress the fast reaction. Examples include various ketones, quinones, and amines, which presumably react preferentially with and destroy free radicals (e.g., $^\bullet CCl_3$ and $^\bullet Cl$). The latter are considered to play an important part in the reaction mechanism [20]. A proposed sequence of intermediate reactions, for example, is the following:

$$CCl_4 \rightarrow {}^\bullet CCl_3 + {}^\bullet Cl$$
$$Al + {}^\bullet Cl \rightarrow AlCl$$
Initiation

$$AlCl + CCl_4 \rightarrow AlCl_2 + {}^\bullet CCl_3$$
$$AlCl_2 + CCl_4 \rightarrow AlCl_3 + {}^\bullet CCl_3$$
$$AlCl_3 + CCl_4 \rightarrow CCl_3^+[AlCl_4]^-$$
Propagation
$$CCl_3^+[AlCl_4]^- \rightarrow AlCl_3 + {}^\bullet CCl_3 + {}^\bullet Cl$$

$$2\,{}^\bullet CCl_3 \rightarrow C_2Cl_6$$
$${}^\bullet Cl + {}^\bullet CCl_3 \rightarrow CCl_4$$
Termination

Subsequent investigations by others have confirmed the presence of $^\bullet CCl_3$ and various of the proposed intermediate reacting species [21,22].

Water itself may react with and destroy initiating species until the water is consumed by chemical reaction or until sufficient $AlCl_3$ forms. Hence, the induction time is extended when water is present in the solvent. By the same token, in anhydrous CCl_4, vacuum treatment of aluminum specimens results in a shorter induction time of 5 min compared to 55 min not treated, probably by reducing the H_2O content of the oxide film. Substances like $AlCl_3$ decrease the induction time by forming a complex with CCl_4 that later dissociates into free radicals taking part in the chain reaction. Similarly, alloying elements like magnesium are considered to react preferentially with free radicals, accounting for the observed improved corrosion resistance of magnesium–aluminum alloys to CCl_4 compared to pure aluminum. The behavior of alloying constituents in general (often opposite to their effect in aqueous environments), the observed lack of effect of galvanic coupling or of an impressed nominal voltage on the rate, along with the same observed corrosion rate in the vapor phase as in the boiling liquid are a strong arguments that the reaction does not follow an electrochemical mechanism. In addition to analytical evidence for $\cdot CCl_3$, (the species that probably accounts for the red color of CCl_4 reacting with Al), the ease with which many added organic substances suppress the fast reaction (free radicals are very reactive) also supports the free-radical mechanism.

In summary, aluminum (type 1100) is resistant to the following:

1. Hot or cold NH_4OH.
2. Hot or cold acetic acid. Highest rates of attack are for the boiling 1–2% acid. Ninety-nine percent acetic acid is not appreciably corrosive, but removal of the last 0.5% H_2O increases attack over a hundredfold. Formic acid and Cl^- contamination also increase attack. Aluminum is resistant to citric, tartaric, and malic acids.
3. Fatty acids. Aluminum equipment is used for distillation of fatty acids.
4. Nitric acid, >80% up to about 50°C (120°F) [23] (see Fig. 21.4).
5. Distilled water.
6. Atmospheric exposure. Excellent resistance to rural, urban, and industrial atmospheres; lesser resistance to marine atmospheres.
7. Sulfur, sulfur atmospheres, and H_2S. Aluminum is used for mining of sulfur.
8. Fluorinated refrigerant gases, such as Freon, but *not* to methyl chloride or bromide.

Aluminum is *not* resistant to the following:

1. Strong acids, such as HCl and HBr (dilute or concentrated), H_2SO_4 (satisfactory for special applications at room temperature below 10%), HF, $HClO_4$, H_3PO_4, and formic, oxalic, and trichloroacetic acids.
2. Alkalies. Lime and fresh concrete are corrosive, as well as strong alkalies; for example, NaOH and the very alkaline organic amines. Corrosion by soap solutions can be inhibited by adding a few tenths percent of sodium silicate (not effective for strong alkalies).

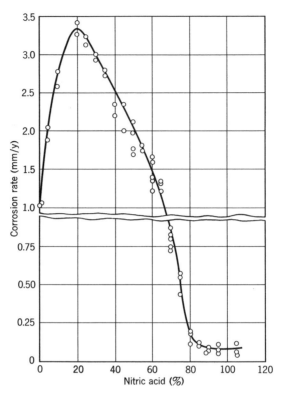

Figure 21.4. Corrosion rates of commercially pure aluminum (1100) in nitric acid, room temperature [23]. (Copyright NACE International 1961.)

3. Mercury and mercury salts.
4. Seawater. Pitting occurs at crevices and surface deposits, especially when trace amounts of heavy metal ions are present.
5. Waters containing heavy metal ions (e.g., mine waters or waters previously passing through copper, brass, or ferrous piping).
6. Chlorinated solvents.
7. Anhydrous ethyl, propyl, or butyl alcohols at elevated temperatures. A trace of water acts as an inhibitor [23].
8. Contact with wet woods, in particular beech wood [24]. Any wood impregnated with copper preservatives is especially damaging.

(For behavior in soils, see Section 10.3.)

21.1.5 Galvanic Coupling

Although cadmium has the nearest potential to that of aluminum in many environments, and cadmium-plated steel screws, bolts, trim, and so on, can be used in direct contact with aluminum, nevertheless, cadmium is now recognized as a biocumulative poison with toxic effects similar to those of mercury and lead; cadmium plating is forbidden on most or all parts in many countries [25]. Tin coatings are reported to be satisfactory. Zinc is somewhat further removed in potential, but it is also usually satisfactory. Zinc is anodic to, and, hence, cathodically protects, aluminum against initiation of pitting in neutral or slightly acidic media (see Section 13.7.1). In alkalies, however, the polarity reverses, and zinc accelerates the corrosion of aluminum. Magnesium is anodic to aluminum, but the potential difference and the resultant current flow when the metals are coupled (e.g., in seawater) is so high that aluminum may be overprotected cathodically with resultant damage to aluminum. Aluminum is damaged in this respect to a lesser extent when alloyed with magnesium. High-purity aluminum, it is reported, can be coupled to magnesium without damage to either metal [26], because galvanic currents are reduced in the absence of iron, copper, and nickel impurities, which act as efficient cathodes.

It is imperative that aluminum never be coupled to copper or copper alloys, because of the resulting damage to aluminum. In this connection, it is also important to avoid contact of aluminum with rain water that has washed over copper flashing or gutters because of the damage caused by small amounts of Cu^{2+} dissolved in such water.

In rural areas, coupling of steel to aluminum is usually satisfactory, but in seacoast areas attack of aluminum is accelerated. In fresh waters, aluminum may be anodic or cathodic to steel, depending on small composition differences of the water.

21.2 ALUMINUM ALLOYS

The usual alloying additions to aluminum in order to improve physical properties include Cu, Si, Mg, Zn, and Mn. Of these, manganese may actually improve the corrosion resistance of wrought and cast alloys. One reason is that the compound $MnAl_6$ forms and takes iron into solid solution. The compound $(MnFe)Al_6$ settles to the bottom of the melt, in this way reducing the harmful influence on corrosion of small quantities of alloyed iron present as an impurity [27]. No such incorporation occurs in the case of cobalt, copper, and nickel, so that manganese additions would not be expected to counteract the harmful effects of these elements on corrosion behavior.

The Duralumin alloys (e.g., types 2017 and 2024) contain several percent copper, deriving their improved strength from the precipitation of $CuAl_2$ along slip planes and grain boundaries. Copper is in solid solution above the homogenizing temperature of about 480°C (900°F) and, on quenching, remains in solution. Precipitation takes place slowly at room temperature with progressive strengthening of the alloy. Should the alloy be quenched from solid–solution temperatures into boiling water, or, after quenching should it be heated (artificially aged) above 120°C (250°F), the compound $CuAl_2$ forms preferentially along the grain boundaries. This results in a depletion of copper in the alloy adjacent to the intermetallic compound, accounting for grain boundaries anodic to the grains, and a marked susceptibility to *intergranular corrosion*. Prolonged heating (overaging) restores uniform composition alloy at grains and grain interfaces, thereby eliminating susceptibility to this type of corrosion, but at some sacrifice of physical properties. In practice, the alloy is quenched from about 490°C (920°F), followed by room-temperature aging.

Exfoliation is a related type of anodic path corrosion in which attack of rolled or extruded aluminum alloy results in surface blisters followed by separation of elongated slivers or laminae of metal. It occurs in various types of aluminum alloys in addition to the copper-bearing series. Proper heat treatment may alleviate such attack.

Exfoliation is commonly experienced on exposure of susceptible aluminum alloys to marine atmospheres. Simulation in the laboratory is accomplished by controlled intermittent spray with 5% NaCl containing added acetic acid to pH 3 at 35–50°C (95–120°F) [28]. In practice, severe exfoliation corrosion has been experienced in water irrigation piping constructed of type 6061 alloy [29]. The cause in this instance was ascribed to pipe fabrication methods and presence of excess amounts of impurity elements (e.g., Fe, Cu, and Mn).

21.2.1 Stress-Corrosion Cracking

Pure aluminum is immune to stress-corrosion cracking (S.C.C.). Should a Duralumin alloy, on the other hand, be stressed in tension in the presence of moisture, it may crack along the grain boundaries. Sensitizing the alloy by heat treatment, as described previously, makes it more susceptible to this type of failure. In aging tests at 160–205°C (320–400°F), maximum susceptibility was observed at times somewhat short of maximum tensile strength [30]. Hence, in heat-treatment procedures, it is better practice to aim at a slightly overaged rather than an underaged alloy.

Magnesium alloyed with aluminum increases susceptibility to S.C.C., especially when magnesium is added in amounts >4.5%. To avoid such failures, slow cooling (50°C/h) from the homogenizing temperature is necessary supposedly in order to coagulate the β phase (Al_3Mg_2), a process that is promoted by addition of 0.2% Cr to the alloy [31]. Microstructures that are resistant to S.C.C. are those either with no precipitate along grain boundaries or with precipitate distributed as uniformly as possible within grains [32].

For structural components in automotive applications, the Aluminum Association guidelines [33, 34] recommend alloys with a maximum of 3% Mg where there is exposure for long periods to temperatures greater than about 75°C (167°F). If an alloy of higher Mg content is being considered, the thermal exposure of the part during its lifetime should be established, and realistic, full-component testing should be carried out as part of a detailed materials assessment to evaluate and qualify the specific alloy under consideration. Following these guidelines has led to successful use of, for example, alloy 5182, a 4.5% Mg alloy, in structural components of the Honda NSX with no known record of field failures caused by SCC [35, 36].

Edeleanu [37] showed that cathodic protection stopped the growth of cracks that had already progressed into the alloy immersed in 3% NaCl solution. On aging the alloy at low temperatures, maximum susceptibility to S.C.C. occurred before maximum hardness values were reached. This behavior paralleled that of a Duralumin alloy cited previously. Accordingly, Edeleanu proposed that the susceptible material along the grain boundary which caused cracking was not the equilibrium β phase responsible for hardness, but instead was made up of magnesium atoms segregating at the grain boundary before the intermetallic compound formed. On this basis, susceptibility to S.C.C. decreased on continued aging of the alloy because the separated β phase consumed the original segregated grain-boundary material responsible for susceptibility. A similar mechanism probably applies to the copper–aluminum alloy series.

High concentrations of zinc in aluminum (4–20%) also induce susceptibility to cracking of the stressed alloys in the presence of moisture. Traces of H_2O, for example, contained in the surface oxide film are sufficient to cause cracking; carefully baked-out specimens in dry air do not fail [38]. Oxygen is not necessary, nor is a liquid aqueous phase required. These conditions, plus the susceptibility of a high-strength aluminum alloy (7075) to failure in organic solvents [39], suggest that a possible mechanism of failure may be one of stress-sorption cracking caused by adsorbed water or organic molecules on appropriate defect sites of the strained alloy. Nevertheless, the mechanism of S.C.C. of aluminum alloys is complex, involving several time-dependent interactions that are difficult to evaluate independently, such as the roles of anodic dissolution, hydrogen embrittlement, and grain boundary precipitates in the initiation and propagation of cracks [40–42]. Understanding has been sufficient, however, to permit new aluminum alloys to be developed with high strength combined with improved resistance to corrosion, including S.C.C., and such alloys are now used, for example, on Boeing 777 aircraft [41].

Many high-strength aluminum alloys are available (some are listed in Table 21.1); specific composition ranges and heat treatments for these alloys are usually chosen with the intent of minimizing susceptibility to S.C.C. Solution heat-treatment temperature affects stress-corrosion susceptibility by altering the grain-boundary composition as well as the alloy metallurgical microstructure [42, 43]. Service temperatures—especially those above room temperature—that can cause artificial aging sometimes induce susceptibility followed by premature failure in the presence of moisture or sodium chloride solutions. Susceptibility of any of the wrought alloys is greatest when stressed at right angles to the rolling direction (in the short transverse direction), probably because more grain-boundary area of elongated grains along which cracks propagate comes into play.

As mentioned earlier, cladding of alloys can serve to cathodically protect them from either intergranular corrosion or S.C.C. Compressive surface stresses are effective for avoiding S.C.C.; hence, practical structures are sometimes shot peened.

REFERENCES

1. J. Draley and W. Ruther, *Corrosion* **12**, 441t, 480t (1956).
2. T. Pearson and H. Phillips, Corrosion of pure aluminum, *Met. Rev. (London)* **2**, 305, (1957).
3. H. Kaesche, *Z. Phys. Chem. N.F.* **34**, 87, (1962).
4. H. Böhni and H. Uhlig, *J. Electrochem. Soc.* **116**, 906, (1969).
5. C. Southwell, A. Alexander, and C. Hummer, Jr., *Mater. Prot.* **4**, 30, (1965).
6. R. Groover, T. Lennox, Jr., and M. Peterson, *Mater. Prot.* **8**, 25, (1969).
7. D. MacLennan, A. Macmillan, and J. Greenblatt, The corrosion of aluminum and aluminum alloys in high temperature water, in *Proceedings of the First International Congress on Metallic Corrosion*, Butterworths, London, 1961, p. 429.
8. R. Hart and J. Maurice, *Corrosion* **21**, 222, (1965).
9. A. Hamstead, G. Elder, and J. Canterbury, *Corrosion* **14**, 189t (1958).
10. E. Ghali, Aluminum and aluminum alloys, in *Uhlig's Corrosion Handbook*, 2nd edition, R. W. Revie, editor, Wiley, New York, 2000, pp. 687–690.
11. B. J. Little and J. S. Lee, *Microbiologically Influenced Corrosion*, Wiley, New York, 2007, pp. 195–201.
12. J.-D. Gu, T. E. Ford, and R. Mitchell, Microbiological corrosion of metals, in *Uhlig's Corrosion Handbook*, 2nd edition, R. W. Revie, editor, Wiley, New York, 2000, pp. 923–924.
13. J. Elphick, in *Microbial Aspects of Metallurgy*, J. D. A. Miller, editor, American Elsevier, New York, 1970, pp. 157–172.
14. H. Hedrick, *Materials Prot.* **9**(1), 27 (1970).
15. P. B. Park, in *Microbial Aspects of the Deterioration of Materials*, edited by R. Gilbert and D. Lovelock, Academic Press, London, 1975, pp. 105–126.
16. R. Salvarezza, M. de Mele, and H. Videla, *Corrosion* **39**, 26, (1983).
17. C. J. McNamara, T. D. Perry, N. Wolf, R. Mitchell, R. Leard, and J. Dante, Corrosion of aluminum alloy 2024 by jet fuel degrading microorganisms, CORROSION/2003, Paper No. 03568, NACE International, Houston, TX, 2003.
18. C. J. McNamara, T. D. Perry IV, R. Leard, K. Bearce, J. Dante, and R. Mitchell, *Biofouling* **221**(5/6), 257 (2005).
19. M. Stern and H. Uhlig, *J. Electrochem. Soc.* **99**, 381, 389, (1952).
20. M. Stern and H. Uhlig, *J. Electrochem. Soc.* **100**, 543, (1953).
21. W. Archer and M. Harter, *Corrosion* **34**, 159, (1978).
22. R. Ellialtioglu, H. White, L. Godwin, and T. Wolfram, *J. Chem. Phys.* **72**(10), 5291 (1980).
23. E. Cook, R. Horst, and W. Binger, *Corrosion* **17**, 25t (1961).
24. R. Farmer, *Metallurgia* **68**, 161, (1963).
25. T. Mooney, Electroplated coatings, in *ASM Handbook*, Vol. 13A, *Corrosion: Fundamentals, Testing, and Protection*, ASM International, Materials Park, OH, 2003, p. 778.
26. M. Bothwell, *J. Electrochem. Soc.* **106**, 1014, (1959).
27. M. Whitaker, *Metal Ind.* **80**, 1, (1952).

28. B. Lifka and D. Sprowls, *Corrosion* **22**, 7, (1966).

29. J. Zahavi and Y. Yahalom, *J. Electrochem. Soc.* **129**, 1181, (1982).

30. W. Robertson, *Trans. Am. Inst. Min. Metall. Eng.* **166**, 216 (1946); see also U. Evans, *The Corrosion and Oxidation of Metals*, Arnold, London, 1960, p. 674.

31. P. Brenner, *Metal Progr.* **65**, 112, (1954).

32. J. G. Kaufman, Corrosion of aluminum and aluminum alloys, in *ASM Handbook*, Vol. 13B, *Corrosion: Materials*, ASM International, Materials Park, OH, 2003, pp. 105–106.

33. *Aluminum: The Corrosion Resistant Automotive Material*, Publication AT7, Aluminum Association, 2001.

34. *Aluminum for Automotive Body Sheet Panels*, Publication AT3, Aluminum Association, 1998.

35. G. Courval, Corrosion of aluminum components in the automotive industry, in *ASM Handbook*, Vol. 13C, *Corrosion: Environments and Industries*, ASM International, Materials Park, OH, 2006, p. 545.

36. Y. Komatsu et al., Application of all aluminum automotive body for Honda NSX, Technical Paper 910548, Society of Automotive Engineers, 1991.

37. C. Edeleanu, *J. Inst. Metals* **80**, 187, (1951).

38. G. Wasserman, *Z. Metall.* **34**, 297, (1942).

39. H. Paxton and R. Procter, in *Fundamental Aspects of Stress Corrosion Cracking*, R. Staehle et al., editor, National Association of Corrosion Engineers, Houston, TX, 1969, p. 509.

40. N. J. H. Holroyd, Environment-induced cracking of high-strength aluminum alloys, in *Proceedings of the First International Conference on Environment-Induced Cracking of Metals*, 1988, NACE-10, R. P. Gangloff and M. B. Ives, editors, National Association of Corrosion Engineers, Houston, TX, 1990, pp. 311–345.

41. V. S. Agarwala, Corrosion in the military, in *ASM Handbook*, Vol. 13C, *Corrosion: Environments and Industries*, ASM International, Materials Park, OH, 2006, pp. 132–133.

42. J. G. Kaufman, Corrosion of aluminum and aluminum alloys, in *ASM Handbook*, Vol. 13B, *Corrosion: Materials*, ASM International, Materials Park, OH, 2005, pp. 105–106.

43. C. Shastry, M. Levy, and A. Joshi, *Corros. Sci.* **21**, 673, (1981).

GENERAL REFERENCES

V. S. Agarwala and G. M. Ugiansky, editors, *New Methods for Corrosion Testing of Aluminum Alloys*, STP 1134, ASTM, Philadelphia, 1992.

E. Ghali, Aluminum and aluminum alloys, in *Uhlig's Corrosion Handbook*, 2nd edition, R. W. Revie, editor, Wiley, New York, 2000, pp. 677–716.

N. J. H. Holroyd, Environment-induced cracking of high-strength aluminum alloys, in *Proceedings of the First International Conference on Environment-Induced Cracking of Metals*, 1988, NACE-10, R. P. Gangloff and M. B. Ives, editors, National Association of Corrosion Engineers, Houston, TX, 1990, pp. 311–345.

J. G. Kaufman, Corrosion of aluminum and aluminum alloys, in *ASM Handbook*, Vol. 13B, *Corrosion: Materials*, ASM International, Materials Park, OH, 2003, pp. 95–124.

J. Snodgrass and J. Moran, Corrosion resistance of aluminum alloys, in *ASM Handbook*, Vol. 13A, *Corrosion: Fundamentals, Testing, and Protection*, ASM International, Materials Park, OH, 2003, pp. 689–691.

Stress-corrosion cracking of aluminum alloys, in *Stress-Corrosion Cracking*, R. H. Jones, editor, ASM International, Materials Park, OH, 1992, pp. 233–249.

22

MAGNESIUM AND MAGNESIUM ALLOYS

22.1 INTRODUCTION

Although magnesium and its alloys have long been perceived as rapidly corroding, it is possible, by appropriate alloy selection combined with good engineering design and correctly applied surface treatment and protection schemes, to use these materials successfully and obtain numerous benefits, including light weight, high strength/weight ratio, high stiffness/weight ratio, and ease of machinability [1]. In the original Volkswagen Beetle, magnesium alloys were successfully used for the engine and for the transmission casings. Today, magnesium alloys are used in consumer applications (e.g., cases for laptop computers and cameras), in the automotive industry (transmission casings, engine blocks), and in the aerospace industry (helicopter transmission and rotor housings). The corrosion behavior of magnesium and its alloys and the principles behind successful use of these materials are discussed in this chapter.

22.2 MAGNESIUM

$$Mg^{2+} + 2e^- \rightarrow Mg \qquad \phi° = -2.34\,V$$

Corrosion and Corrosion Control, by R. Winston Revie and Herbert H. Uhlig
Copyright © 2008 John Wiley & Sons, Inc.

Several investigators have obtained evidence that magnesium dissolves anodically to a large extent as Mg^+ and that this ion in solution reduces water to form Mg^{2+} plus H_2 (see Section 13.6).

Magnesium has the lowest density ($1.7 g/cm^3$) of all structural metals, having about one-fourth the density of iron and one-third that of aluminum. This very low density makes magnesium and magnesium alloys useful wherever low weight is important, if not required.

Magnesium is also the most active metal in the Emf Series that is used for structural purposes. The potential–pH diagram for magnesium and water (Fig. 4.4, Section 4.5) indicates the theoretical domains of corrosion, passivity, and immunity. Magnesium corrodes over a wide range of potential and pH. At high pH, in the region identified as passive, the magnesium hydroxide corrosion product film is only partially protective. This low-density film allows electrolyte to contact the underlying metal. In addition, the film ruptures because of the higher molar volume of magnesium hydroxide compared to metallic magnesium, resulting in the constant exposure of fresh metal [2].

Corrosion resistance of magnesium depends on purity of the metal even more than the corrosion resistance of aluminum depends on its purity. Distilled magnesium corrodes in seawater, for example, at the rate of 0.25 mm/y (0.01 ipy), which is about twice the rate for iron, but commercial magnesium corrodes at a rate 100–500 times higher, with visible hydrogen evolution. The impurities in commercial magnesium largely responsible for the higher rate are iron and, to a lesser extent, nickel and copper. Their effect becomes marked above a critical concentration called the tolerance limit [3]. For iron, the tolerance limit is 0.017%; similarly, for nickel and copper, it is 0.0005% and 0.1%, respectively (Fig. 22.1). Additions of manganese or zinc to the metal raise these limits to higher values. Manganese has been reported to surround iron particles and to render them inactive as local cathodes [4]. An additional mechanism may involve formation of an insoluble intermetallic compound that includes iron in its structure. This is heavier than magnesium; hence, similar to the situation for aluminum (Section 21.2), it settles to the bottom of the melt, in this way reducing the Fe content of the cast metal.

22.3 MAGNESIUM ALLOYS

Because unalloyed magnesium is not used extensively for structural applications, it is the corrosion resistance of magnesium alloys that is of primary interest. To enhance strength and resistance to corrosion, magnesium is alloyed with aluminum, lithium, zinc, rhenium, thorium, and silver, with minor additions of cerium, manganese, and zirconium sometimes being used as well.

There are two major alloying systems:

1. Alloys containing 2–10% aluminum with minor additions of zinc and manganese; these alloys are used up to 95–120°C (200–250°F), above

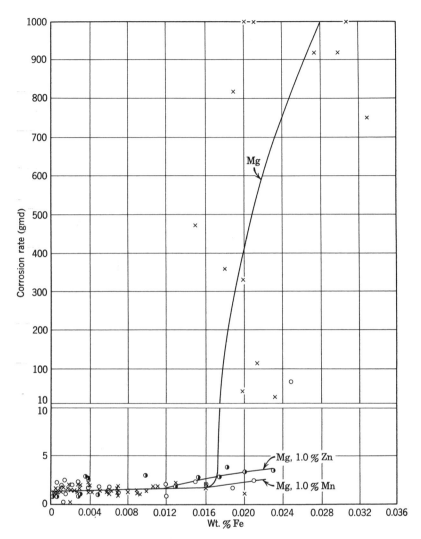

Figure 22.1. Corrosion of magnesium in 3% NaCl, alternate immersion, 16 weeks, showing tolerance limit for iron and beneficial effect of alloyed zinc and manganese [Figure 4 from J. Hanawalt, C. Nelson, and J. Peloubet, Corrosion studies of magnesium and its alloys, *Trans. AIME, Inst. Metals Div.* **147**, 281 (1942)].

which temperature range both corrosion resistance and strength deteriorate.

2. Alloys containing various elements (rare earths, zinc, thorium, and silver), but not aluminum, and containing zirconium, which provides grain refinement and improved mechanical properties; these alloys provide improved elevated temperature properties compared to those in the first group.

Some alloys in both systems that are commonly available are listed in Table 22.1. In the ASTM system for designating magnesium alloys, the first two letters indicate the principal alloying elements: A, aluminum; E, rare earths; H, thorium; K, zirconium; M, manganese; Q, silver; Z, zinc. The letter corresponding to the element present in greater concentration is listed first; if the elements are in equal concentration, they are listed alphabetically. The letters are followed by numbers that indicate the nominal compositions of these alloying elements, rounded off to the nearest whole number; for example, AZ91 indicates the alloy Mg–9Al–1Zn. In addition, letters that are sometimes appended to the alloy designation are assigned chronologically and usually indicate alloy improvements in purity [5].

In saline solutions, the corrosion rate is controlled by the concentration and distribution of the critical elements, iron, nickel, and copper, which create cathodic sites of low hydrogen overpotential. For this reason, these elements are usually controlled at low impurity levels.

Alloy purity, however, does not prevent galvanic corrosion that occurs if the magnesium alloy is coupled, for example, to a steel bolt. Design to prevent galvanic corrosion is, therefore, essential. In addition to the anode/cathode area ratio, the conductivity and composition of the medium in which a couple is immersed are controlling factors in the rate of galvanic corrosion. All commonly used metals cause galvanic corrosion of magnesium in saline solutions, but zinc plating the cathode (e.g., iron or steel) has been found to reduce the galvanic corrosion of the magnesium to one-tenth the rate. In addition, a reduction in the conductivity (e.g., changing from 3% NaCl to tap water), results in an even greater reduction in the galvanic corrosion rate.

22.3.1 Stress-Corrosion Cracking

The magnesium alloys with greatest susceptibility to stress-corrosion cracking (S.C.C.) are the Mg–Al alloys, and susceptibility increases with Al concentration. Magnesium–zinc alloys have intermediate susceptibility, and alloys that contain neither aluminum nor zinc are the most resistant [7].

Stress-corrosion cracking of magnesium alloys occurs in many environments, including distilled water, seawater, and atmospheric environments in rural, urban, industrial, and coastal areas. In water, dissolved oxygen accelerates S.C.C. and deaeration retards it. S.C.C. in magnesium alloys is usually transgranular with secondary cracks, or branching, and initiation usually occurs at pits. Although several mechanisms for S.C.C. of magnesium alloys have been proposed, hydrogen embrittlement is most likely a factor [8].

Intergranular S.C.C. of magnesium alloys has also been reported, attributed to a grain-boundary phase of $Mg_{17}Al_{12}$ that causes galvanic attack of the adjacent matrix [8].

As in most systems in which S.C.C. occurs, both applied and residual stresses are important. Welded components of magnesium alloys are normally stress-relieved after welding to reduce the residual stresses that develop during welding.

TABLE 22.1. Nominal Compositions of Some Magnesium Alloys[a]

Alloy Number		Element (%)							Product Form[b]
ASTM	UNS	Al	Zn	Mn	Ag	Zr	Th	Re	
M1	M15100	—	—	1.5	—	—	—	—	W
AM50	M10500	5	—	0.3	—	—	—	—	C
AM60	M10600	6	—	0.2	—	—	—	—	C
AZ31	M11310	3	1	0.2	—	—	—	—	W
AZ61	M11610	6	1	0.2	—	—	—	—	W
AZ63	M11630	6	3	0.2	—	—	—	—	C
AZ80	M11800	8	0.5	0.2	—	—	—	—	C, W
AZ91	M11910	9	1	0.2	—	—	—	—	C
EZ33	M12331	—	2.5	—	—	0.5	—	2.5	C
ZM21	—	—	2	1	—	—	—	—	W
HK31	M13310	—	0.1	—	—	0.5	3	—	C, W
HZ32	M13320	—	2	—	—	0.5	3	—	C
QE22	M18220	—	—	—	2.5	0.5	—	2	C
QH21	M18210	—	—	—	2.5	0.5	1	1	C
ZE41	M16410	—	4.5	—	—	0.5	—	1.5	C
ZE63	M16630	—	5.5	—	—	0.5	—	2.5	C
ZK40	M16400	—	4.0	—	—	0.5	—	—	C, W
ZK60	M16600	—	6.0	—	—	0.5	—	—	C, W

[a]Adapted from Ref. 6.
[b]C, cast alloy; W, wrought alloy.

Resistance to S.C.C. may also be increased by using shot peening and other processes that produce compressive residual stresses at the surface [8].

22.3.2 Coatings

Coatings provide an important strategy for protecting magnesium alloys from corrosion. In light-truck applications, heads of bolts made from AZ91D were protected from galvanic corrosion by using a nylon coating that was electrostatically applied. Molded plastic caps were also found to be effective. Zinc alloy coatings (e.g., 80Sn–20Zn) electroplated on magnesium and given a chromate treatment were found to decrease the corrosion rate of magnesium by more than 90% in salt spray testing [9, 10].

Various protective anodized coatings are available as produced in electrolytes composed mainly of fluorides, phosphates, or chromates [11]. Anodizing treatments result in porous coatings that provide excellent bases for subsequent painting. Anodized magnesium castings are used successfully, for example, in helicopter drive system components. In general, in aerospace applications where extended component life and low maintenance costs are required, all surfaces of magnesium alloys exposed to corrosive environments must be coated.

Driven by the need to reduce both gasoline consumption and greenhouse gas production, the automotive industry is reducing vehicle weight to the extent possible, and one approach being used is increasing the magnesium content of automobiles. Large heavy automobile parts are particularly attractive candidates for replacement by magnesium. The extent of corrosion protection required depends on the severity of the environment; for example, for underbody and wheel applications, comprehensive protection schemes are essential, including protection against galvanic corrosion. Underbody wax-type coatings may provide additional protection. One of the more recent developments in the automotive industry is the application of Alloy AM50 for the front-end support assembly for light-duty Ford trucks [12]. Magnesium alloys are also being used for air cleaner covers, engine compartment grills, retractable headlight assemblies, clutch and brake pedal supports, and clutch and transmission housings. Fasteners that are specially designed using nylon or plastic washers, sleeves, and so on, are used to help control galvanic corrosion of magnesium [9].

22.4 SUMMARY

Limitations in the use of magnesium alloys for structural parts are as follows:

1. Sensitivity to stress-corrosion cracking in moist air.
2. Tendency of magnesium and its alloys to corrode when galvanically coupled to other metals. Aluminum alloys containing some magnesium (e.g., types 5050, 5052, and 5056) are least affected by alkalies generated by a magnesium–aluminum couple and, hence, can be applied in contact with magnesium. Pure aluminum is also satisfactory. In general, however, magnesium must be insulated from all dissimilar metals; insulating washers must be used under bolt or screw heads in order to extend the path of electrolyte resistance and thereby minimize galvanic action.

Magnesium, in brief, is resistant to the following:

1. Atmospheric exposure. The controlled purity 3% Al, 1.5% Zn–Mg alloy exposed for eight years to the tropical marine atmosphere at Panama lost 2.4 mm (0.094 in.) per year, compared to 0.1 mm (0.004 in.) per year in the temperate marine atmosphere of Kure Beach, North Carolina [13]. Most alloys must be stress relieved, avoiding subsequent high applied stress in order to avoid stress-corrosion cracking in moist atmospheres.
2. Distilled water (same precautions as for 1).
3. >2% HF. A protective film of magnesium fluoride forms. Pitting may occur at the water–air interface. The corrosion rate of 8% Al, 0.2% Mn alloy in 5% HF is 2.3 gmd; in 48% HF, it is 0.05 gmd.
4. Alkalies. Unlike aluminum, magnesium is resistant to alkalies. The rate in 48% NaOH–4% NaCl is 0.2 gmd (high-purity Mg). Above 60°C (140°F), the rate increases appreciably.

Magnesium is *not* resistant to the following:

1. Waters containing traces of heavy metal ions.
2. Seawater.
3. Inorganic or organic acids and acid salts (e.g., NH_4 salts).
4. Methanol (anhydrous); magnesium methylate forms. The reaction can be inhibited by $(NH_4)_2S$, H_2CrO_4, turpentine, or dimethylglyoxime (DMG). The higher alcohols are resisted satisfactorily.
5. Leaded gasoline mixtures.
6. Freon (CCl_2F_2) plus water. The anhydrous gas is not corrosive.

REFERENCES

1. E. Ghali, Magnesium and magnesium alloys, in *Uhlig's Corrosion Handbook*, 2nd edition, R. W. Revie, editor, Wiley, New York, 2000, p. 793.
2. B. A. Shaw and R. C. Wolfe, Corrosion of magnesium and magnesium-base alloys, in *ASM Handbook*, Vol. 13B, *Corrosion: Materials*, ASM International, Materials Park, OH, 2003, p. 205.
3. J. Hanawalt, C. Nelson, and J. Peloubet, *Min. Metall. Eng., Inst. Metals Div.* **147**, 273 (1942).
4. H. Robinson and P. George, *Corrosion* **10**, 182 (1954).
5. Ref. 1, p. 796.
6. B. A. Shaw, Corrosion resistance of magnesium alloys, in *ASM Handbook*, Vol. 13A, *Corrosion: Fundamentals, Testing, and Protection*, ASM International, Materials Park, OH, 2003, p. 693.
7. Ref. 1, p. 813.
8. W. K. Miller, Stress-corrosion cracking of magnesium alloys, in *Stress-Corrosion Cracking*, R. H. Jones, editor, ASM International, Materials Park, OH, 1992, pp. 251–263.
9. D. L. Hawke and T. Ruden, Galvanic compatibility of coated steel fasteners with magnesium, paper no. 950429, International Congress and Exposition, Detroit, Michigan, SAE International, Warrendale, PA, 1995.
10. Ref. 2, pp. 218–219.
11. E. Emley, *Principles of Magnesium Technology*, Pergamon Press, New York, 1966, pp. 692–705.
12. Ref. 2, pp. 222–223.
13. C. Southwell, A. Alexander, and C. Hummer, Jr., *Mater. Prot.* **4**(12), 30 (1965).

GENERAL REFERENCES

M. M. Avedesian and H. Baker, editors, *Magnesium and Magnesium Alloys*, ASM Specialty Handbook, ASM International, Materials Park, OH, 1999.

E. Ghali, Magnesium and magnesium alloys, in *Uhlig's Corrosion Handbook*, 2nd edition, R. W. Revie, editor, Wiley, New York, 2000, pp. 793–830.

R. Hodkinson and J. Fenton, *Lightweight Electric/Hybrid Vehicle Design*, SAE International, Butterworth-Heinemann, Oxford, 2001.

Alan A. Luo, editor, *Magnesium Technology 2004*, TMS, Warrendale, PA, 2004.

W. K. Miller, Stress-corrosion cracking of magnesium alloys, in *Stress-Corrosion Cracking*, R. H. Jones, editor, ASM International, Materials Park, OH, 1992, pp. 251–263.

N. R. Neelameggham, H. I. Kaplan, and B. R. Powell, *Magnesium Technology 2005*, TMS, Warrendale, PA, 2005.

M. O. Pekguleryuz and L. W. F. Mackenzie, *Magnesium Technology in the Global Age*, Canadian Institute of Mining, Metallurgy and Petroleum, Montreal, Canada, 2006.

B. A. Shaw, Corrosion resistance of magnesium alloys, in *ASM Handbook*, Vol. 13A, *Corrosion: Fundamentals, Testing, and Protection*, ASM International, Materials Park, OH, 2003, pp. 692–696.

23

NICKEL AND NICKEL ALLOYS

23.1 INTRODUCTION

Nickel has many applications in industry where corrosion resistance is required. Nickel is an excellent base on which to design alloys because, first, the element itself is moderately corrosion resistant, and, second, it can be alloyed with many other elements while maintaining its ductile face-centered-cubic structure. As a result, over the past 100 years, a wide range of binary, ternary, and other nickel-base alloys were developed to meet the increasingly demanding requirements in many industries for resistance to aqueous corrosion and to high-temperature oxidation conditions. These new nickel-base alloys were designed by developing and applying fundamental knowledge on the role of individual alloying elements and by improving thermomechanical processes to optimize alloy metallurgy. Today, nickel-base alloys are used in chemical process, petrochemical, pulp and paper, energy conversion, power production, supercritical water, waste incineration, pharmaceutical, and many other industries. Alloy selection from the large number of alloys available depends on the specific properties that are required. Tables 23.1 and 23.2 list the major alloying elements and the effects that they have on properties of alloys intended for aqueous corrosion and for high-temperature application, respectively [1].

Corrosion and Corrosion Control, by R. Winston Revie and Herbert H. Uhlig
Copyright © 2008 John Wiley & Sons, Inc.

TABLE 23.1. Alloying Elements and Their Major Effects in Alloys for Aqueous Corrosion Resistance

Alloying Element	Main Features	Other Features
Ni	Provides matrix for metallurgical compatibility with alloying elements. Improves thermal stability and fabricability.	Enhances corrosion resistance in mild reducing media and in alkali media. Improves chloride stress-corrosion cracking (S.C.C.).
Cr	Provides resistance to oxidizing corrosive media.	Enhances localized corrosion resistance.
Mo	Provides resistance to reducing (nonoxidizing) corrosive media.	Enhances localized corrosion resistance and chloride S.C.C. Provides solid solution strengthening.
W	Behaves similarly to Mo, but less effective. Detrimental to thermal stability in high Ni–Cr–Mo alloys.	Provides solid solution strengthening.
N	Austenitic stabilizer; economical substitute for nickel.	Enhances localized corrosion resistance, thermal stability, and mechanical properties.
Cu	Improves resistance to seawater and sulfuric acid; detrimental to thermal stability in higher Ni–Cr–Mo alloys.	Enhances resistance to environments containing H_2SO_4 and HF.
Ti, Nb, Ta	Carbon stabilizer.	Improves HAZ and intergranular corrosion resistance.
Si	High silicon (>4%) improves resistance to oxidizing mineral acids, sulfuric acid, and nitric acid.	Detrimental in certain corrosive environments.
Fe	Provides matrix for metallurgical compatibility with various alloying elements.	Reduces cost by replacing nickel and enhances scrap utilization.

Because of the increased cost of these alloys compared to stainless steels, they are used when stainless steels are not suitable or when product purity and/or safety are critically important [1].

23.2 NICKEL

$$Ni^{2+} + 2e^- \rightarrow Ni \qquad \phi^\circ = -0.250\,V$$

Nickel is active in the Emf Series with respect to hydrogen, but noble with respect to iron. The theoretical potential–pH domains of corrosion, immunity, and passivation of nickel at 25°C are shown in Fig. 23.1 [2], which shows, for example,

TABLE 23.2. Alloying Elements and Their Major Effects in High Temperature Alloys

Alloying Element	Main Features	Other Features
Cr	Improves oxidation resistance; detrimental to nitriding and fluorination resistance	Improves sulfidation and carburization resistance.
Si	Improves oxidation, nitriding, sulfidation and carburizing resistance; detrimental to nonoxidizing chlorination resistance.	Synergistically acts with chromium to improve high temperature degradation.
Al	Independently and synergistically with Cr improves oxidation resistance; detrimental to nitriding resistance.	Helps improve sulfidation resistance; improves age hardening effects.
Mo	Improves high temperature strength; improves creep strength; detrimental to oxidation resistance at higher temperatures.	Reduces chlorination resistance.
W	Similar to molybdenum.	
Nb	Increases short-term creep strength; may be beneficial in carburizing; detrimental to nitriding resistance.	
C	Improves strength; helps nitriding resistance; beneficial to carburization resistance; oxidation resistance adversely affected.	
Ti	Improves strengthening by age hardening.	Detrimental to nitriding resistance.
Mn	Slight positive effect on high temperature strength and creep; detrimental to oxidation resistance; increases solubility of nitrogen.	
Co	Reduces rate of sulfur diffusion; hence, improves sulfidation resistance; improves solid solution strength.	Increases solubility of carbon in nickel-base alloys, increasing resistance to carburization.
Ni	Improves carburization, nitriding, and chlorination resistance; detrimental to sulfidation resistance.	
Y, Re	Improves adherence and spalling resistance of oxide layer, and hence, improves oxidation, sulfidation, carburization resistance.	

that nickel can be cathodically protected in acid or neutral solution by controlling the potential below about $-0.4\,V$ versus S.H.E.

Nickel does not react rapidly with dilute nonoxidizing acids (e.g., H_2SO_4 and HCl) unless dissolved oxygen is present. It is thermodynamically inert to (will not corrode in) deaerated water at room temperature, in which $Ni(OH)_2$ forms as a corrosion product. It is passive in contact with many aerated aqueous solutions, but the passive film is not as stable as, for example, that on chromium (for

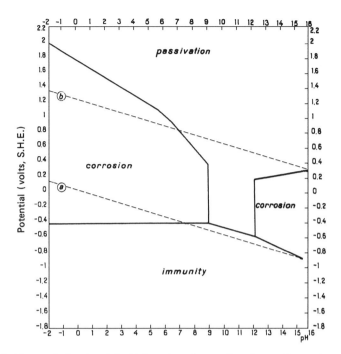

Figure 23.1. Theoretical potential–pH domains of corrosion, immunity, and passivation for nickel at 25°C [2]. (M. Pourbaix, *Atlas of Electrochemical Equilibria in Aqueous Solutions*, 2nd English edition, p. 334. Copyright ANCE International 1974 and CEBELCOR.)

nickel, Flade potential $\phi_F^\circ = 0.2\,\text{V}$) [3]. It corrodes by pitting when exposed to seawater (passive–active cells).

Nickel is outstandingly resistant to hot or cold alkalies. Only silver and possibly zirconium are more resistant. Nickel exposed to boiling 50% NaOH corrodes at the rate of 0.06 gmd (0.0001 ipy). It also resists fused NaOH, low-carbon nickel being preferred for this application in order to avoid intergranular attack of the stressed metal; stress relief anneal, 5 min at 875°C (1100°F) is advisable. Nickel is attacked by aerated aqueous ammonia solutions, a soluble complex, $Ni(NH_3)_6^{2+}$, forming as a corrosion product. It is also attacked by strong hypochlorite solutions with formation of pits.

Nickel is not subject to stress-corrosion cracking except in high concentrations of alkali or in fused alkali.

Nickel has excellent resistance to oxidation in air up to 800–875°C (1500–1600°F) and is often used in service at still higher temperatures. When subject to alternate oxidizing and reducing atmospheres at elevated temperatures, it tends to oxidize along grain boundaries. It also fails along grain boundaries in the presence of sulfur-containing environments above about 315°C (600°F). Fused salts contaminated with sulfur, or with sulfates in the presence of organic or other chemically reducing impurities, may damage nickel or high-nickel alloys in this manner.

In summary, nickel is resistant to the following:

1. Alkalies, hot or cold, including fused alkalies.
2. Dilute nonoxiding inorganic and organic acids. Resistance is improved if acids are deaerated.
3. The atmosphere. In industrial atmospheres, a nonprotective film forms, composed of basic nickel sulfate (fogging). Fogging is minimized by a thin chromium electroplate over nickel. There is good resistance to oxidation in air at elevated temperatures.

Nickel is *not* resistant to the following:

1. Oxidizing acids (e.g., HNO_3).
2. Oxidizing salts (e.g., $FeCl_3$, $CuCl_2$, and $K_2Cr_2O_7$).
3. Aerated ammonium hydroxide.
4. Alkaline hypochlorites.
5. Seawater.
6. Sulfur or sulfur-containing reducing environments, $>315°C$ ($>600°F$).

23.3 NICKEL ALLOYS

23.3.1 General Behavior

Adding *copper* to nickel improves, to a moderate degree, resistance under reducing conditions (e.g., nonoxidizing acids, including hydrofluoric acid). Also, in line with the resistance of copper to pitting, the tendency of nickel–copper alloys to form pits in seawater is less pronounced than for nickel, and pits tend to be shallow rather than deep. Above about 60–70 at.% Cu (62–72 wt.% Cu), the alloys lose the passive characteristics of nickel and tend to behave more like copper (see Section 6.8.1) retaining, however, appreciably improved resistance to impingement attack. Hence, 10–30% Ni, bal. Cu alloys (cupro-nickels) do not pit in stagnant seawater and are resistant to corrosion in rapidly flowing seawater. Such alloys containing a few tenths to 1.75% Fe, which still further improves resistance to impingement attack, are used for seawater condenser tubes. The 70% Ni–Cu alloy (Monel), on the other hand, pits in stagnant seawater and, in seawater, is best used under rapidly moving, aerated conditions that ensure uniform passivity. It does not pit under conditions that provide cathodic protection, such as when the alloy is coupled to a more active metal (e.g., iron).

Additions of *chromium* to nickel impart resistance to oxidizing conditions (e.g., HNO_3 and H_2CrO_4) by supporting the passivation process. The critical minimum chromium content [4] obtained from critical current densities for anodic passivation in sulfuric acid is 14 wt.% Cr. These alloys are more sensitive than nickel to attack by Cl^- and by HCl, and deep pits form when the alloys are exposed to stagnant seawater. Chromium also imparts to nickel improved resistance to oxidation at elevated temperatures. One important commercial alloy contains 20% Cr, 80% Ni (see Section 11.13.4).

Alloying nickel with *molybdenum* provides, to a marked degree, improved resistance under reducing conditions, such as in hydrochloric acid. The corrosion potentials of such alloys in acids, whether aerated or deaerated, tend to be more active than their Flade potentials [5]; hence, the alloys are not passive in the sense of Definition 1 (Section 6.1). For example, the corrosion potentials of nickel alloys containing 3–22.8% Mo in 5% H_2SO_4, hydrogen-saturated, all lie within 2 mV of a platinized platinum electrode in the same solution [5]. Notwithstanding an active corrosion potential, the alloy containing 15% Mo, for example, corrodes at 1/12 the rate of nickel in deaerated 10% HCl at 70°C, and the rate decreases still further with increasing molybdenum content (1/100 the rate for nickel at 25% Mo). Polarization measurements show that molybdenum alloyed with nickel has little effect on hydrogen overpotential, but increases anodic polarization instead; hence, the corrosion rate of the alloy is anodically controlled. The mechanism of increased anodic polarization is related most likely to a sluggish hydration of metal ions imparted by molybdenum, or alternatively to a porous diffusion-barrier film of molybdenum oxide, rather than to formation of a passive film typical of chromium or the passive chromium–nickel alloys.

As an alloying element in nickel, *tungsten* behaves similarly to molybdenum, but is less effective [6].

Because the binary nickel–molybdenum alloys have poor physical properties (low ductility, poor workability), other elements, for example, iron, are added to form ternary or multicomponent alloys. These are also difficult to work, but they mark an improvement over the binary alloys. Resistance of such alloys to hydrochloric and sulfuric acids is better than that of nickel, but it is not improved with respect to oxidizing media (e.g., HNO_3). Since the Ni–Mo–Fe alloys have active corrosion potentials and do not, therefore, establish passive–active cells, they do not pit in the strong acid media to which they are usually exposed in practice.

By alloying nickel with both molybdenum *and* chromium, an alloy is obtained resistant to oxidizing media imparted by alloyed chromium, as well as to reducing media imparted by molybdenum. One such alloy, which also contains a few percent iron and tungsten (Alloy C), is immune to pitting and crevice corrosion in seawater (10-year exposure) and does not tarnish appreciably when exposed to marine atmospheres. Alloys of this kind, however, despite improved resistance to Cl^-, corrode more rapidly in hydrochloric acid than do the nickel–molybdenum alloys that do not contain chromium.

Nominal compositions of some commercial nickel-base alloys containing copper, molybdenum, or chromium are given in Table 23.3. The Ni–Cu alloys are readily rolled and fabricated, whereas the Ni–Cr alloys are less readily, and the Ni–Mo–Fe and Ni–Mo–Cr alloys are difficult to work or fabricate.

Some commercial Cr–Ni–Fe–Mo alloys corresponding in composition to high-nickel stainless steels also contain a few percent copper. They are designed to resist, among other media, sulfuric acid over a wide range of concentrations.

TABLE 23.3. Typical Compositions (%) of Commercial Nickel-Base Alloys

Alloy System	Common Name of Alloy	UNS No.	Ni	Mo	Co[a]	Cr	W	Fe	Si[a]	Mn[a]	C[a]	Other
Ni–Cu	400	N04400	66.5	—	—	—	—	2.5[a]	0.5	0.2	0.3	Cu: 31.5
Ni–Cr	625	N06625	61	9	1.0	21.5	—	5.0[a]	0.5	0.5	0.1	Nb + Ta: 3.6 Al: 0.4; Ti: 0.4
Ni–Mo	B	N10001	Bal.	29.5	2.5	1.0[a]	—	6.0[a]	1.0	1.0	0.12	Cu: 0.5[a]
	B-2	N10665	Bal.	28	1.0	1.0[a]	0.5[a]	2.0[a]	0.1	1.0	0.02	Cu: 0.2[a]; Al: 0.5[a]
	B-3	N10675	65[b]	28.5	3.0	1.5	3.0[a]	1.5	0.1	3.0	0.01	
Ni–Cr–Fe	600	N06600	Bal.	—	—	15.5	—	8	0.5	0.5	0.15	Cu: 0.2
	800H	N08810	32	—	—	21	—	Bal.	0.5		0.08	Al: 0.4; Ti: 0.38
	690	N06690	58.0[b]	—	—	30	—	10	0.5	0.5	0.05	Cu: 0.5[a]
Ni–Cr–Fe– Mo–Cu	G	N06007	Bal.	6.5	2.5	22.0	1.0[a]	19.5	1.0	2.0	0.05	Cu: 2.0; Nb: 2.0
	G-3	N06985	Bal.	7	5.0	22.0	1.5[a]	19.5	1.0	1.0	0.015	Cu: 2.0
Ni–Cr–Mo	C	N10002	Bal.	16	2.5	16	4	6	1.0	1.0	0.08	V: 0.35[a]
	C-22	N06022	Bal	13	2.5	22	3	3	0.08	0.5	0.015	V: 0.35[a]
	C-276	N10276	Bal.	16	2.5	16	4	5	0.08	1.0	0.02	V: 0.35[a]
	59	N06059	Bal.	16	—	23	—	1.5[a]	0.1	0.5	0.01	Al: 0.25
Ni–Fe–Cr	825	N08825	42	3	—	21.5	—	22.0[b]	0.2	0.5	0.05	Al: 0.2[a]; Ti: 0.9

[a]Maximum. [b]Minimum.

Source: Adapted from Metals Handbook, Desk Edition, 2nd edition, ASM International, Materials Park, OH, 1998, p. 610. For detailed alloy specifications, see, for example, Worldwide Guide to Equivalent Nonferrous Metals and Alloys, 4th edition, ASM International, Materials Park, OH, 2001.

The function of alloyed copper is the same as that of alloyed palladium in titanium mentioned in Section 6.4, namely, to accelerate the cathodic reaction (H^+ or O_2 reduction) to the point where the anodic current density reaches or exceeds the critical value for anodic passivation.

23.3.2 Ni–Cu System: Alloy 400—70% Ni, 30% Cu

Because nickel and copper in all proportions form solid solutions, many nickel–copper and copper–nickel alloys are possible. The first commercial nickel–copper alloy was Alloy 400 (commonly known as Monel), which is still widely used today. With about 31% copper in a nickel matrix, its corrosion behavior is similar to that of nickel in many ways, but improved relative to nickel in other ways [7]. Since Alloy 400 is resistant to high-velocity seawater, it is often used for valve trim and pump shafts. It is also used for industrial hot freshwater tanks and for equipment in the chemical industry. It resists boiling sulfuric acid in concentrations less than 20%, with the corrosion rate being <0.20 mm/y (<0.008 ipy)(23-h test) [8]. It is outstandingly resistant to unaerated HF at all concentrations and temperatures up to boiling. [Rate in N_2-saturated 35% HF, 120°C (248°F) is 0.025 mm/y (0.001 ipy); when air saturated, 3.8 mm/y (0.15 ipy) [9].] Resistance to alkalies is good, *except* in hot concentrated caustic solutions or aerated NH_4OH.

Alloy 400 is not resistant to oxidizing media (e.g., HNO_3, $FeCl_3$, $CuSO_4$, and H_2CrO_4) nor to wet Cl_2, Br_2, SO_2, and NH_3. The alloy is susceptible to S.C.C. in moist aerated hydrofluoric acid and in hydrofluorosilicic acid vapor, but susceptibility can be minimized by deaeration of the environment and by stress relief annealing the alloy component [10].

23.3.3 Ni–Cr–Fe System: Alloy 600—76% Ni, 16% Cr, 7% Fe

Alloy 600 resists oxidizing aqueous media [e.g., mine waters, $Fe_2(SO_4)_3$, $CuSO_4$, and HNO_3]. In nitric acid, resistance is best above 20% HNO_3, including red fuming acid, but is not as good, in general, as the stainless steels. Resistance to alkalies is good, except in concentrated hot caustic solutions. It is also resistant to all concentrations of NH_4OH at room temperature. Like the stainless steels, this alloy tends to pit in seawater and also in oxidizing metal chlorides, such as $FeCl_3$.

Alloy 600 is used extensively where oxidation resistance at elevated temperatures is required (see Section 11.13.4).

Pressurized water reactors of nuclear-power plants use mill-annealed Alloy 600 tubes for heat exchangers. Typically, reactor coolant circulates through the tubes at 315°C (600°F) exiting 30–35°C lower in temperature. Boiler water in contact with the outside of the tubes is all-volatile treated (minimum dissolved solids and dissolved O_2, slightly alkaline using NH_3). Thinning and intergranular S.C.C. of tubes tend to occur at inlet-end sections above the tube sheet in crevice and sludge buildup areas [11]. Washings from such areas test alkaline and are high in sodium. Hence, laboratory tests of stressed alloy exposed to hot NaOH solutions (290–365°C) are used as an accelerated measure of resistance to S.C.C. in operational steam units. Such tests show that heat treatment of Alloy 600 at either 650°C for 4 h or at 700°C for 16 h or more considerably improves resistance to S.C.C. in NaOH solutions [12]. The improved resistance also extends to an environment of relatively pure water at 345°C (650°F) [12]. Through heat treatment, discontinuous carbide precipitates form at grain boundaries accompanied at 650°C (but not at 700°C) by chromium depletion. The latter situation accounts for intergranular corrosion of the alloy by 25% HNO_3 and by polythionic acids. But whether or not chromium depletion occurs is not related to the cause of intergranular S.C.C.

Accordingly, Alloy 600 fails in 10% NaOH at 315°C (600°F) independent of carbon content (0.006–0.046%) [13]; and an alloy (18% Cr, 77% Ni) approximating the Alloy 600 composition, but containing only 0.002% C, remains susceptible in water at 350°C (660°F) [14]. The long incubation time typically required for cracks to initiate in pure water (several months) supports the view that specifically damaging elements must reach a critical concentration by slow diffusion to grain boundaries in order for the alloy to become susceptible. Both phosphorus and boron have been suspected [13, 15] (see also Section 19.2.3.3). Alloy 690, however, has been found to provide satisfactory service, with no reported failures by intergranular S.C.C. [16].

Alloy 600 is resistant to S.C.C. in boiling $MgCl_2$ solutions. The extent of S.C.C. of the alloy in water at 350°C is unaffected by, and remains intergranular in, the presence of 0.1% NaCl [14].

23.3.4 Ni–Mo System: Alloy B—60% Ni, 30% Mo, 5% Fe

The original Alloy B (N10001) was invented in the 1920s. Alloy B and commercial alloys of similar composition are resistant to hydrochloric acid of all concentrations and temperatures up to the boiling point. The rate of corrosion is 0.23 mm/y (0.009 ipy) in boiling 10% HCl; 0.5 mm/y (0.02 ipy) in boiling 20% HCl; and 0.05 mm/y (0.002 ipy) in 37% HCl at 65°C (150°F) [17]. Resistance to boiling sulfuric acid is good up to 60% H_2SO_4 (<0.2 mm/y, <0.007 ipy). In phosphoric acid, rate of attack is low for all concentrations and temperatures, with the highest rate for the pure acid applying to the boiling 86% H_3PO_4 (0.8 mm/y, 0.03 ipy). In various organic acids, hot or cold, resistance is also good.

Because of the lack of chromium, Alloy B is not resistant to oxidizing conditions (for example, HNO_3) or to oxidizing metal chlorides (such as $FeCl_3$) [18]. Contamination of nonoxidizing acids with oxidizing ions (such as Fe^{3+} or Cu^{2+}) causes a large increase in corrosion rate [19]. Nickel–molybdenum alloys must not be used where oxidizing conditions exist.

Because of the high carbon content of Alloy B, it is subject to intergranular attack in nonoxidizing acids (e.g., acetic, formic, and hydrochloric) when heated in the range 500–700°C (930–1300°F), as may occur in the heat-affected zone (HAZ) of a weld. Heat treatment to avoid such attack consists of annealing at 1150–1175°C (2100–2150°F), followed by rapid cooling in air or water. The low-carbon alloy, B-2, is less susceptible to intergranular attack and HAZ corrosion, but it is difficult to fabricate and can crack during manufacturing operations. These embrittlement problems were resolved by developing Alloy B-3, with carefully controlled additions of iron, chromium, and manganese [18, 19].

23.3.5 Ni–Cr–Fe–Mo–Cu System: Alloy G—Ni, 22% Cr, 20% Fe, 6.5% Mo, 2% Cu

The high nickel content in these alloys provides corrosion resistance in reducing environments. The chromium content contributes to strength, aqueous corrosion resistance in oxidizing environments, and oxidation resistance. Alloy G was developed during the 1960s. Alloy G-3 is an improvement on Alloy G, but with greater resistance to HAZ attack and better weldability. The lower carbon content results in slower kinetics of carbide precipitation, and the higher molybdenum content improves the localized corrosion resistance. As a result, Alloy G-3 has replaced Alloy G in most applications. Alloy G-3 is widely used for downhole tubulars in the oil and gas industry [20, 21].

23.3.6 Ni–Cr–Mo System: Alloy C—54% Ni, 15% Cr, 16% Mo, 4% W, 5% Fe

The Ni–Cr–Mo alloys are widely used within the chemical process industry. These alloys are resistant to both oxidizing and reducing environments. They resist chloride-induced pitting, crevice corrosion, and S.C.C. In addition, they are easily formed and are weldable.

Because of its chromium content, Alloy C is resistant to such oxidizing media as HNO_3, HNO_3–H_2SO_4 mixed acids, H_2CrO_4, $Fe_2(SO_4)_3$, $FeCl_3$, and $CuCl_2$. Corrosion rates fall below 0.5 mm/y (0.02 ipy) in <50% HNO_3 at 65°C (150°F), but are higher above this concentration. In boiling >15% HNO_3, rates are high. Resistance is excellent to wet or dry Cl_2 at room temperature; pitting in wet Cl_2 may occur at higher temperatures. Alloy C also resists wet or dry SO_2 up to about 70°C (155°F)[17]. Alloy C has outstanding resistance to pitting or crevice corrosion in seawater. Resistance to acetic acid is excellent. In boiling 40% formic acid, the rate is 0.25 mm/y (0.01 ipy).

Alloy C has good resistance to hydrochloric acid at room temperature [0.025 mm/y (0.001 ipy) for 37% HCl], but not at higher temperatures [5 mm/y (0.2 ipy) for 20% HCl, 65°C (150°F)]. It is recommended for boiling sulfuric acid up to 10% acid (0.25 mm/y) and for boiling phosphoric acid up to at least 50% acid (0.25 mm/y).

When heated in the temperature range 500–700°C (930–1300°F), the alloy tends to corrode intergranularly; for this reason, weld HAZs undergo severe intergranular corrosion, which can be avoided by a final heat treatment at 1210–1240°C (2210–2260°F), followed by rapid cooling in air or water [22].

Alloy C-276, introduced in the 1960s, contains less carbon and is less sensitive to intergranular corrosion resulting from intermediate heat treatment. For this reason, Alloy C-276 can be used in most applications in the as-welded condition without severe intergranular attack.

Alloy C-22 and Alloy 59 were developed for applications where resistance to corrosion under highly oxidizing conditions is required. The high chromium content in these alloys, 22% and 23% for C-22 and 59, respectively, imparts excellent corrosion resistance of these alloys to nitric acid [23]. In addition, these alloys have been found to have superior crevice corrosion resistance in seawater [24].

23.3.7 Ni–Fe–Cr System: Alloy 825—Ni, 31% Fe, 22% Cr

The Ni–Fe–Cr system bridges the gap between the high-nickel austenitic stainless steels and the Ni–Cr–Fe alloy system. Alloy 825 is an upgrade from the 300 series stainless steels and is used to obtain resistance to localized corrosion and S.C.C. This alloy was developed from Alloy 800 by increasing the nickel content and by adding molybdenum (3%), copper (2%), and titanium (0.9%) for improved corrosion resistance in many media [20, 21].

REFERENCES

1. D. C. Agarwal, Nickel and nickel alloys, in *Uhlig's Corrosion Handbook*, 2nd edition, R. W. Revie, editor, Wiley, New York, 2000, p. 831.

2. M. Pourbaix, *Atlas of Electrochemical Equilibria in Aqueous Solutions*, National Association of Corrosion Engineers, Houston, TX, and CEBELCOR, Brussels, 1974, p. 334.

3. J. Osterwald and H. Uhlig, *J. Electrochem. Soc.* **106**, 515 (1961).

4. P. Bond and H. Uhlig, *J. Electrochem. Soc.* **107**, 488 (1960).

5. H. Uhlig, P. Bond, and H. Feller, *J. Electrochem. Soc.* **110**, 650 (1963).

6. Ref. 1, p. 834.

7. Ibid., p. 835.

8. W. Z. Friend, in *Corrosion Handbook*, H. H. Uhlig, editor, Wiley, New York, 1948, p. 271.

9. Ibid., p. 273.

10. Ref. 1, p. 836.

11. G. Airey and F. Pement, *Corrosion* **39**, 46 (1983).

12. H. Domian et al., *Corrosion*, **33**, 26 (1977).

13. J. Crum, *Corrosion* **38**, 40 (1982).

14. H. Coriou et al., in Proceedings of Conference Fundamental Aspects of Stress Corrosion Cracking, R. Staehle et al., editor, National Association of Corrosion Engineers, Houston, TX, 1969, p. 352.

15. N. Pessall et al., *Corrosion* **35**, 100 (1979).

16. P. M. Scott and P. Combrade, Corrosion in pressurized water reactors, in *ASM Handbook*, Vol. 13C, *Corrosion: Environments and Industries*, ASM International, Materials Park, OH, 2006, pp. 367, 369.

17. Manufacturer's literature.

18. Ref. 1, p. 837.

19. P. Crook, Corrosion of nickel and nickel-base alloys, in *ASM Handbook*, Vol. 13B, *Corrosion: Materials*, ASM International, Materials Park, OH, 2005, p. 230.

20. Ref. 1, p. 840.

21. Ref. 19, p. 231.

22. C. Samans, A. Meyer, and G. Tisinai, *Corrosion* **22**, 336 (1966).

23. Ref. 19, p. 236.

24. Ref. 19, p. 238.

GENERAL REFERENCES

D. C. Agarwal, Nickel and nickel alloys, in *Uhlig's Corrosion Handbook*, 2nd edition, R. W. Revie, editor, Wiley, New York, 2000, pp. 831–851.

P. Crook, Corrosion of nickel and nickel-base alloys, in *ASM Handbook*, Vol. 13B, *Corrosion: Materials*, ASM International, Materials Park, OH, 2005, pp. 228–251.

A. J. Sedriks, Corrosion resistance of stainless steels and nickel alloys, in *ASM Handbook*, Vol. 13A, *Corrosion: Fundamentals, Testing, and Protection*, ASM International, Materials Park, OH, 2006, pp. 697–702.

Corrosion in the Nuclear Industry

B. M. Gordon, Introduction to corrosion in the nuclear power industry, in *ASM Handbook*, Vol. 13C, *Corrosion: Environments and Industries*, ASM International, Materials Park, OH, 2006, pp. 339–340.

F. P. Ford, B. M. Gordon, and R. M. Horn, Corrosion in boiling water reactors, in *ASM Handbook*, Vol. 13C, *Corrosion: Environments and Industries*, ASM International, Materials Park, OH, 2006, pp. 341–361.

P. M. Scott and P. Combrade, Corrosion in pressurized water reactors, in *ASM Handbook*, Vol. 13C, *Corrosion: Environments and Industries*, ASM International, Materials Park, OH, 2006, pp. 362–385.

G. Was, J. Busby, and P. L. andresen, Effect of irradiation on stress-corrosion cracking and corrosion in light water reactors, in *ASM Handbook*, Vol. 13C, *Corrosion: Environments and Industries*, ASM International, Materials Park, OH, 2006. pp. 386–414.

Environmental Degradation of Materials in Nuclear Power Systems—Water Reactors, Proceedings, Eighth International Symposium, American Nuclear Society, La Grange Park, IL, 1997.

24

COBALT AND COBALT ALLOYS

24.1 INTRODUCTION

$$Co^{2+} + 2e^- \rightarrow Co \qquad \phi^\circ = -0.277\,V$$

The standard potential of cobalt is close to that of nickel, being 27 mV more active (Table 3.2, Section 3.8). Like nickel, the metal is rapidly attacked by oxidizing acids and salts (for example, HNO_3 and $FeCl_3$); it resists alkalies, hot or cold, although not as well as does nickel. Cobalt is also corroded by aerated aqueous ammonia solutions, forming a soluble complex—for example, $Co(NH_3)_6^{2+}$.

The theoretical potential–pH domains of corrosion, immunity, and passivation of cobalt in aqueous solution are presented in Fig. 24.1 [1]. As this figure indicates, cobalt can be cathodically protected in acid or neutral solution by polarizing to potentials active to –0.5 V versus S.H.E. [1].

At elevated temperatures, cobalt is relatively resistant to oxidation in air. It resists hot reducing sulfur-bearing atmospheres better than nickel does.

Corrosion and Corrosion Control, by R. Winston Revie and Herbert H. Uhlig
Copyright © 2008 John Wiley & Sons, Inc.

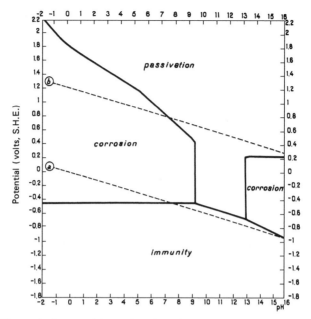

Figure 24.1. Theoretical potential–pH domains of corrosion, immunity, and passivation of cobalt at 25°C [1]. (M. Pourbaix, *Atlas of Electrochemical Equilibria in Aqueous Solutions*, 2nd English edition, p. 326. Copyright NACE International 1974 and CEBELCOR.)

24.2 COBALT ALLOYS

Being less plentiful and more expensive than nickel, cobalt is usually alloyed with chromium for applications where the alloys have practical advantages over similar nickel- or iron-base alloys. The cobalt-base alloys, for example, are better resistant to fretting corrosion, to erosion by high-velocity liquids, and to cavitation damage.

Chromium-bearing cobalt alloys can be divided into three groups [2]:

1. Alloys for wear resistance
2. Alloys for high-temperature use
3. Alloys for resistance to aqueous corrosion and wear

Cobalt alloys were developed in the early 1900s by Elwood Haynes of Kokomo, Indiana, for resisting corrosive environments and for high-temperature strength and hardness. The alloys are used in cutting tools operating in aggressive chemical media, steam valves and valve seats, pressure gauges, bushings, nozzles, pressure seats, and implant materials for the human body.

At temperatures up to 417°C, cobalt has a hexagonal close packed (hcp) structure, whereas above 417°C, it is face-centered cubic (fcc). Alloying elements that are fcc stabilizers (such as nickel, iron, and carbon) lower the transformation temperature. On the other hand, chromium, molybdenum, and tungsten, hcp stabilizers, raise the transformation temperature. Alloys of cobalt that contain both fcc and hcp stabilizers have transformation temperatures that are a complex function of the alloy composition [3]. Applying stress to alloys in a metastable fcc structure at ambient temperature can partially transform them to hcp.

Cobalt can be anodically passivated in $1N$ H_2SO_4, with the required minimum current density of $5000 \, A/m^2$ ($500 \, mA/cm^2$) being 14 times higher than that for nickel [4]. Alloying cobalt with chromium reduces the current density, with the 10% Cr alloy requiring only $10 \, A/m^2$ ($1 \, mA/cm^2$) to become passive. The alloy containing 10–12% Cr is negligibly attacked by 10% HNO_3, hot or cold; but in 10% H_2SO_4 or HCl, passivity is lost, and corrosion rates become excessively high. Alloying of Co–Cr alloys with molybdenum or tungsten reduces attack by H_2SO_4 or HCl, but not by HNO_3.

Some commercial cobalt-alloy compositions are listed in Table 24.1. They typically contain chromium as well as molybdenum and/or tungsten, making them relatively resistant to both reducing and oxidizing conditions. High resistance to abrasion results from the precipitation of carbides, which can be carbides of chromium, molybdenum, and tungsten [3].

The alloys resist pitting in $FeCl_3$ at room temperature, except Alloy 6B because of its high carbon content of 1.2%. At a lower carbon content—for example, 0.4%—the alloy is resistant. Vitallium, having a very noble critical pitting potential is also resistant to pitting in dilute NaCl solutions, a property that extends to Alloy 25 and MP35N as well.

Substantial additions of cobalt to chromium plus molybdenum (or tungsten) alloys are detrimental to S.C.C. behavior, resembling the effect of added iron rather than the beneficial effect of added nickel. Accordingly, the MP35N alloy resists S.C.C. in $MgCl_2$ solution at 153–154°C, but by replacing most nickel with cobalt (and the incidental reduction of molybdenum to 6% and increase of chromium to 30%), as in Vitallium, susceptibility results [5]. Alloy 25 is also susceptible. This susceptibility does not include Vitallium exposed to saline solutions at 37°C (body temperature) in which the alloy is resistant. The situation is analogous to the observed resistance of 18–8 (types 304 and 316) stainless steels to S.C.C. in aerated chlorides at temperatures below 60–80°C, but not above.

In boiling 50% NaOH, all of the stressed cobalt alloys fail by S.C.C., or sometimes by relatively rapid uniform dissolution. When severely cold-worked and cathodically polarized at room temperature in 5% H_2SO_4 plus As_2O_3, the stressed alloys are susceptible to failure by hydrogen cracking. Similar failures are also expected when the alloys are coupled in the same acid to a more active metal like iron, with the resultant galvanic action paralleling cathodic polarization. Failures in 5% NaCl–0.1% acetic acid, saturated with H_2S (NACE solution [6]), which simulates deep sour gas well environments, depend on temperature and the prevailing uniform corrosion rate to produce hydrogen. At room temperature, failures in this solution by hydrogen cracking (also sometimes called sulfide stress cracking) usually occur only for the cold-worked alloys that are subsequently heat-treated. This heat treatment improves strength, but may also

TABLE 24.1. Nominal Compositions (%) of Cobalt Alloys

Alloys for	Common Name	UNS No.	Cr	Mo	W	Ni	Fe	C	Mn	Si	Co	Other
Wear resistance	6B	R30016	30	1.5^a	4	2.5	3^a	1	1.4	0.7	bal.	—
High temperatures	25	R30605	20	—	15	10	3^a	0.1	1.5	0.4^a	bal.	—
Aqueous corrosion + wear resistance	21	R30021	27	5.5	—	2.75	3^a	0.25	1^a	1^a	bal.	B 0.007^a
	MP35N	R30035	20	9.75	—	35	1^a	0.025^a	0.15^a	0.15^a	bal.	Ti 1^a
	Ultimet	R31233	26	5	2	9	3	0.06	0.8	0.3	bal.	N 0.08
	Vitallium	—	30	6	—	—	—	0.5^a	0.75^a	—	bal.	—

[a]Maximum.
Source: Data from ASM Handbook, Vol. 13B, Corrosion: Materials, ASM International, Materials Park, OH, 2005, pp. 165, 172.

increase uniform corrosion rates in acids sufficient to generate hydrogen in amounts necessary to induce cracking. Failure by hydrogen cracking usually diminishes as the temperature is raised (less hydrogen enters the metal and more escapes as gas), but chloride S.C.C. may displace hydrogen cracking as the failure mechanism in the higher-temperature range. In this event, coupling of the alloys to a more active metal, as in cathodic protection, prevents cracking.

The advantage of cobalt-base alloys to reduce fretting corrosion contributes to the advantage of using Vitallium in the human body. Waterhouse [7] showed that a Vitallium screw mounted in a metal plate in saline solution and subjected to variable stress so as to cause slight rubbing of the screw head was damaged less than stainless steel. Also, in laboratory cavitation-erosion tests in distilled water, cobalt alloys exhibited superior resistance [8]. Vitallium and Alloys 6B and 25 lost only 1/3 to 1/14 the weight loss of similar specimens of nickel-base C-276 and iron-base type 304 stainless steel. Similarly, in high-velocity (244 m/s) hot brines common to geothermal wells, Alloy 25 and MP35N resisted corrosion-erosion better than C-276 and much better than 26% Cr–1% Mo stainless steel [9].

REFERENCES

1. M. Pourbaix, *Atlas of Electrochemical Equilibria in Aqueous Solutions*, 2nd English edition, National Association of Corrosion Engineers, Houston, TX, and CEBELCOR, Brussels, 1974, p. 326.
2. P. Crook and W. L. Silence, Cobalt alloys, in *Uhlig's Corrosion Handbook*, 2nd edition, R. W. Revie, editor, Wiley, New York, 2000, p. 717.
3. Ref. 2, p. 718.
4. A. Bond and H. Uhlig, *J. Electrochem. Soc.* **107**, 488 (1960).
5. H. Uhlig and A. Asphahani, *Mater. Perf.* **18**(11), 9 (1979).
6. Standard Test Method, TM0284-2003, *Evaluation of Pipeline and Pressure Vessel Steels for Resistance to Hydrogen-Induced Cracking*, NACE International, Houston, TX.
7. R. Waterhouse, *Fretting Corrosion*, Pergamon Press, New York, 1972, p. 59.
8. P. Crook and H. Richards, *Proceedings of the Conference on Production and Applications of Less Common Metals*, The Metals Society, London, 1982.
9. *The LLL Geothermal Energy Program Status Report*, A. Austin et al., editors, Lawrence Livermore Laboratory, Livermore, California, April 1977.

GENERAL REFERENCES

Corrosion of cobalt and cobalt-base alloys, in *ASM Handbook*, Vol. 13B, *Corrosion: Materials*, ASM International, Materials Park, OH, 2005, pp. 164–176.

P. Crook and W. L. Silence, Cobalt Alloys, in *Uhlig's Corrosion Handbook*, 2nd edition, R. W. Revie, editor, Wiley, New York, 2000, pp. 717–728.

J. R. Davis, editor, *Nickel, Cobalt, and Their Alloys*, ASM Specialty Handbook, ASM International, Materials Park, OH, 2000.

25

TITANIUM

25.1 TITANIUM

$$Ti^{2+} + 2e \rightarrow Ti \qquad \phi^{\circ} = -1.63\,V$$

Titanium is very active in the Emf Series, about 1.2 V more active than iron (Table 3.2, Section 3.8). The theoretical potential–pH domains of corrosion, immunity, and passivation of titanium in aqueous solution are presented in Fig. 25.1 [1]. In this figure, passivation is attributed to a film of TiO_2. The excellent corrosion resistance of titanium in many environments results from a hard, tightly adherent oxide film that forms instantaneously in the presence of an oxygen source [2]. This film, typically less than 10 nm thick, is very chemically resistant and is attacked by very few substances, including hot concentrated HCl, H_2SO_4, NaOH, and HF [3].

Titanium is a metal of relatively high melting point [m.p. = 1668°C (3053°F); density = 4.5 g/cm³]. It was developed for the aircraft industry because of its high strength-to-weight ratio and excellent corrosion resistance; it is now widely used in chemical process equipment.

It has been reported that, in the active state, titanium may oxidize to form Ti^{3+} ions in solution [4]. The metal is readily passivated in aerated aqueous

Corrosion and Corrosion Control, by R. Winston Revie and Herbert H. Uhlig
Copyright © 2008 John Wiley & Sons, Inc.

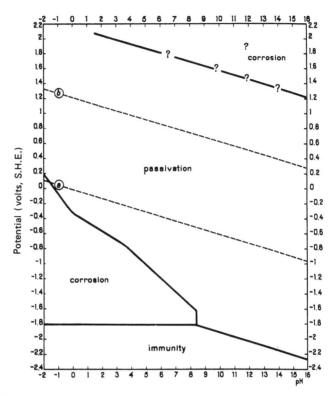

Figure 25.1. Theoretical potential–pH domains of corrosion, immunity, and passivation of titanium at 25°C, assuming passivation by TiO_2 [1]. (M. Pourbaix, *Atlas of Electrochemical Equilibria in Aqueous Solutions*, 2nd English edition, p. 219. Copyright NACE International 1974 and CEBELCOR.)

solutions, including dilute acids and alkalies. In the passive state, titanium is covered with a nonstoichiometric oxide film [4], the average composition of which is TiO_2. Semiconducting properties of the passive film are generally attributed to oxygen anion vacancies and Ti^{3+} interstitials that function as donor states and give *n*-type semiconducting characteristics to the oxide. The galvanic potential of titanium in seawater is near the noble value for stainless steels (see Fig. 3.3, Section 3.8). The Flade potential region (see Section 6.2) is relatively active ($\phi_F^\circ \approx -0.05\,V$), indicating stable passivity [5, 6]. Only in strong acids or alkalies does passivity break down, accompanied by appreciable corrosion.

Titanium is markedly resistant to oxidizing media in the presence of Cl^-, for example, heavy metal chlorides ($FeCl_3$), aqua regia at room temperature, or wet chlorine. It is better resistant to nitric acid at elevated temperatures than are the stainless steels. It resists alkalies at room temperature, but is attacked by hot concentrated or fused sodium hydroxide. Although titanium resists attack by hot dilute NaOH or by dilute H_2O_2, combining the two produces high initial

TABLE 25.1. Corrosion Rates of Commercial Titanium in Alkaline-Peroxide Solutions [8]

60°C, Test Period 1.5 h		gmd	mm/y
% NaOH	% H_2O_2		
2	1	696	56
2	1 + 0.5% $(NaPO_3)_6$	169	14
0.5	1 + 0.5% $(NaPO_3)_6$	22	1.8
1	0.35 + 1% Na_2SiO_3 + 0.05% $(NaPO_3)_6$	0.0	0.0

corrosion rates of more than 50 mm/y (2 ipy) [7, 8]. Some data are given in Table 25.1, which also shows the inhibiting properties of added sodium hexametaphosphate and sodium silicate.

On exposure of the metal for 1 h or more, stressed or unstressed, to fuming HNO_3 containing 2.5–28% NO_2 and not more than 1.25% H_2O, a dark substance forms over the surface (97.5% Ti metal by X-ray analysis) [9]. In the dry state, this surface film is pyrophoric when scratched, and it explodes when abraded in contact with concentrated HNO_3. The 8% Mn–Ti alloy is especially sensitive in this regard.

Titanium, presumably as sponge, in contact with liquid oxygen is reported to be sensitive to detonation by impact [10].

25.2 TITANIUM ALLOYS

Some commonly used titanium alloys are listed in Table 25.2. Group I alloys are the commercially pure grades, whereas Group II alloys contain palladium for improved corrosion resistance in reducing media. Groups I and II contain a single phase, the α phase, which has a hexagonal close-packed (hcp) structure. Groups III, IV, and V contain increasing concentrations of alloying elements and of the second phase, the β, body-centered cubic phase. Beta alloys are used predominantly in the aerospace industry because of their high strength/density ratios [2].

Although corrosion resistance is poor in boiling HCl or H_2SO_4 (114 mm/y in boiling 10% HCl), resistance is improved by orders of magnitude in the presence of small amounts of Cu^{2+} or Fe^{3+} (0.15 mm/y in 10% HCl, boiling, containing 0.02 mole Cu^{2+} or Fe^{3+} per liter) [11]. Small amounts of nickel, either in the environment or alloyed with titanium, also improve corrosion resistance; for example, 0.1% Ni in titanium or 0.2 ppm Ni^{2+} added to the solution are reported [12] to establish passivity in boiling 3.5% NaCl acidified to pH 1. In this solution, it is furthermore reported that the corrosion rate is lowest for the basal plane of the hexagonal close-packed titanium lattice. Small alloying additions of palladium, platinum, or ruthenium also effectively reduce the corrosion rate in boiling 10% HCl (2.5 mm/y for 0.1% Pd alloy, Fig. 25.2) [13, 14]. Ion implanting titanium

TABLE 25.2. Some Commonly Used Titanium Alloys [2]

Common Alloy Designation	UNS Number	ASTM Grade	Nominal Composition (%)
	Group I Commercially Pure Titanium		
Grade 2	R50400	2	0.12 O
	Group II Low Alloy Content Titanium with Pd Additions		
Grade 2, Pd	R52400	7	0.12 O, 0.15 Pd
Grade 2, low Pd	R52402	16	0.12 O, 0.05 Pd
	Group III Other Alpha and Near-Alpha Alloys		
Grade 12	R53400	12	0.3 Mo, 0.8 Ni
Ti 8–1–1	R54810	—	8 Al, 1 V, 1 Mo
	Group IV Alpha–Beta Alloys		
Ti 6–4	R56400	5	6 Al, 4 V
	Group V Beta Alloys		
Beta C	R58640	19	3 Al, 8 V, 6 Cr, 4 Zr, 4 Mo

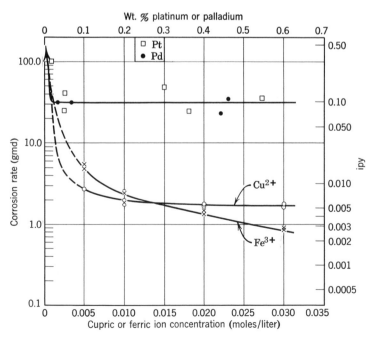

Figure 25.2. Corrosion of titanium in boiling 10% HCl as a function of Fe^{3+} and Cu^{2+} concentration and alloyed palladium or platinum.

surfaces with palladium reduces the corrosion rate [15] by a factor of 1000 in boiling $1\,M$ H_2SO_4; surface alloying with palladium by an electrospark technique or by electrochemical deposition with subsequent annealing are similarly effective [16]. These alloying elements or oxidizing ions accelerate the cathodic reaction (Cu^{2+}, Fe^{3+}, Ni^{2+} reduction, or increased rate of H^+ discharge) to a level at which the corresponding anodic current density reaches or exceeds the critical value for anodic passivation (see Section 6.4). Anodic passivation (or anodic protection) is also practical in HCl, H_2SO_4, and some other acids using a small impressed current [17].

When titanium is cathodically polarized or coupled to a more active metal in an acid like HCl, a surface film of titanium hydride may form. However, because of restricted diffusion of hydrogen in titanium at room temperature, penetration of hydrogen into the metal accompanied by metal embrittlement occurs only above about 80°C (175°F) [18].

25.3 PITTING AND CREVICE CORROSION

Titanium is characterized by marked resistance to pitting and crevice corrosion in seawater. Only very slight uniform weight loss occurs over periods of exposure of several years (<0.0025 mm/y). No pitting or appreciable weight loss occurred when titanium was buried in a variety of soils for eight years [19]. Resistance to pitting is accounted for by a very noble critical pitting potential of about 12–14 V in dilute chloride solutions at room temperature [20, 21]. At high Cl^- concentrations and elevated temperatures, the critical potential becomes much more active (Fig. 25.3); hence, pitting corrosion is observed in hot concentrated $CaCl_2$ and similar solutions [22]. Addition of 1% Mo to titanium is effective in shifting the critical potential to more noble values above 125°C (260°F), with accompanying greater resistance to pitting in this temperature range.

Initiation of pitting in titanium is more pronounced in Br^- and I^- than in Cl^- solutions. The critical pitting potential in $1\,M$ Br^- and $1\,M$ I^- solutions at room temperature are 0.9 and 1.8 V, respectively [23]. Mansfeld [24], following earlier Russian investigations, showed that titanium loses passivity in anhydrous $1\,N$ HCl in CH_3OH, but that passive behavior as indicated by polarization curves is restored when >0.6% H_2O is present, and that the pitting potential becomes increasingly noble as more H_2O is added.

Titanium does not undergo crevice corrosion in seawater at room temperature, but instances have been recorded in hot seawater, such as under an asbestos gasket at 95–120°C (200–250°F) [25]. Attack of this kind, it is reported, does not occur at temperatures below about 95°C (200°F); also, such attack is more commonly observed in acid and neutral rather than in alkaline chloride solutions. In laboratory tests, Griess [26] showed that titanium undergoes crevice attack in $1\,M$ NaCl containing dissolved oxygen at 150°C (300°F). For this type of corrosion, Cl^- was not necessary; crevice attack was also observed in hot solutions of I^-, Br^-, and SO_4^{2-}. Susceptibility was related to acid anodic corrosion products

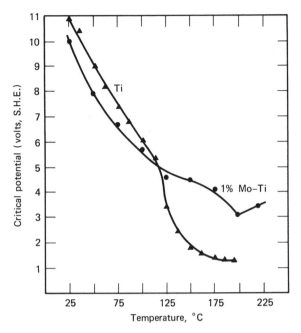

Figure 25.3. Critical pitting potentials of commercial titanium and 1% Mo–Ti alloy in 1*M* NaCl as a function of temperature [22].

accumulating in the crevice which reached a pH as low as 1.0. Hence, the 0.1% Pd–Ti alloy, which resists breakdown of passivity in acids, was found to resist crevice corrosion better than did titanium [26, 27]. In deaerated NaCl in which differential aeration cells are not established, crevice corrosion is suppressed. Various case histories of crevice and pitting corrosion observed in practice have been described [28].

Titanium tubing can be used successfully for seawater-cooled heat exchangers [29], where its resistance to pitting and corrosion-erosion make it a useful material.

25.4 INTERGRANULAR CORROSION AND STRESS-CORROSION CRACKING

Intergranular corrosion of titanium (and various of its alloys) occurs in fuming nitric acid at room temperature (3- to 16-h tests). Addition of 1% NaBr acts as an inhibitor [30]. Similar corrosion of commercially pure titanium occurs in methanol solutions containing Br_2, Cl_2, or I_2; or Br^-, Cl^-, or I^- [31]. Water acts as an inhibitor.

The 6% Al, 4% V–Ti alloy has failed by stress-corrosion cracking (S.C.C.) in liquid N_2O_4 within 40 h at 40°C (105°F) [32]. A slight excess of NO (or the presence of H_2O) inhibits such failure.

Various titanium alloys, including 8% Al, 1% Mo, 1% V–Ti (8–1–1) heated in air in contact with moist sodium chloride (e.g., from fingerprints) at 260°C (500°F) or higher, undergo S.C.C. (or intergranular corrosion?) usually along grain boundaries [33–35]. Pure titanium is resistant to this type of failure.

B. F. Brown et al. [36] showed that titanium alloys in seawater, despite otherwise excellent corrosion resistance, tend to undergo transgranular S.C.C. when a sharp surface flaw is present. It has been observed that aqueous solutions of Cl^-, Br^-, and I^- are unique in causing S.C.C. of commercial titanium high in oxygen content (0.2–0.4% O) and of various alloys including the 8–1–1 alloy. On the other hand, F^-, SO_4^{2-}, OH^-, S^{2-}, NO_3^-, and ClO_4^- do not cause failure and, to the contrary, may inhibit crack propagation in some alloys susceptible to S.C.C. in distilled water (e.g., about 100 ppm KNO_3 is effective) [37, 38]. Various anions of the group just listed also act as inhibitors of S.C.C. in the presence of halide ions; in this respect, their behavior simulates the effect of extraneous anions in the case of austenitic stainless steels (see Section 19.2.6).

Beck [38] found that cathodic polarization of the 8–1–1 alloy to –0.76 V (S.H.E.) prevents failure in the presence of Cl^-, Br^-, or I^-. Leckie [39] reported protection of 7% Al, 2% Nb, 1% Ta–Ti alloy in 3% NaCl at –1.1 V (S.H.E.) and also at –1.3 V, at which hydrogen evolution was copious. Successful anodic protection against S.C.C. of the 8–1–1 alloy was reported [38] in the presence of Cl^- but not in the presence of Br^- or I^-.

Failure of the 8–1–1 alloy also occurs in pure methanol (CH_3OH). Interestingly, adding a small amount of Cl^- to distilled water or to methanol did not increase velocity of crack propagation, but less potassium nitrate (10 ppm) was needed to inhibit cracking in methanol as compared to water [37]. The stressed alloy was also found to be sensitive to cracking in the pure nonaqueous solvents CCl_4 and CH_2Cl_2.

The 8–1–1 alloy is a mixture of mostly α (hexagonal close packed) and some β phase (body-centered cubic), with observed cracks proceeding across grains of the α alloy but with the β phase failing by ductile fracture. Heat-treatment procedures and composition changes (e.g., lowering the aluminum content) that favor the β phase increase resistance to S.C.C. However, the composition of the phase is also a determining factor, because the β phase of several other titanium alloys is found to be susceptible to S.C.C. [40]. The sensitive effect of alloy structure, the specificity of the environment, and the effects of extraneous anions and applied potential are similar to effects observed in the stainless steels (see Section 8.3), suggesting similarities in the mechanisms. Nevertheless, the mechanisms of cracking of titanium alloys are still being discussed, with debate continuing to take place regarding the relative importance of anodic dissolution and hydrogen reduction processes.

In brief, titanium is resistant to the following:

1. Seawater, 0.0008 mm/y, 4.5-year exposure, including high-velocity conditions (0.005 mm/y, 42 m/s) [41]. Known to resist pitting and crevice attack for five years and longer. High oxygen concentration in titanium may induce susceptibility to S.C.C.
2. Wet chlorine (>0.9% H_2O, less in flowing gas) [42]. In dry Cl_2, titanium may ignite.
3. Nitric acid, all concentrations and temperatures up to boiling, but not to fuming HNO_3.
4. Oxidizing solutions, hot or cold (e.g., $CuCl_2$, $FeCl_3$, $CuSO_4$, and $K_2Cr_2O_7$).
5. Hypochlorites.

Titanium is *not* resistant to the following:

1. Aqueous hydrogen fluoride.
2. Fluorine.
3. Hydrochloric and sulfuric acids, except when dilute, or in moderate concentrations when inhibited by oxidizing metal ions (e.g., Fe^{3+} and Cu^{2+}) or other oxidizing substances (e.g., $K_2Cr_2O_7$ and $NaNO_3$) or when alloyed with platinum or palladium.
4. Oxalic, >10% formic, anhydrous acetic [43] acids.
5. Boiling $CaCl_2$, >55% [44].
6. Concentrated hot alkalies, or dilute alkalies plus H_2O_2.
7. Molten salts (e.g., NaCl, LiCl, and fluorides).
8. High-temperature exposure to air, nitrogen, or hydrogen. Oxidation in air occurs above 450°C (840°F) with formation of titanium oxides and nitrides. Ignition temperature decreases with increased air pressure sometimes causing localized combustion of titanium-alloy blades in the compressor section of gas-turbine engines [45]. Titanium hydride forms rapidly above 250°C (480°F); it forms at lower temperatures when hydrogen is cathodically discharged. Absorption of oxygen, nitrogen, or hydrogen at elevated temperatures leads to embrittlement.

REFERENCES

1. M. Pourbaix, *Atlas of Electrochemical Equilibria in Aqueous Solutions*, 2nd English edition, National Association of Corrosion Engineers, Houston, TX, and CEBELCOR, Brussels, 1974, p. 219.
2. J. Been and J. S. Grauman, Titanium and titanium alloys, in *Uhlig's Corrosion Handbook*, 2nd edition, R. W. Revie, editor, Wiley, New York, 2000, p. 864.
3. R. W. Schutz, Corrosion of titanium and titanium alloys, in *ASM Handbook*, Vol. 13B, *Corrosion: Materials*, ASM International, Materials Park, OH, 2005, p. 253.

4. E. J. Kelly, *Modern Aspects Electrochem.* **14**, 319 (1982).

5. M. Stern and H. Wissenberg, *J. Electrochem. Soc.* **106**, 755 (1959).

6. N. Thomas and K. Nobe, *J. Electrochem. Soc.* **116**, 1748 (1969).

7. T. Fennel and N. Stalter, TAPPI, *J. Tech. Assoc. Pulp and Paper Ind.* **51**(1), 62A (1968).

8. T. Sigalovskaya et al., *Zashchita Metallov.* **12**(4), 363 (1976).

9. J. Rittenhouse, *Trans. Am. Soc. Metals* **51**, 871 (1951).

10. T.M.L. Report No. 88, Battelle Memorial Institute, Columbus, OH, November 1957; *Metals Handbook*, 8th edition, Vol. 1, American Society for Metals, Cleveland, OH, 1961, p. 1152.

11. J. Cobb and H. Uhlig, *J. Electrochem. Soc.* **99**, 13 (1952).

12. J. Green and R. Latanision, *Corrosion* **29**, 386 (1973).

13. M. Stern and H. Wissenberg, *J. Electrochem. Soc.* **106**, 759 (1959); M. Stern and C. Bishop, *Trans. Am. Soc. Metals* **52**, 239 (1960).

14. A. Sedriks, *Corrosion* **31**, 60 (1975).

15. E. McCafferty and G. Hubler, *J. Electrochem. Soc.* **125**, 1892 (1975).

16. N. Tomashov, G. Chernova, and T. Fedoseeva, *Corrosion* **36**, 201 (1980).

17. J. Cotton, *Chem. Ind. (London)* **1958**, 68.

18. L. Covington, *Corrosion* **35**, 378 (1979).

19. B. Sanderson and M. Romanoff, *Mater. Prot.* **8**, 29 (April 1969).

20. C. Hall, Jr., and N. Hackerman, *J. Phys. Chem.* **57**, 262 (1953).

21. I. Dugdale and J. Cotton, *Corros. Sci.* **4**, 397 (1964).

22. F. Posey and E. Bohlmann, *Second Symposium On Fresh Water from the Sea*, Athens, Greece, May 1967.

23. T. Beck, in *Localized Corrosion*, R. Staehle et al., editors, National Association of Corrosion Engineers, Houston, TX, 1974, p. 644.

24. F. Mansfeld, *J. Electrochem. Soc.* **118**, 1412 (1971).

25. J. Jackson and W. Boyd, in *Applications Related Phenomena in Titanium Alloys*, Special Technical Publication No. 432, American Society for Testing and Materials, Philadelphia, PA, 1968, pp. 218–226. See also K. Shiobara and S. Morioka, in *Proceedings of the Fifth International Congress on Metallic Corrosion*, National Association of Corrosion Engineers, Houston, TX, 1974, pp. 313–317.

26. J. Griess, Jr., *Corrosion* **24**, 96 (1968).

27. A. Takamura, *Corrosion* **23**, 306 (1967).

28. *Mater. Prot.* **6**(10), 22 (1967).

29. J. McMaster, *Mater. Perf.* **18**(4), 28 (1979); K. Suzuki and Y. Nakamoto, *Mater. Perf.* **20**(6), 23 (1981).

30. G. Kiefer and W. Harple, *Metals Progr.* **63**, 2, 74 (1953).

31. A. Sedriks and J. Green, *Corrosion* **25**, 324 (1969).

32. J. Jackson and W. Boyd, *Corrosion of Ti*, DMIC Memo 218, Battelle Memorial Institute, Columbus, OH, September 1966.

33. *Stress Corrosion Cracking of Titanium*, Special Technical Publication No. 397, American Society for Testing and Materials, Philadelphia, 1966. Also S. Rideout et al. in Ref. 40, p. 650.

34. R. Newcomber, H. Tourkakis, and H. Turner, *Corrosion* **21**, 307 (1965).
35. H. Gray, *Corrosion* **25**, 337 (1969); H. Gray and J. Johnston, *Metall. Trans.* **1**, 3101 (1970).
36. B. F. Brown et al., *Marine Corrosion Studies* (Third Interim Report of Progress), NRL Memo, Report 1634, Naval Research Laboratory, Washington, D.C., July 1965.
37. T. Beck and M. Blackburn, *Am. Inst. Aeronaut. Astron. J.* **6**, 326 (1968).
38. T. Beck, *J. Electrochem. Soc.* **114**, 551 (1967).
39. H. Leckie, *Corrosion* **23**, 187 (1967).
40. R. Adams and E. Von Tiesenhausen, in *Proceedings of Conference Fundamental Aspects of Stress Corrosion Cracking*, R. Staehle et al., editors, National Association of Corrosion Engineers, Houston, TX, 1969, p. 691.
41. T. May, International Nickel Co., private communication, 1962.
42. E. Mellway and M. Kleinman, *Corrosion* **23**, 88 (1967).
43. K. Risch, *Chem.-Ing.-Tech.* **39**, 385 (1967).
44. P. Gegner and W. Wilson, *Corrosion* **15**, 341t (1959).
45. T. Strobridge et al., Report NBS 1R-79-1616, National Bureau of Standards, Boulder, Colorado, July 1979 (Metals and Ceramic Inf. Center, CAB, June 1980, Issue No. 84, Battelle Columbus Labs, Columbus, OH).

GENERAL REFERENCES

J. Been and J. S. Grauman, Titanium and titanium alloys, in *Uhlig's Corrosion Handbook*, 2nd edition, R. W. Revie, editor, Wiley, New York, 2000, pp. 863–885.

R. Boyer, G. Welsch, and E. W. Collins, editors, *Materials Properties Handbook, Titanium Alloys*, ASM International, Materials Park, OH, 1994.

E. Kelly, Electrochemical behavior of titanium, *Mod. Aspects Electrochem.* **14**, 319 (1982).

R. W. Schulz, Corrosion of titanium and titanium alloys, in *ASM Handbook*, Vol. 13B, *Corrosion: Materials*, ASM International, Materials Park, OH, 2005, pp. 252–299.

PROBLEM

1. Using the polarization curve for titanium in sulfuric acid, show the effect of alloying with palladium or platinum and explain the reduction of corrosion rate in the alloyed material compared to pure titanium.

26

ZIRCONIUM

26.1 INTRODUCTION

Zirconium is an active metal in the Emf Series, normally exhibiting very stable passivity. As illustrated in Fig. 26.1, zirconium is passive over a very wide range of potential and pH [1].

The metal [m.p. = 1852°C (3366°F); density = 6.45 g/cm^3] reacts readily at elevated temperatures with O_2, N_2, and H_2. At temperatures below ~860°C, α-zirconium, hexagonal close-packed (hcp) structure, is the stable form, which transforms to body-centered cubic (bcc), β-zirconium, at higher temperatures.

An unusual property is the high solid solubility of the metal for oxygen, with the metal dissolving up to 29 at.% (6.7 wt.%) according to the oxygen–zirconium phase diagram [2]. Similarly, α-zirconium absorbs nitrogen in solid solution up to 25 at.% (4.8 wt.%). It reacts with air to form both zirconium oxides and nitrides. The reaction is slow enough, however, to permit hot rolling at 600–750°C (1100–1400°F) [3].

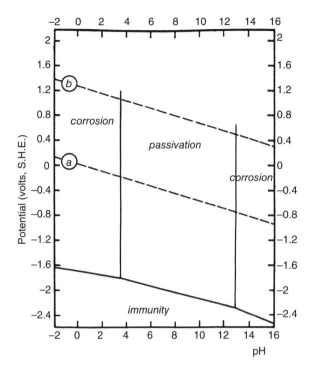

Figure 26.1. Theoretical conditions of corrosion, immunity, and passivation of zirconium at 25 °C, assuming passivation by $ZrO_2 \cdot 2H_2O$ [1]. (M. Pourbaix, *Atlas of Electrochemical Equilibria in Aqueous Solutions*, 2nd English edition, p. 227. Copyright NACE International 1974 and CEBELCOR.)

26.2 ZIRCONIUM ALLOYS

Because of good corrosion resistance in both acids and bases, zirconium alloys are widely used in chemical plants. Commercial zirconium, as used primarily for corrosion resistance in the chemical industry [4], contains up to 4.5% hafnium, which is difficult to separate because of the similar chemical properties of zirconium and hafnium. The presence or absence of hafnium has no effect on the corrosion resistance, which is controlled by a very stable oxide. At ambient temperature, this passive oxide is 2–5 nm thick [4]. The pure metal low in hafnium (0.02% max) has a low thermal neutron capture, making it useful for nuclear-power applications.

The outstanding corrosion property of zirconium is its resistance to alkalies at all concentrations up to the boiling point. It also resists fused sodium hydroxide. In this respect, it is distinguished from tantalum and, to a lesser extent, titanium, which are attacked by hot alkalies. Zirconium is resistant to hydrochloric and nitric acids at all concentrations and to <70% H_2SO_4 up to boiling temperatures. In HCl and similar media, the metal must be low in carbon (<0.06%) for optimum resistance. In boiling 20% HCl, a transition or "breakaway" point is observed in the corrosion rate (see below) after a specific time of exposure. The final rate, which is higher than the initial rate, is usually less than 0.11 mm/y (0.0045 ipy) [5]. Zirconium is not resistant to oxidizing metal chlorides (e.g., $FeCl_3$; pitting occurs), nor is it resistant to HF or fluosilicic acid.

Critical pitting potentials of 0.38 V (S.H.E.) in $1N$ NaCl and 0.45 V in $0.1N$ NaCl [6] indicate that the metal is vulnerable to pitting in seawater. It undergoes intergranular S.C.C. in anhydrous methyl or ethyl alcohol containing HCl, but not when a small amount of water is added [7]. This behavior, similar to that of commercial titanium, suggests that stress may not be necessary and that the failure is perhaps better described as intergranular corrosion.

Although zirconium alloys do have good resistance to S.C.C., they are susceptible in many environments. In general, weldments should be heat-treated to reduce residual stresses. Unalloyed zirconium has been reported to be resistant to S.C.C. because of its low yield strength, whereas the higher-strength alloys are more susceptible. Cracking can be intergranular, transgranular, or mixed intergranular/transgranular [8]. In Zr–2.5% Nb, not only S.C.C., but also a delayed hydride cracking process, can occur at highly stressed locations [8–11].

In the nuclear industry, zirconium alloys are used as nuclear fuel cladding and structural fuel assembly components (see Section 8.5). The intense radiation within the reactor core accelerates degradation by increasing the rates of corrosion and hydriding. Zircaloy-2 (Zr, 1.5% Sn, 0.1% Fe, 0.1% Cr, 0.05% Ni), an alloy used in nuclear reactors, is subject to S.C.C. in chlorides at 25 °C at potentials noble to the breakdown potential of the air-formed oxide film [0.34 V (S.H.E.) in 5% NaCl] [12]. Stress-corrosion cracking may also occur in $FeCl_3$ and $CuCl_2$ solutions [13]. In slow strain rate tests, commercially pure zirconium, Zircaloy-2, and Zircaloy-4 (Zr, 1.5% Sn, 0.2% Fe, 0.1% Cr) undergo S.C.C. in >20% HNO_3 at 25 °C with the maximum cracking rate in 70–90% HNO_3. Unlike the situation for titanium, the presence of NO_2 does not have a significantly damaging effect [14]. Constant strain (U-bend, C-ring) tests of commercial zirconium and some zirconium alloys in 70% HNO_3 up to the boiling point showed high resistance to SCC, but not necessarily immunity [15]. Both zirconium and Zircaloy-2 are subject to SCC in iodine vapor (a major fission product of uranium) at 300–350 °C [16, 17]. Cold work and irradiation hardening have been reported to increase susceptibility.

To help improve the corrosion resistance of Zircaloy, several new zirconium alloys have been developed, such as Zirlo (Zr–1.0% Nb–1.0% Sn–0.1% Fe). Notwithstanding the progress so far, materials reliability does have a significant effect on the economics of nuclear power plants, and there is considerable incentive to develop a full understanding of the mechanisms of corrosion of zirconium alloys in reactors and to develop alloys that are resistant to both irradiation and corrosion in reactors [18].

26.3 BEHAVIOR IN HOT WATER AND STEAM

The good resistance of zirconium to deaerated hot water and steam is of special importance in nuclear-power applications. The metal or its alloys can be exposed in general for prolonged periods without pronounced attack at temperatures below about 425 °C (800°F). The rate of attack is characteristically low at first; but after a certain time of exposure, ranging from minutes to years depending on temperature, the rate suddenly increases. This "transition" phenomenon is reported to occur for pure or impure zirconium after a weight gain on the order of 3.5–5.0 g/m², and similar additional accelerated oxidation may occur at still higher weight gains [19]. It occurs at lower temperatures if the zirconium is contaminated with nitrogen (>0.005%) or with carbon (>0.04%) [20]. The damaging effect of nitrogen in this respect is offset by alloying additions of 1.5–2.5% tin in combination with lesser amounts of iron, nickel, and chromium.

Marker experiments indicate that oxidation proceeds by diffusion of oxygen ions toward the metal–oxide interface (anion defect lattice). It has been suggested on this basis that trivalent nitrogen ions in the ZrO_2 lattice increase anion-defect concentration, thereby increasing the diffusion rate of oxygen ions. But were this the mechanism, oxidation in oxygen would also be affected, which is not the case. Adding to the complexity is the observation that alloyed tin appreciably shortens corrosion life of zirconium in water; but tin in the presence of small amounts of alloyed iron, nickel, or, to a lesser extent, chromium again increases corrosion resistance, with the combination overcoming the detrimental effect of alloyed nitrogen.

The mechanism accounting for the transition phenomenon is not well understood. It has been explained on the basis of cracks forming in the oxide because of stresses accumulating as the oxide thickens. However, an increased corrosion rate does not occur when the metal oxidizes in oxygen except for much longer times and much thicker oxide films. Hydrogen formed by the decomposition of H_2O during reaction appears to exert a dominant role, especially that portion which dissolves in the metal, causing higher oxidation rates [19]. X-ray data for the oxides that form in H_2O show a monoclinic modification of ZrO_2 either before or after the breakaway time, but with some evidence that the initial oxide is a tetragonal modification [20].

The oxidation behavior of Zircaloy-2 in water and steam is shown in Fig. 26.2.

In brief, zirconium is resistant to the following:

1. Alkalies, all concentrations up to boiling point, including fused caustic.
2. Hydrochloric acid, all concentration up to boiling point. Embrittlement of metal and higher corrosion rates occur above boiling temperatures under pressure.
3. Nitric acid, all concentrations up to boiling point, including red fuming acid (S.C.C. may occur under slow strain rate conditions [14]).
4. Sulfuric acid, <70%, boiling.
5. Phosphoric acid, <55% H_3PO_4, boiling.
6. Boiling formic, acetic, lactic, or citric acids.

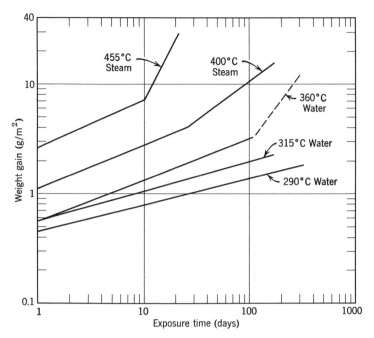

Figure 26.2. Corrosion of Zircaloy-2 in high-temperature water and steam, showing transition points [19].

Zirconium is *not* resistant to the following:

1. Oxidizing metal chlorides (e.g., $FeCl_3$ and $CuCl_2$).
2. Hydrofluoric acid and fluosilicic acid.
3. Wet chlorine.
4. Oxygen, nitrogen, and hydrogen at elevated temperatures.
5. Aqua regia.
6. Trichloroacetic or oxalic acids, boiling.
7. Boiling $CaCl_2$, >55% [21].
8. Carbon tetrachloride, 200 °C (explodes) [22].

REFERENCES

1. M. Pourbaix, *Atlas of Electrochemical Equilibria in Aqueous Solutions*, 2nd English edition, National Association of Corrosion Engineers, Houston, TX, and CEBELCOR, Brussels, 1974, p. 227.
2. M. Hansen, *Constitution of Binary Alloys*, McGraw-Hill, New York, 1958, p. 1079.
3. W. Hurford, in *The Metallurgy of Zirconium*, B. Lustman and F. Kerze, Jr., editors, McGraw-Hill, New York, 1955, p. 263.

4. B. Cox, Zirconium alloy corrosion, in *Uhlig's Corrosion Handbook*, 2nd edition, R. W. Revie, editor, Wiley, New York, 2000, pp. 905–906.

5. W. Kuhn, *Corrosion* **15**, 103t (1959).

6. J. Kolotyrkin, *Corrosion* **19**, 261t (1963).

7. K. Mori, A. Takamura, and T. Shimose, *Corrosion* **22**, 29 (1966).

8. Ref. 4, p. 908.

9. B. A. Cheadle, C. E. Coleman, and J. F. R. Ambler, Prevention of delayed hydroide cracking in zirconium alloys, ASTM STP 939, 1987, pp. 224–240.

10. F. H. Huang and W. J. Mills, *Met. Trans. A* **22A**, 2049 (1991).

11. J. W. Mills and F. H. Huang, *Eng. Fracture Mech.* **39**, 39, 241 (1991).

12. B. Cox, *Corrosion* **29**, 157 (1973).

13. B. Cox, *Rev. Coatings Corros.* **1**, 366 (1974).

14. J. Beavers, J. Griess, and W. Boyd, *Corrosion* **37**, 292 (1981).

15. Te-Lin Yau, *Corrosion* **39**, 167 (1983).

16. B. Cox, *Corrosion* **28**, 207 (1972); ibid. 33, 79 (1977); B. Cox and J. C. Wood, in *Corrosion Problems in Energy Conversion and Generation*, C. Tedmon, editor, Electrochemical Society, Princeton, NJ, 1974, p. 275.

17. R. Gangloff et al., *Corrosion* **35**, 316 (1979).

18. J. Busby, Irradiation effects on corrosion of zirconium alloys, in *ASM Handbook*, Vol. 13C, *Corrosion: Environments and Industries*, ASM International, Materials Park, OH, 2006, pp. 406–408. See also C. Lemaignan, Corrosion of zirconium alloy components in light water reactors, in *ASM Handbook*, Vol. 13C, *Corrosion: Environments and Industries*, ASM International, Materials Park, OH 2006, pp. 415–420.

19. B. Cox, *Corrosion* **18**, 33t (1962).

20. D. Thomas, in *The Metallurgy of Zirconium*, B. Lustman and F. Kerze, Jr., editors, McGraw-Hill, New York, 1955, pp. 608–640.

21. P. Gegner and W. Wilson, *Corrosion* **15**, 341t (1959).

22. W. Archer and M. Harter, *Corrosion* **34**, 159 (1978).

GENERAL REFERENCES

B. Cox, Zirconium alloy corrosion, in *Uhlig's Corrosion Handbook*, 2nd edition, R. W. Revie, editor, Wiley, New York, 2000, pp. 905–914.

B. Cox, *J. Nucl. Mater.* **170**, 1 (1990).

B. Cox, Oxidation of zirconium and its alloys, *Advances in Corrosion Science and Technology*, Vol. 5, M. Fontana and R. Staehle, editors, Plenum Press, New York, 1976, p. 173.

IAEA-TECDOC-996 (1998) Waterside corrosion of zirconium alloys in nuclear power plants, International Atomic Energy Agency, Vienna, 1998.

Te-Lin Yau and R. C. Sutherlin, Corrosion of zirconium and zirconium alloys, in *ASM Handbook*, Vol. 13B, *Corrosion: Materials*, ASM International, Materials Park, OH, 2005, pp. 300–324.

27

TANTALUM

27.1 INTRODUCTION

Tantalum [m.p. = 3000 °C (5430°F); density = 16.6 g/cm^3] exhibits the most stable passivity among known metals, as expected from the theoretical potential–pH domains of immunity and passivation shown in Figure 27.1 [1]. Tantalum retains passivity in boiling acids (e.g., HCl, HNO$_3$, or H$_2$SO$_4$) and in moist chlorine or FeCl$_3$ solutions above room temperature. Corrosion resistance of this order suggests a Flade potential much more active than the hydrogen electrode potential, and a low passive current that is insensitive to Cl$^-$.

27.2 CORROSION BEHAVIOR

High corrosion resistance to acids makes tantalum useful for special applications in the chemical industry (e.g., H$_2$SO$_4$ concentrators and HCl absorption systems). Liners of tantalum sheet may average only 0.3 mm (0.013 in.) thick, permitting varied applications of the metal despite high cost.

Tantalum is attacked by alkalies and by hydrofluoric acid. It is readily embrittled by hydrogen at room temperature when the metal is cathodically polarized,

Corrosion and Corrosion Control, by R. Winston Revie and Herbert H. Uhlig
Copyright © 2008 John Wiley & Sons, Inc.

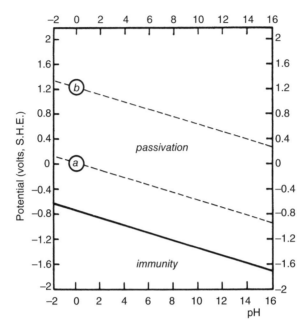

Figure 27.1. Theoretical potential–pH domains of immunity and passivation for tantalum at 25 °C [1]. (M. Pourbaix, *Atlas of Electrochemical Equilibria in Aqueous Solutions*, 2nd English edition, p. 254. Copyright NACE International 1974 and CEBELCOR.)

or when coupled in an electrolyte to a metal more active in the Galvanic Series. The damage by cathodic hydrogen can be avoided by coupling tantalum to a very small area of a low-overpotential metal, such as platinum [2]. Hydrogen ions then discharge on the platinum instead of entering the tantalum lattice. The benefit of coupled platinum extends to avoiding embrittlement caused by hydrogen generated through a corrosion reaction. For example, even though tantalum is not embrittled by concentrated hydrochloric acid at boiling temperatures, it becomes embrittled at 190 °C (375°F) under pressure. This damage does not occur if an area ratio of platinum to tantalum is provided on the order of at least 1:10,000. Platinum can be attached to tantalum by riveting, welding, or electrodeposition. Metal embrittled by cathodic hydrogen or by that entering at elevated temperatures can be restored to normal properties only by heating in vacuum.

Fischer [3] reported that tantalum, despite its pronounced passivity in aqueous media, corrodes generally in anhydrous CH_3OH–Br_2 mixtures at room temperature and undergoes pitting corrosion in CH_3OH, 10% Br_2, 4% H_2O. Palit and Elayaperumal [4] stated that tantalum is subject to pitting in CH_3OH + 0.4 vol.% concentrated HCl at potentials slightly noble to the corrosion potential in this solution.

In brief, tantalum is resistant to the following:

1. Hydrochloric acid, all concentrations up to boiling point.
2. Nitric acid, all concentrations up to and above boiling point.

3. Sulfuric acid, all concentrations (except fuming) <175 °C (<350°F). Fuming acid attacks tantalum at room temperature.
4. Chromic acid, hot or cold.
5. Phosphoric acid. Resistant to all concentrations up to and, in some cases, above boiling temperatures. For 85% H_3PO_4 at 225 °C, rate = 0.09 mm/y (0.0035 ipy). Attack occurs at lower temperatures when the acid is contaminated with HF (>4 ppm).
6. Halogen gases [wet or dry Cl_2 up to 150 °C (300°F); Br_2 up to 175 °C (350°F)].
7. Aqua regia.
8. Oxidizing metal chlorides, hot or cold (e.g., $FeCl_3$ and $CuCl_2$).
9. Organic acids: lactic, oxalic, and acetic.

Tantalum is *not* resistant to the following:

1. Alkalies. Embrittlement occurs, for example, with 5% NaOH, 100 °C.
2. Hydrofluoric acid and fluorides, trace amounts.
3. Fuming sulfuric acid.
4. Oxygen, nitrogen, or hydrogen at elevated temperatures. Oxidation in air becomes appreciable above 250 °C (500°F). Cathodic hydrogen causes embrittlement at room temperature.
5. Methanol solutions of Br_2 or HCl.

REFERENCES

1. M. Pourbaix, *Atlas of Electrochemical Equilibria in Aqueous Solutions*, 2nd English edition, National Association of Corrosion Engineers, Houston, TX, and CEBELCOR, Brussels, 1974, p. 254.
2. C. Bishop and M. Stern, *Corrosion* **17**, 379t (1961).
3. W. Fischer, *Tech. Mitt. Krupp.* **22**, 125 (1964).
4. G. Palit and K. Elayaperumal, *Corros. Sci.* **18**, 169 (1978).

GENERAL REFERENCE

Corrosion of tantalum and tantalum alloys, in *ASM Handbook*, Vol. 13B, *Corrosion: Materials*, ASM International, Materials Park, OH, 2005, pp. 337–353.

28

LEAD

28.1 INTRODUCTION

$$Pb^{2+} + 2e^- \rightarrow Pb \qquad \phi° = -0.126\,V$$

Lead is a relatively active metal in the Emf Series, becoming passive in many corrosive media that form insoluble lead compounds (e.g., H_2SO_4, HF, H_3PO_4, and H_2CrO_4) by reason of thick diffusion-barrier coatings (Definition 2, Section 6.1). In these acids, corrosion resistance is good provided relative velocity of metal and acid is below the value that causes erosion of protective films. Lead is used, therefore, in the chemical industry as lining and piping. The metal is, however, corroded by dilute nitric acid and by several dilute aerated organic acids (e.g., acetic and formic acids).

The theoretical potential–pH domains for corrosion, immunity, and passivation are shown in Figure 28.1 [1]. As shown in the diagram, in acid or neutral solution, lead can be cathodically protected by controlling the potential to less than –0.3 V. In alkaline solution, potential control for cathodic protection is below –0.4 to –0.8 V, depending on the pH [1].

Corrosion and Corrosion Control, by R. Winston Revie and Herbert H. Uhlig
Copyright © 2008 John Wiley & Sons, Inc.

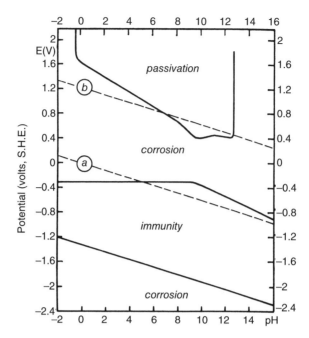

Figure 28.1. Theoretical potential–pH domains of corrosion, immunity, and passivation for lead at 25 °C [1]. (M. Pourbaix, *Atlas of Electrochemical Equilibria in Aqueous Solutions*, 2nd English edition, p. 490. Copyright NACE International 1974 and CEBELCOR.)

28.2 CORROSION BEHAVIOR OF LEAD AND LEAD ALLOYS

Being amphoteric, lead is corroded by alkalies at moderate or high rates, depending on aeration, temperature, and concentration. Nevertheless, lead resists corrosion in many environments because the products of corrosion are insoluble and form self-healing protective films. Because of these protective films, the corrosion rate of lead is usually under anodic control [2].

It is attacked, for example, by calcium hydroxide solutions at room temperature, including waters that have been in contact with fresh Portland cement. Organic acids harmful to lead can be leached from wood structures, particularly damp timber including western cedar, oak, and Douglas fir. This cause of corrosion can be prevented by fully drying the wood, treating it to prevent contact with moist air, or inserting a moisture barrier between wood and lead [3].

An alloy of 2% Ag–Pb is used as a corrosion-resistant anode in impressed current systems for cathodic protection of structures in seawater (see Section 13.6) [4]. Alloying with 6–12% Sb increases strength [only at temperatures <120 °C (<250°F)] of the otherwise weak metal, but corrosion resistance of the alloy in some media is below that of pure lead.

Lead is resistant to seawater. It is also durable for use in contact with fresh waters; however, the toxic properties of trace amounts of lead salts make it mandatory to exclude its use, and the use of its alloys, for soft potable waters, carbonated beverages, and all food products. The U.S. Environmental Protection Agency (EPA) Lead and Copper Rule, implemented in 1991, limits lead in potable water to a maximum of 15 ppb [5]. The rate of corrosion in aerated distilled water is high (approximately 9 gmd) and the rate increases with concentration of dissolved oxygen [6]. In the absence of dissolved oxygen, the corrosion rate in waters or dilute acids is either negligible or very low.

Lead is resistant to atmospheric exposures, particularly to industrial atmospheres in which a protective film of lead sulfate forms. Buried underground, the corrosion rate may exceed that of steel in some soils (e.g., those containing organic acids), but in soils high in sulfates the rate is low. Soluble silicates, which are components of many soils and natural waters, also act as effective corrosion inhibitors.

Lead is used in solder alloys (typically with 5% tin, and a small amount of silver to increase strength) for joints of automobile radiators [7].

In applications where thermal cycling occurs, the high coefficient of expansion ($30 \times 10^{-6}/°C$) of lead may cause intergranular cracking due to fatigue or corrosion fatigue.

28.2.1 Lead–Acid Battery

Lead–acid batteries are a principal use of lead. Following are the reactions that take place during discharge of a lead–acid battery [8]:

$$\text{At the anode:} \quad Pb \rightarrow Pb^{2+} + 2e^- \tag{28.1}$$

$$\text{At the cathode:} \quad PbO_2 + 4H^+ + 2e^- \rightarrow Pb^{2+} + 2H_2O \tag{28.2}$$

$$\text{The net reaction:} \quad Pb + PbO_2 + 4H^+ \rightarrow 2Pb^{2+} + 2H_2O \tag{28.3}$$

Corrosion of lead in Eq. (28.1) is part of the overall electrochemical reaction by which electrical energy is generated using the lead–acid battery. In addition to this positive and beneficial aspect of corrosion in energy generation, corrosion can also lead to shorter battery lifetime and reduced capacity to produce energy. For example, corrosion during battery storage reduces the quantity of lead available for energy production and may also produce corrosion products that inhibit the corrosion that takes place by Eq. (28.1). Passivation of the lead anode increases the anodic overvoltage, decreases the current, and causes a deterioration of battery performance. The lead–acid battery uses lead grids at both the anode and the cathode. Corrosion of lead at the cathode has been reported to cause loss of capacity of batteries [9]. For corrosion by Eq. (28.1), which results in energy production, the overvoltage should be as low as possible. For the destructive forms of corrosion, which lead to reduced battery lifetime and capacity, the overvoltage should be as high as possible [10].

Various alloys of lead are used as support grids for positive and negative plates of lead–acid batteries. Alloying elements are added to improve mechanical properties, but anodic corrosion of these alloys can limit battery lifetime [11, 12]. Lead–antimony alloys, containing typically about 5% antimony, have been used for many years [13]. These alloys develop an adherent protective corrosion product layer; however, the deposition of antimony on the negative plate lowers the hydrogen overpotential [13, 14]. Furthermore, on overcharge, toxic stibine (SbH_3) is produced. For these reasons, alloys of lead with calcium, tin, and aluminum are also being used [13, 14]. Other possible alloying elements that have been studied include magnesium, titanium, and bismuth [14, 15]. Lead–calcium–tin alloys used as grid material for maintenance-free batteries have been reported to be susceptible to intergranular corrosion at large grain sizes [16].

28.3 SUMMARY

In summary, lead is resistant to the following:

1. Many strong acids.
   a. H_2SO_4, <96%, room temperature; the rate for <80% H_2SO_4, boiling, is <2 mm/y (<0.08 ipy); for 20% H_2SO_4, boiling, the rate is 0.08 mm/y (0.003 ipy) [17].
   b. Commercial H_3PO_4 (hot or cold).
   c. H_2CrO_4 (as used in chromium plating).
   d. HF <60–65% (room temperature).
   e. H_2SO_3.
2. Atmospheric exposures, particularly industrial.
3. Seawater (0.01 mm/y, 0.0005 ipy).
4. Chlorine, wet or dry, <100 °C (but only dry Br_2 and at lower temperatures), SO_2, SO_3, H_2S.

Lead is not resistant to the following:

1. HNO_3, <70%.
2. HCl.
3. Concentrated H_2SO_4 (>96%, room temperature).
4. Alkalies. For some chemical applications, the rate in caustic alkalies is considered tolerable.
5. HF, gaseous.
6. Many aerated organic acids.

REFERENCES

1. M. Pourbaix, *Atlas of Electrochemical Equilibria in Aqueous Solutions*, 2nd English edition, National Association of Corrosion Engineers, Houston, TX, and CEBELCOR, Brussels, 1974, p. 490.

2. F. E. Goodwin, Lead and lead alloys, in *Uhlig's Corrosion Handbook*, 2nd edition, R. W. Revie, editor, Wiley, New York, 2000, p. 768.

3. Ref. 2, p. 775.

4. R. H. Heidersbach, Cathodic protection, in *ASM Handbook*, Vol. 13A, *Corrosion: Fundamentals, Testing, and Protection*, ASM International, Materials Park, OH, 2003, p. 860.

5. Environmental Protection Agency website: http://www.epa.gov/safewater/lcmr/

6. Ref. 2, p. 770.

7. S. J. Alhassan, Corrosion of lead and lead alloys, in *ASM Handbook*, Vol. 13B, *Corrosion: Materials*, ASM International, Materials Park, OH, 2003, p. 203.

8. M. Pourbaix, *Lectures on Electrochemical Corrosion*, third English edition, J. A. S. Green, translator, R. W. Staehle, editor, NACE International, Houston, TX, 1995, p. 91.

9. D. Pavlov, *J. Power Sources* **46**, 171 (1993); ibid., **48**, 179 (1994); ibid., **53**, 9 (1995).

10. C. F. Windisch, Anodes for batteries, in *ASM Handbook*, Vol. 13A, *Corrosion: Fundamentals, Testing, and Protection*, ASM International, Materials Park, OH, 2003, pp. 170–177.

11. P. Ruetschi, in *Lead–Acid Batteries: A Reference and Data Book*, Elsevier, Lausanne, Switzerland, 1977, pp. 19–21.

12. C. Dacres, S. Reamer, R. Sutula, and I. Angres, *J. Electrochem. Soc.* **128**, 2060 (1981).

13. N. Hampson, S. Kelly, and K. Peters, *J. Appl. Electrochem.* **19**, 91 (1980).

14. J. Bialacki, N. Hampson, and K. Peters, *J. Electrochem. Soc.* **130**, 1797 (1983).

15. N. Hampson, S. Kelly, and K. Peters, *J. Electrochem. Soc.* **127**, 1456 (1980).

16. M. V. Rose and J. A. Young, in *Proceedings of the Fifth International Conference on Lead*, Metal Bulletin Limited, London, 1976, p. 37.

17. Ref. 2, pp. 771–775.

GENERAL REFERENCES

S. J. Alhassan, Corrosion of lead and lead alloys, in *ASM Handbook*, Vol. 13B, *Corrosion: Materials*, ASM International, Materials Park, OH, 2003, pp. 195–204.

F. E. Goodwin, Lead and lead alloys, in *Uhlig's Corrosion Handbook*, 2nd edition, R. W. Revie, editor, Wiley, New York, 2000, pp. 767–792.

D. Pavlov, *Essentials of Lead–Acid Batteries*, Society for Advancement of Electrochemical Science and Technology, Karaikudi, India, 2006.

C. F. Windisch, Anodes for batteries, in *ASM Handbook*, Vol. 13A, *Corrosion: Fundamentals, Testing, and Protection*, ASM International, Materials Park, OH, 2003, pp. 170–177.

29

APPENDIX

29.1 ACTIVITY AND ACTIVITY COEFFICIENTS OF STRONG ELECTROLYTES

Let μ be the partial molal free energy, $\mu°$ the partial molal free energy in the standard state ($a = 1$), a the activity, M the molality, and γ the activity coefficient.

By definition,

$$\mu = \mu° + RT \ln a$$

$$a = \gamma M$$

where $\gamma \to 1$ as $M \to 0$.

For a binary electrolyte of molality M, assuming total dissociation,

$$\mu = \mu° + RT \ln a_- + RT \ln a_+$$
$$= \mu° + RT \ln \gamma_- M_- + RT \ln \gamma_+ M_+$$
$$= \mu° + 2RT \ln \gamma_\pm M$$

Corrosion and Corrosion Control, by R. Winston Revie and Herbert H. Uhlig
Copyright © 2008 John Wiley & Sons, Inc.

where $M_- = M_+ = M$, and $\gamma_\pm$ is the mean ion activity coefficient (it is impossible to measure the activities of individual ions). For an electrolyte such as lanthanum sulfate of molality M, we have

$$La_2(SO_4)_3 \rightarrow 2La^{3+} + 3SO_4^{2-}$$

$$\mu = \mu^\circ + 2RT \ln \gamma_\pm M_{La^{3+}} + 3RT \ln \gamma_\pm M_{SO_4^{2-}}$$

Since $M_{La^{3+}} = 2M$ and $M_{SO_4^{2-}} = 3M$,

$$\mu = \mu^\circ + RT \ln \gamma_\pm^5 (2M)^2 (3M)^3$$
$$= \mu^\circ + 5RT \ln \gamma_\pm M (2^2 \times 3^3)^{1/5}$$

For the general case in which one mole of electrolyte dissociates into ν_1 moles of cation and ν_2 moles of anion, the mean activity is given by

$$a_\pm = \gamma_\pm M(\nu_1^{\nu_1} \nu_2^{\nu_2})^{1/(\nu_1+\nu_2)}$$

(See also: C. W. Davies, *Electrochemistry*, George Newnes, Ltd., London, 1967; J. O'M. Bockris and A. K. N. Reddy, *Modern Electrochemistry*, 2nd edition, Vol. 1: *Ionics*, Plenum, New York, 1998, pp. 251–257.) Activity coefficients are listed in Table 29.1.

Example. Calculate the emf of the following cell:

$$Zn; ZnCl_2(0.01M); Pt(Cl_2, 0.5 \, atm)$$

$$\gamma_\pm \text{ for } 0.01M \text{ ZnCl}_2 = 0.71 \text{ (Table 29.1)}$$

$$Zn^{2+} + 2e^- \rightarrow Zn \qquad \phi^\circ = -0.763 \, V \qquad (29.1)$$

$$Cl_2 + 2e^- \rightarrow 2Cl^- \qquad \phi^\circ = 1.360 \, V \qquad (29.2)$$

Tentative reaction: (29.2)– (29.1)

$$Zn + Cl_2 \rightarrow Zn^{2+} + 2Cl^- \qquad (29.3)$$

The corresponding Nernst equation is

$$E = 0.763 + 1.360 - \frac{0.0592}{2} \log \frac{(Zn^{2+})(Cl^-)^2}{p_{Cl_2}}$$

$$= 2.123 - \frac{0.0592}{2} \log \frac{(0.01)(0.71)[(0.02)(0.71)]^2}{0.5}$$

$$= 2.29 \, V$$

TABLE 29.1. Activity Coefficients of Strong Electrolytes (M = molality)

$M \to$	0.001	0.002	0.005	0.01	0.02	0.05	0.1	0.2	0.5	1.0	2.0	3.0	4.0
HCl	0.966	0.952	0.928	0.904	0.875	0.830	0.796	0.767	0.758	0.809	1.01	1.32	1.76
HBr	0.966	—	0.929	0.906	0.879	0.838	0.805	0.782	0.790	0.871	1.17	1.67	—
HNO$_3$	0.965	0.951	0.927	0.902	0.871	0.823	0.785	0.748	0.715	0.720	0.783	0.876	0.982
HClO$_4$	—	—	—	—	—	—	—	—	—	0.81	1.04	1.42	2.02
HIO$_3$	0.96	0.94	0.91	0.86	0.80	0.69	0.58	0.46	0.29	0.19	0.10	0.073	0.060
H$_2$SO$_4$	0.830	0.757	0.639	0.544	0.453	0.340	0.265	0.209	0.154	0.130	0.124	0.141	0.171
NaOH	—	—	—	—	—	0.82	—	0.73	0.69	0.68	0.70	0.77	0.89
KOH	—	—	0.92	0.90	0.86	0.82	0.80	0.76	0.73	0.76	0.89	1.08	1.35
CsOH	—	—	—	0.92	0.88	0.83	0.80	—	0.74	0.78	—	—	—
Ba(OH)$_2$	—	0.853	0.773	0.712	0.627	0.526	0.443	0.370	—	—	—	—	—
AgNO$_3$	—	—	0.92	0.90	0.86	0.79	0.72	0.64	0.51	0.40	0.28	—	—
Al(NO$_3$)$_3$	—	—	—	—	—	—	0.20	0.16	0.14	0.19	0.45	1.0	1.2
BaCl$_2$	0.88	—	0.77	0.72	—	0.56	0.49	0.44	0.39	0.39	0.44 (1.8M)	—	—
Ba(NO$_3$)$_2$	0.88	0.84	0.77	0.71	0.63	0.52	0.43	0.34	—	—	—	—	—
Ba(IO$_3$)$_2$	0.83	0.79	0.71	0.64	0.55	—	—	—	—	—	—	—	—
CaCl$_2$	0.89	0.85	0.785	0.725	0.66	0.57	0.515	0.48	0.52	0.71	—	—	—
Ca(NO$_3$)$_2$	0.88	0.84	0.77	0.71	0.64	0.54	0.48	0.42	0.38	0.35	0.35	0.37	0.42
CdCl$_2$	0.76	0.68	0.57	0.47	0.38	0.28	0.21	0.15	0.09	0.06	—	—	—
CdI$_2$	0.76	0.65	0.49	0.38	0.28	0.17	0.11	0.068	0.038	0.025	0.018	—	—
CdSO$_4$	0.73	0.64	0.50	0.40	0.31	0.21	0.17	0.11	0.067	0.045	0.035	0.036	—
CsF	0.98	0.97	0.96	0.95	0.94	0.91	0.89	0.87	0.85	0.87	—	—	—
CsCl	—	—	0.92	0.90	0.86	0.79	0.75	0.69	0.60	0.54	0.49	0.48	0.47
CsBr	—	—	0.93	0.90	0.86	0.79	0.75	0.69	0.60	0.53	0.48	0.46	0.46
CsI	—	—	—	—	—	—	0.75	0.69	0.60	0.53	0.47	0.43	—
CsNO$_3$	—	—	—	—	—	—	0.73	0.65	0.52	0.42	—	—	—

TABLE 29.1. Continued

$M \rightarrow$	0.001	0.002	0.005	0.01	0.02	0.05	0.1	0.2	0.5	1.0	2.0	3.0	4.0
CsAc	0.89						0.79	0.77	0.76	0.80	0.95	1.15	
$CuCl_2$	0.89	0.85	0.78	0.72	0.66	0.58	0.52	0.47	0.42	0.43	0.51	0.59	
$CuSO_4$	0.74		0.53	0.41	0.31	0.21	0.16	0.11	0.068	0.047			
$FeCl_2$			0.80	0.75	0.70	0.62	0.58	0.55	0.59	0.67			
$In_2(SO_4)_3$				0.142	0.092	0.054	0.035	0.022					
KF	0.965	0.96	0.95	0.93	0.92	0.88	0.85	0.81	0.74	0.71	0.70		
KCl	0.965	0.952	0.927	0.901	0.872	0.815	0.769	0.719	0.651	0.606	0.576	0.571	0.579
KBr	0.965	0.952	0.927	0.903	0.88	0.822	0.777	0.728	0.665	0.625	0.602	0.603	0.622
KI		0.951	0.927	0.905		0.84	0.80	0.76	0.71	0.68	0.69	0.72	0.75
$K_4Fe(CN)_6$						0.19	0.14	0.11	0.067				
$KClO_3$	0.967	0.955	0.932	0.907	0.875	0.813	0.755						
K_2CO_3	0.89	0.86	0.81	0.74	0.68	0.58	0.50	0.43	0.36	0.33	0.33	0.39	0.49
$KClO_4$	0.965	0.951	0.924	0.895	0.857	0.788							
K_2SO_4	0.89		0.78	0.71	0.64	0.52	0.43	0.36					
$LaCl_3$				0.57	0.49	0.39	0.32	0.28	0.27	0.36			
$La(NO_3)_3$	0.963					0.39	0.33	0.27					
LiCl	0.966	0.948	0.921	0.89	0.86	0.82	0.78	0.75	0.73	0.76	0.91	1.18	1.46
LiBr		0.954	0.932	0.909	0.882	0.842	0.810	0.784	0.783	0.848	1.06	1.35	
LiI	0.966						0.81	0.80	0.81	0.89	1.19	1.70	
$LiNO_3$	0.967	0.953	0.930	0.904	0.878	0.834	0.798	0.765	0.743	0.76	0.84	0.97	
$LiClO_3$	0.967	0.955	0.933	0.911	0.884	0.842	0.810	0.782	0.77	0.81			
$LiClO_4$		0.956	0.935	0.915	0.890	0.853	0.825	0.805	0.82	0.91			
$MgCl_2$	0.88	0.84	0.77	0.71	0.64	0.55	0.56	0.53	0.52	0.62	1.05	2.1	
$Mg(NO_3)_2$							0.51	0.46	0.44	0.50	0.69	0.93	
$MgSO_4$				0.40	0.32	0.22	0.18	0.13	0.088	0.064	0.055	0.064	0.079
$MnSO_4$							0.25	0.17	0.11	0.073	0.058	0.062	

$M \rightarrow$	0.001	0.002	0.005	0.01	0.02	0.05	0.1	0.2	0.5	1.0	2.0	3.0	4.0
$NiSO_4$	0.961	0.944	0.911				0.18	0.13	0.075	0.051	0.041		
NH_4Cl	0.964	0.949	0.901	0.88	0.84	0.79	0.74	0.69	0.62	0.57			
NH_4Br	0.962	0.946	0.917	0.87	0.83	0.78	0.73	0.68	0.62	0.57			
NH_4I	0.959	0.942	0.912	0.89	0.86	0.80	0.76	0.71	0.65	0.60			
NH_4NO_3				0.88	0.84	0.78	0.73	0.66	0.56	0.47			
$(NH_4)_2SO_4$	0.874	0.821	0.726	0.67	0.59	0.48	0.40	0.32	0.22	0.16			
NaF	0.966	0.953	0.93	0.90	0.87	0.81	0.75	0.69	0.62				
$NaCl$	0.966	0.955	0.929	0.904	0.875	0.823	0.780	0.730	0.68	0.66	0.67	0.71	0.78
$NaBr$	0.97	0.96	0.934	0.914	0.887	0.844	0.800	0.740	0.695	0.686	0.734	0.826	0.934
NaI	0.966	0.953	0.94	0.91	0.89	0.86	0.83	0.81	0.78	0.80	0.95		
$NaNO_3$	0.97	0.95	0.93	0.90	0.87	0.82	0.77	0.70	0.62	0.55	0.48	0.44	0.41
Na_2SO_4	0.887	0.847	0.778	0.714	0.641	0.53	0.45	0.36	0.27	0.20			
$NaClO_4$			0.93	0.90	0.87	0.82	0.77	0.72	0.64	0.58			
$PbCl_2$	0.86	0.80	0.70	0.61	0.50	0.46	0.37	0.27	0.17	0.11			
$Pb(NO_3)_2$	0.88	0.84	0.76	0.69	0.60								
$RbCl$			0.93	0.90			0.76	0.71	0.63	0.58	0.54	0.54	0.54
$RbBr$							0.76	0.70	0.63	0.58	0.53	0.52	0.51
RbI							0.76	0.70	0.63	0.57	0.53	0.52	0.51
$RbNO_3$							0.73	0.65	0.53	0.43	0.32	0.25	0.21
$RbAc$							0.73	0.65	0.52	0.42			
$TlCl$	0.96	0.95	0.93	0.90		0.77	0.70	0.60	0.53	0.51			
$TlNO_3$						0.79	0.73	0.65	0.59				
$TlClO_4$						0.80	0.74	0.68					
$TlAc$						0.56							
$ZnCl_2$	0.88	0.84	0.77	0.71	0.64		0.50	0.45	0.38	0.33	0.44	0.40	0.38
$ZnSO_4$	0.70	0.61	0.48	0.39			0.15	0.11	0.065	0.045	0.036	0.04	

Sourse: Oxidation Potentials, W. Latimer, Prentice–Hall, Englewood Cliffs, NJ, 1952 (with permission).

[Reaction (29.3) is spontaneous (ΔG is negative); zinc is anode (−), and platinum is cathode (+)].

29.2 DERIVATION OF STERN–GEARY EQUATION FOR CALCULATING CORROSION RATES FROM POLARIZATION DATA OBTAINED AT LOW CURRENT DENSITIES

Assume that the corrosion current, I_{corr}, occurs at a value within the Tafel region for both anodic and cathodic reactions. Also assume that concentration polarization and IR drop are negligible. Current relations are shown in Fig. 29.1 for the corroding metal polarized as anode by means of an external current to a potential ϕ. Increasing anodic current, I_a, is accompanied by decreasing cathodic current, I_c, because of the relation,

$$I_{appl} = I_a - I_c \qquad (29.4)$$

Similarly, for cathodic polarization, for which I_{appl} changes sign, we have

$$-I_{appl} = I_c - I_a \qquad (29.5)$$

For anodic polarization, we obtain

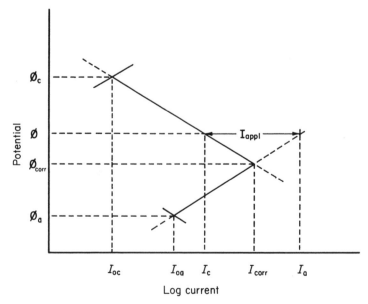

Figure 29.1. Polarization diagram for corroding metal polarized anodically from ϕ_{corr} to ϕ.

$$\phi - \phi_{\text{corr}} = \Delta\phi = \beta_a \log \frac{I_a}{i_{oa} A_a} - \beta_a \log \frac{I_{\text{corr}}}{i_{oa} A_a}$$
$$= \beta_a \log \frac{I_a}{I_{\text{corr}}} \tag{29.6}$$

where A_a is the fraction of area that is anode, and i_{0a} is the exchange current density for the anodic reaction. Note that i_{0a} in the Tafel equation refers to current per unit local area (not total area); hence, I_{corr}/A_a is the current density to which i_{0a} applies. Likewise, if the metal is polarized an equal amount in the cathodic direction, we have

$$\Delta\phi = -\beta_c \log \frac{I_c}{I_{\text{corr}}} \tag{29.7}$$

or

$$I_c = I_{\text{corr}} 10^{-\Delta\phi/\beta_c} \quad \text{and} \quad I_a = I_{\text{corr}} 10^{\Delta\phi/\beta_a}$$

then

$$I_{\text{appl}} = I_{\text{corr}} (10^{\Delta\phi/\beta_a} - 10^{-\Delta\phi/\beta_c}) \tag{29.8}$$

Expressed as a series, we have

$$10^x = 1 + 2.3x + \frac{(2.3x)^2}{2!} + \cdots$$

If $\Delta\phi/\beta c$ and $\Delta\phi/\beta a$ are small, higher terms can be neglected and (29.8) can be approximated by

$$I_{\text{appl}} = 2.3 I_{\text{corr}} \Delta\phi \left(\frac{1}{\beta_c} + \frac{1}{\beta_a} \right) \tag{29.9}$$

or

$$I_{\text{corr}} = \frac{I_{\text{appl}}}{2.3\Delta\phi} \left(\frac{\beta_a \beta_c}{\beta_a + \beta_c} \right) = \frac{1}{2.3R} \left(\frac{\beta_a \beta_c}{\beta_a + \beta_c} \right) \tag{29.10}$$

where R is the polarization resistance, determined experimentally by applying both anodic and cathodic currents to the electrode and measuring the polarization. The value of R used in Eq. (29.10) is the slope of the ϕ versus I graph at the corrosion potential.

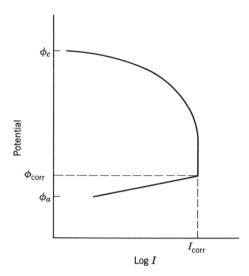

Figure 29.2. Polarization diagram for metal corroding under control by oxygen depolarization.

Equation (29.10) is the Stern–Geary equation. Should the cathodic reaction be controlled by concentration polarization, as occurs in corrosion reactions controlled by oxygen depolarization, the corrosion current equals the limiting diffusion current (Fig. 29.2). This situation is equivalent to a large or infinite value of β_c in (29.10). Under these conditions, (29.10) becomes

$$I_{corr} = \frac{\beta_a I_{appl}}{2.3\Delta\phi} = \frac{\beta_a}{2.3R} \qquad (29.11)$$

If a noncorroding metal (minimum local action) that polarizes only slightly (high value of i_{0a}) is polarized anodically at moderate current densities, then i_{0a} can be substituted for I_{corr} in Fig. 29.1, and

$$\Delta\phi = \frac{I_{appl}}{2.3i_{0a}} \frac{\beta_c\beta_a}{\beta_a + \beta_c} \qquad (29.12)$$

indicating, as observed, that anodic polarization of many metals at low current densities is a linear function of applied current.

29.2.1 The General Equation

If ϕ_{corr} falls outside the Tafel region, for example, when it approaches either the reversible anode or cathode potential, "back-reactions" become appreciable and (29.10) is subject to error. Mansfeld and Oldham [1] considered this case and formulated the required modifications. The derivation of the more exact form of (29.10) starts with modifying (29.4). Because I_{appl} changes sign at potentials above compared to below ϕ_{corr}, choosing I_{appl} and I_a with the same sign—that is, anodic polarization—results in

$$I_{appl} = I_a - I_{ar} - (I_c - I_{cr}) \tag{29.13}$$

where I_{ar} and I_{cr} are the back-reactions for the anodic and cathodic reactions, respectively.

Introducing the Tafel equation expressed in natural logarithms, $\eta = b \ln(I/I_0)$ and letting $b_a = \beta_a/2.3$, $b_c = \beta_c/2.3$, $b_{ar} = \beta_{ar}/2.3$, and $b_{cr} = \beta_{cr}/2.3$, we have

$$I_{appl} = I_{oa} \exp\left(\frac{\phi - \phi_a}{b_a}\right) - I_{oa} \exp\left(-\frac{\phi - \phi_a}{b_{ar}}\right) - I_{oc} \exp\left(-\frac{\phi - \phi_c}{b_c}\right) + I_{oc} \exp\left(\frac{\phi - \phi_c}{b_{cr}}\right) \tag{29.14}$$

According to electrode kinetics, we obtain

$$\frac{1}{b_a} + \frac{1}{b_{ar}} = \frac{n_a F}{RT} \quad \text{and} \quad \frac{1}{b_c} + \frac{1}{b_{cr}} = \frac{n_c F}{RT}$$

where n_a and n_c are constants related to the transfer coefficient. Using these relations to eliminate b_{ar} and b_{cr}, it follows that

$$\begin{aligned} I_{appl} = &I_{oa} \exp\left(\frac{\phi - \phi_a}{b_a}\right)\left[1 - \exp\left(\frac{n_a F(\phi_a - \phi)}{RT}\right)\right] \\ &- I_{oc} \exp\left(\frac{\phi_c - \phi}{b_c}\right)\left[1 - \exp\left(\frac{n_c F(\phi - \phi_c)}{RT}\right)\right] \end{aligned} \tag{29.15}$$

Differentiating (29.15) with respect to ϕ, we get

$$\begin{aligned} \frac{\partial I_{appl}}{\partial \phi} = &\frac{I_{oa}}{b_a} \exp\left(\frac{\phi - \phi_a}{b_a}\right)\left[1 + \left(\frac{b_a n_a F}{RT} - 1\right)\exp\left(\frac{n_a F(\phi_a - \phi)}{RT}\right)\right] \\ &+ \frac{I_{oc}}{b_c} \exp\left(\frac{\phi_c - \phi}{b_c}\right)\left[1 + \left(\frac{b_c n_c F}{RT} - 1\right)\exp\left(\frac{n_c F(\phi - \phi_c)}{RT}\right)\right] \end{aligned} \tag{29.16}$$

At the corrosion potential, $I_{appl} = 0$ and $I_a = I_c = I_{corr}$.
Letting $\Delta\phi_a = \phi_{corr} - \phi_a$ and $\Delta\phi_c = \phi_c - \phi_{corr}$, (29.15) becomes

$$I_{corr} = I_{oa} \exp\left(\frac{\Delta\phi_a}{b_a}\right)\left[1 - \exp\left(\frac{-n_a F \Delta\phi_a}{RT}\right)\right] \tag{29.17}$$

$$= I_{oc} \exp\left(\frac{\Delta\phi_c}{b_c}\right)\left[1 - \exp\left(\frac{-n_c F \Delta\phi_c}{RT}\right)\right] \tag{29.18}$$

and (29.16) becomes

$$\frac{1}{R} = \left(\frac{\partial I_{appl}}{\partial \phi}\right)_{\phi=\phi_{corr}} = \frac{I_{oa}}{b_a}\exp\left(\frac{\Delta\phi_a}{b_a}\right)\left[1+\left(\frac{b_a n_a F}{RT}-1\right)\exp\left(\frac{-n_a F \Delta\phi_a}{RT}\right)\right]$$
$$+ \frac{I_{oc}}{b_c}\exp\left(\frac{\Delta\phi_c}{b_c}\right)\left[1+\left(\frac{b_c n_c F}{RT}-1\right)\exp\left(\frac{-n_c F \Delta\phi_c}{RT}\right)\right] \tag{29.19}$$

Substituting (29.17) into the first part of (29.19) and (29.18) into the second part of (29.19) gives

$$\frac{1}{R} = I_{corr}\left\{\frac{1}{b_a}+\frac{1}{b_c}+\frac{n_a F}{RT}\left[\exp\left(\frac{n_a F \Delta\phi_a}{RT}\right)-1\right]^{-1}+\frac{n_c F}{RT}\left[\exp\left(\frac{n_c F \Delta\phi_c}{RT}\right)-1\right]^{-1}\right\} \tag{29.20}$$

Equation (29.20) was developed by Mansfeld and Oldham [1] and is equivalent to the one derived originally by Wagner and Traud [2]. If the correction terms involving $\Delta\phi_a$ and $\Delta\phi_c$ can be neglected, (29.20) becomes the Stern–Geary equation. On the other hand, if either $\Delta\phi_a$ or $\Delta\phi_c$ approaches zero, the corresponding correction term approaches infinity.

Small values of $nF\Delta\phi/RT$ allow the exponential terms to be expanded, simplifying (29.20) to

$$\frac{1}{R} = I_{corr}\left(\frac{1}{b_a}+\frac{1}{b_c}+\frac{1}{\Delta\phi_a}+\frac{1}{\Delta\phi_c}\right) \tag{29.21}$$

If $\Delta\phi_c = 0.005\,V$, $T = 298\,°K$, and n_c is assumed to be 0.5, then $F/RT = 39\,V^{-1}$ and $n_c F\Delta\phi_c/RT = 0.1$. The corresponding correction term from (29.20) is $190\,V^{-1}$, whereas the approximated correction term in (29.21) is $1/\Delta\phi_c = 200\,V^{-1}$. These correction terms are large compared to $1/b_a + 1/b_c = 46\,V^{-1}$ (assuming $\beta_a = \beta_c = 0.1\,V$). Should $\Delta\phi_c = 0.1\,V$, the corresponding correction term from (29.20) is $3.2\,V^{-1}$. If at the same time, $\Delta\phi_a = 0.5\,V$ and $n_a = 0.5$, the correction term as given by (29.20) equals $0.001\,V^{-1}$, which is negligible.

Additional approximations [1] that could be used if the corrosion potential is close to one of the reversible potentials are as follows. If $n_a\Delta\phi_a < 2RT/F$ (i.e., the corrosion potential is close to the reversible potential of the metal),

$$\frac{1}{R} = I_{corr}\left(\frac{1}{b_a}+\frac{1}{b_c}+\frac{1}{\Delta\phi_a}-\frac{n_a F}{2RT}\right) \tag{29.22}$$

If $n_c\Delta\phi_c < 2RT/F$ (i.e., the corrosion potential is close to the reversible potential of the reduction reaction), we obtain

$$\frac{1}{R} = I_{corr}\left(\frac{1}{b_a}+\frac{1}{b_c}+\frac{1}{\Delta\phi_c}-\frac{n_c F}{2RT}\right) \tag{29.23}$$

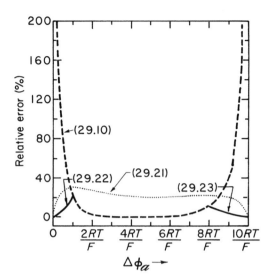

Figure 29.3. Relative errors in corrosion currents calculated by use of (29.10), (29.21), (29.22), and (29.23), instead of the exact equation (29.20) [1] (with permission from Pergamon Press).

The errors that result from each of these approximations to (29.20) are presented in Fig. 29.3 [1], where $n_a = 2$, $n_c = 1$, $b_a = RT/F$, $b_c = 2RT/F$, and $\phi_c - \phi_a = \Delta\phi_a + \Delta\phi_c = 10\,RT/F$ were assumed.

29.3 DERIVATION OF EQUATION EXPRESSING THE SATURATION INDEX OF A NATURAL WATER

In accord with the discussion of saturation index in Section 7.2.6.1, we have

$$K'_s = (Ca^{2+})(CO_3^{2-}) \tag{29.24}$$

$$K'_2 = \frac{(H^+)(CO_3^{2-})}{(HCO_3^-)} \tag{29.25}$$

Assume that salts of weak acids other than carbonic acid are absent. Then when a water is titrated, equivalents of added acid equal equivalents of carbonate and bicarbonate, plus OH⁻ or minus H⁺, depending on pH of the water:

$$(alk) + (H^+) \rightarrow 2(CO_3^{2-}) + (HCO_3^-) + (OH^-) \tag{29.26}$$

where (alk) = alkalinity, or titratable equivalents of base per liter (by titrating with acid) obtained by using methyl orange as indicator. Concentrations of (H^+) and (OH^-) are small between pH values 4.5 and 10.3 and may be neglected.

From Eq. (29.25), we obtain

$$(CO_3^{2-}) = \frac{K_2'(HCO_3^-)}{(H^+)} \tag{29.27}$$

From Eq. (29.26),

$$(CO_3^{2-}) = \frac{(alk) - (HCO_3^-)}{2} \tag{29.28}$$

Therefore,

$$(HCO_3^-) = \frac{(alk)}{1 + 2K_2'/(H^+)} \tag{29.29}$$

Equation (29.29) was derived by Langelier [3] on the assumption that K_s' and K_2' are based on concentrations (moles/liter) rather than on activities. For example, referring to (29.24), if K_s is the true activity product, then $K_s = K_s'\gamma_\pm^2$, where $\gamma_\pm$ refers to the mean ion activity coefficient for $CaCO_3$. The activity coefficient was approximated by Langelier using the Debye–Hückel theory, $-\log\gamma = 0.5\,z^2\mu^{1/2}$, where μ is the ionic strength and z is the valence. Hence, concentrations of CO_3^{2-} and HCO_3^- obtained by titration can be equated to corresponding concentrations of these species in K_s' and K_2'. Accordingly, K_s' and K_2' vary not only with temperature, but also with total dissolved solids because of the effect of ionic strength of a solution on activities of specific ions.

Substituting (29.29) into (29.27), we obtain

$$(CO_3^{2-}) = \frac{K_2'}{(H^+)} \frac{(alk)}{1 + 2K_2'/(H^+)} \tag{29.30}$$

and substituting (29.30) into (29.24), we get

$$(Ca^{2+}) \frac{K_2'}{(H^+)} \frac{(alk)}{1 + 2K_2'/(H^+)} = K_s' \tag{29.31}$$

Taking logarithms of both sides and using the notation $\log 1/\alpha = p\alpha$, we obtain

TABLE 29.2. Values of $\log\left(1+\dfrac{2K_2'}{(H^+)_s}\right)$ as a Function of pH_s

pH_s:	10.3	10.0	9.7	9.4	9.1
$\log\left(1+\dfrac{2K_2'}{(H^+)_s}\right)$:	0.47	0.29	0.17	0.09	0.05

$$pH_s = pK_2' - pK_s' + p(Ca^{2+}) + p(alk) + \log\left(1+\frac{2K_2'}{(H^+)_s}\right) \qquad (29.32)$$

where pH_s is the pH of a given water at which solid $CaCO_3$ is in equilibrium with its saturated solution.

The last term is ordinarily small and can be omitted when $pH_s < 9.5$. Based on the value $K_2' = 4.8 \times 10^{-11}$, typical values as a function of pH_s in the alkaline range at 25°C are given in Table 29.2.

Values of ($pK_2' - pK_s'$) decrease with increasing temperature as follows: 0°C, 2.48; 20°C, 2.04; 25°C, 1.96; 50°C, 1.54. In the presence of other salts (e.g., NaCl, Na_2SO_4, or $MgSO_4$), the increasing ionic strength of the solution depresses the activity of other ions in solution. This effect increases values of ($pK_2' - pK_s'$). For example, at 25°C, at a total dissolved-solids content of 100 ppm, the value is 2.13, and for 500 ppm it is 2.19.

A nomogram for obtaining pH_s of a water at various temperatures and dissolved-solids content was constructed by C. Hoover [4]. A chart for the same purpose as prepared by Powell, Bacon, and Lill [5] is reproduced in Fig. 29.4. To use the chart, we must know the alkalinity of a water and calcium ion concentration calculated as ppm $CaCO_3$, total dissolved solids in ppm, and the temperature.

The saturation index is then the algebraic difference between the measured pH of a water and the computed pH_s:

$$\text{Saturation index} = pH_{measured} - pH_s \qquad (29.33)$$

To calculate the saturation index at above-room temperature, the actual pH of water at the higher temperature should be used. This can be estimated from the room temperature value by using Fig. 29.5 [5], which gives values for two waters of differing alkalinity.

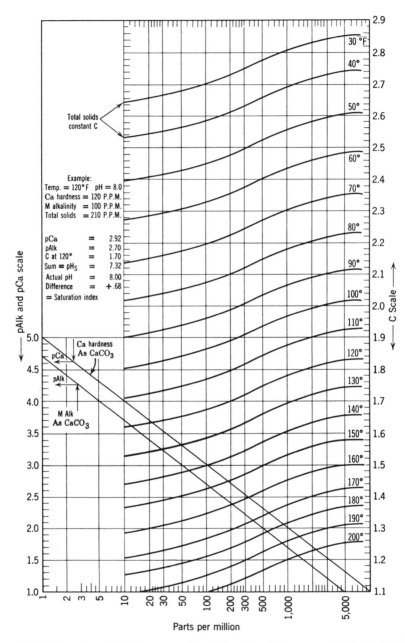

Figure 29.4. Chart for calculating saturation index (Powell, Bacon, and Lill [5]). ("Ca" and "alkalinity" are expressed as ppm $CaCO_3$, and temperature is expressed in °F.)

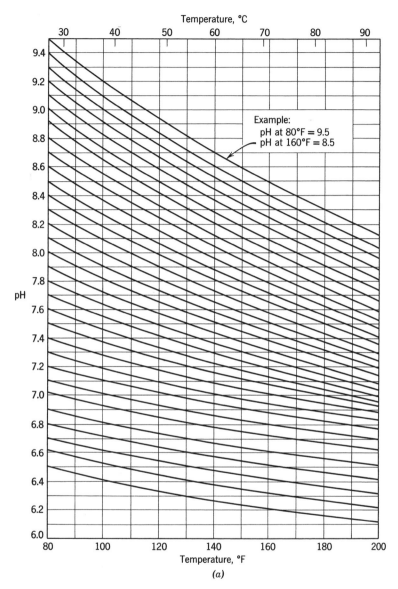

Figure 29.5. Values of pH of water at elevated temperatures (Powell, Bacon, and Lill [5]). (a) For water at 25 ppm alkalinity (methyl orange end point). (b) For water of 100 ppm alkalinity (methyl orange end point).

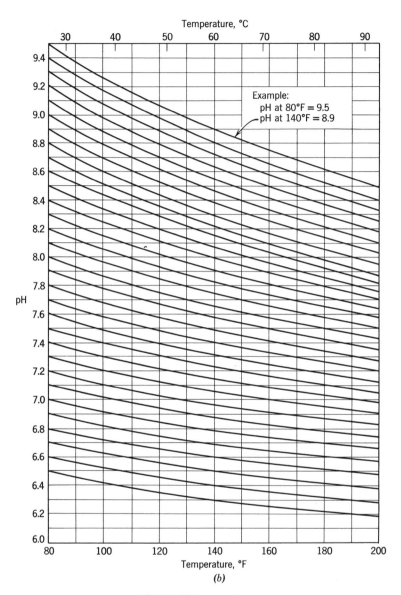

Figure 29.5. *Continued*

29.4 DERIVATION OF POTENTIAL CHANGE ALONG A CATHODICALLY PROTECTED PIPELINE

Assume that current enters a pipe from the soil side through a porous insulating coating, returning to the anode by way of the pipe (Fig. 29.6). Then the change of current, I_x, in the pipe per unit length at x is equal to the total current entering the pipe at x, or

$$\frac{dI_x}{dx} = -2\pi r i_x \qquad (29.34)$$

where r is the radius of the pipe.

By Ohm's law, the potential change along the pipe is given by

$$\frac{dE_x}{dx} = -R_L I_x \qquad (29.35)$$

where R_L is the resistance of metallic pipe per unit length. Combining (29.34) and (29.35), we obtain

$$\frac{d^2 E_x}{dx^2} = R_L(2\pi r i_x) \qquad (29.36)$$

At small values of i_x, polarization of the pipe surface is a linear function of the true current density i'_x at the base of pores in the coating, or

$$E_x = k_1 i'_x \qquad (29.37)$$

Assuming that the resistance z per unit area of coating is inversely proportional to the total cross-sectional area of pores per unit area, true current density increases with z, or

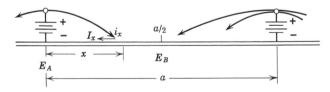

Figure 29.6. Sketch of buried pipe cathodically protected by anodes distance a apart. i_x is the current density at pipe surface at distance x from point of bonding; I_x is the total current in pipe at distance x; E is the difference between measured and corrosion potentials of pipe; r is the radius of pipe; R_L is the resistance of metallic pipe per unit length; and z is the resistance of pipe coating per unit area.

$$i'_x = k_2 z i_x \tag{29.38}$$

Combining (29.37) and (29.38), we get

$$E_x = k z i_x \tag{29.39}$$

where $k = k_1 k_2$.

Substituting (29.39) in (29.36), we obtain

$$\frac{d^2 E_x}{dx^2} = \left(\frac{R_L 2\pi r}{kz}\right) E_x \tag{29.40}$$

For an infinite pipeline, this differential equation has the solution

$$E_x = E_A \exp\left[-\left(\frac{2\pi r R_L}{kz}\right)^{1/2} x\right] \tag{29.41}$$

for the boundary conditions $E_x = 0$ at $x = \infty$, and $E_x = E_A$ at $x = 0$.

On the other hand, if the length of pipeline to be protected is $a/2$, or half the distance to the next point of bonding (see Fig. 29.6), and the potential E_x at $a/2 = E_B$, it follows that current in the pipeline at $a/2 = 0$, or $(dE_x/dx)_{x=a/2} = 0$. Introducing this boundary condition, we obtain

$$E_x = E_B \cosh\left[\left(\frac{2\pi r R_L}{kz}\right)^{1/2}\left(x - \frac{a}{2}\right)\right] \tag{29.42}$$

and

$$E_A = E_B \cosh\left[-\left(\frac{2\pi r R_L}{kz}\right)^{1/2}\frac{a}{2}\right] \tag{29.43}$$

The current in an infinite pipeline at any point x is given by substituting the derivative of (29.41) into (29.35), or

$$I_x = \left(\frac{2\pi r R_L}{kz}\right)^{1/2}\frac{E_A}{R_L}\exp\left[-\left(\frac{2\pi r R_L}{kz}\right)^{1/2} x\right] \tag{29.44}$$

The total current at $x = 0$ (sometimes called drainage current) multiplied by 2, to account for current flowing from either side of the pipeline to the point of bonding, equals

$$I_A = \frac{2E_A}{R_L}\left(\frac{2\pi r R_L}{kz}\right)^{1/2} \tag{29.45}$$

Similarly, for a finite pipe, we have

$$I_A = \left(\frac{2E_B}{R_L}\right)\left(\frac{2\pi r R_L}{kz}\right)^{1/2} \sinh\left[\frac{a}{2}\left(\frac{2\pi r R_L}{kz}\right)^{1/2}\right] \tag{29.46}$$

29.5 DERIVATION OF THE EQUATION FOR POTENTIAL DROP ALONG THE SOIL SURFACE CREATED BY CURRENT ENTERING OR LEAVING A BURIED PIPE

For a deeply buried pipe,

$$i = \frac{j}{2\pi R} \tag{29.47}$$

But for a pipe buried h cm below the soil surface, less current flows in the h direction toward the soil surface than in other directions. The computation for current flow in this situation is approximated by assuming an image pipe located h cm above the soil surface that supplies an equal current j (Fig. 29.7). Then the component of current density along the soil surface equals

$$i_y = \frac{2j}{2\pi R}\frac{y}{R} \tag{29.48}$$

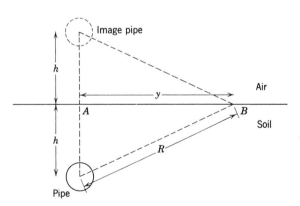

i = current density in soil at distance R from pipe
j = total current entering or leaving pipe per unit length
$\Delta\phi$ = potential difference between A and B at right angle to pipe
ρ = soil resistivity

Figure 29.7. Sketch of buried pipe at which current enters or leaves, causing potential drop along soil surface.

The same result can be obtained by assuming half-cylindrical current distribution in the soil multiplied by the factor y/R to account for zero current flowing to the pipe in the h direction ($y = 0$) and an increasing current density i_y as distance y increases.

By Ohm's law, the potential gradient along the soil surface is

$$\frac{d\phi}{dy} = \frac{j}{\pi R}\frac{y\rho}{R} \tag{29.49}$$

Since $R^2 = h^2 + y^2$, we have

$$\Delta\phi = \int_0^y \frac{\rho j}{\pi R^2}\, y\, dy = \int_0^y \frac{\rho j}{\pi}\frac{y}{h^2 + y^2}\, dy = \frac{\rho j}{2\pi}\ln\frac{h^2 + y^2}{h^2} \tag{29.50}$$

If we take $y = 10$ times distance h, or $y = 10h$,

$$\Delta\phi = \frac{\rho j}{2\pi}2.3\log 101 = 0.734\rho j \tag{29.51}$$

where $\Delta\phi$ is in volts, ρ in Ω-cm, and j in A/cm. [See also, R. Howell, *Corrosion* **8**, 300 (1952).]

29.6 DERIVATION OF THE EQUATION FOR DETERMINING RESISTIVITY OF SOIL BY FOUR-ELECTRODE METHOD

Two metal electrodes, A and B, are at a distance $3a$ apart. Two reference electrodes, C and D, are located distance a apart and also a from A and B, respectively (see Fig. 29.8).

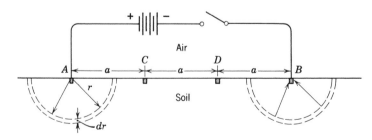

I = total current in battery circuit
r = distance from electrodes A or B
ρ = soil resistivity
$\Delta\phi$ = measured potential difference between electrodes C and D

Figure 29.8. Sketch of four-electrode arrangement for determining soil resistivity.

Assuming hemispherical symmetry at distance r from A or B, current density in the soil $= I/2\pi r^2$. By Ohm's law,

$$\frac{d\phi}{dr} = \frac{I}{2\pi r^2}\rho \tag{29.52}$$

$$\Delta\phi = \int_{r_1}^{r_2} \frac{I\rho}{2\pi r^2}\,dr = \frac{-I\rho}{2\pi}\left(\frac{1}{r_2} - \frac{1}{r_1}\right) \tag{29.53}$$

Because current leaves A, the potential difference $\Delta\phi_1$ between C and D is

$$\Delta\phi_1 = \frac{-I\rho}{2\pi}\left(\frac{1}{2a} - \frac{1}{a}\right) = \frac{I\rho}{4\pi a} \tag{29.54}$$

Similarly, the potential difference $\Delta\phi_2$ between C and D because of current entering electrode B is $I\rho/4\pi a$. Hence,

$$\Delta\phi_{\text{total}} = \Delta\phi_1 + \Delta\phi_2 = \frac{I\rho}{2\pi a} \quad \text{or} \quad \rho = 2\pi a\frac{\Delta\phi}{I} \tag{29.55}$$

[See also S. Ewing and J. Hutchinson, *Corrosion* **9**, 221 (1953).]

29.7 DERIVATION OF THE EQUATION EXPRESSING WEIGHT LOSS BY FRETTING CORROSION

Assume n asperities or contact points per unit area of metal (or oxide) surface which, for mathematical convenience, are circular in shape. The average diameter of asperities is c and the average distance apart is s (Fig. 8.21, Section 8.7.1). In the fretting process, the asperities move over a plane surface of metal at linear velocity υ, with each asperity plowing out a path of clean metal of width averaging c and of length depending on distance of travel. Behind each asperity on the track of clean metal, gas from the atmosphere adsorbs rapidly, followed in time by formation of a thin oxide film. The next asperity, moving in the same path as the first, scrapes the oxide film off and leaves, in turn, a track of clean metal behind. The average time during which oxidation occurs is t. Then

$$t = \frac{s}{\upsilon} \tag{29.56}$$

The corresponding amount of oxide W removed by one asperity plowing out a path l long and c wide depends on the amount of oxide formed in time t. For thin-film oxidation, the logarithmic equation is obeyed (see Section 11.4.1):

$$W = clk \ln\left(\frac{t}{\tau} + 1\right) \qquad (29.57)$$

where τ and k are constants.

Preliminary to oxidation, we can also consider the situation of oxygen adsorbing rapidly as physically adsorbed gas, followed by conversion at a slower rate to chemisorbed oxygen atoms. The chemisorbed oxygen, in turn, reacts with underlying metal to form metal oxide, a reaction that is activated mechanically by asperities moving over the metal surface. Chemisorption limits the amount of oxide that is formed in such a process, the rate of chemisorption following an equation identical in form to that of (29.57) [6]. Hence, whichever process applies, the form of the final equation is essentially the same.*

Substituting (29.56) into (29.57), we obtain

$$W = clk \ln\left(\frac{s}{\upsilon\tau}\right) + 1 \qquad (29.58)$$

Assuming that relative motion of the two surfaces is sinusoidal, $2l$ is the total length of travel in any one cycle and x, the linear displacement from the midpoint of travel at time t, is given by

$$x = \frac{l}{2}\cos\theta \qquad (29.59)$$

and

$$\frac{dx}{dt} = \upsilon = -\frac{l}{2}\sin\theta\frac{d\theta}{dt} \qquad (29.60)$$

If f represents constant linear frequency, this is related to constant angular velocity by the expression

$$\frac{d\theta}{dt} = 2\pi f$$

$$(29.61)$$

Therefore, the average velocity is given by

---

*This argument is not compelling, because the logarithmic term is eventually expanded and only the first term is used. This is equivalent to a linear rate of oxidation or of gas adsorption with time. A linear rate of gas adsorption suggests that the amount of oxygen reaching the clean metal surface as physically adsorbed gas may actually be controlling, rather than its conversion to chemisorbed oxygen atoms. This possibility is given support by the observed increase of fretting weight loss as the temperature is lowered, corresponding to increased rate and extent of physical adsorption at lower temperatures. The rate of chemisorption, on the other hand, usually decreases as the temperature is lowered.

$$\overline{V} = -\frac{\pi l f \int\limits_0^\pi \sin\theta d\theta}{\pi} = 2lf \tag{29.62}$$

Hence, for n contacts or asperities per unit area of interface, weight loss W per cycle caused solely by oxidation is

$$W_{corr} = 2nlck \ln\left(\frac{s}{2lf\tau} + 1\right) \tag{29.63}$$

To this must be added loss of metal by wear because each asperity, on the average, digs below the oxide layer and dissipates metal in an amount proportional to the area of contact of the asperities and the length of travel. The area of asperity, rather than the width, is important now because of the "tearing out" or welding action taking place during mechanical wear, in contrast to scraping off of chemical products from the surface, as discussed previously. A shearing off of asperities without welding also leads to wear that depends on total area of contact. For n circular asperities, weight loss per cycle is given by

$$W_{mechanical} = 2k'n\left(\frac{c}{2}\right)^2 \pi l \tag{29.64}$$

But $n\pi(c/2)^2$, the total area of contact, is equal [7] to the load L divided by the yield pressure p_m. The term p_m is approximated by three times the elastic limit; hence, for mild steel, p_m equals $100\,kg/mm^2$ or $140,000\,psi$.

Therefore,

$$W_{mechanical} = 2k'\frac{lL}{p_m} = k_2 lL \tag{29.65}$$

where k_2 is a constant equal to $2k'/p_m$. The total wear or metal loss per cycle is the sum of the oxidation, or corrosion, term and the mechanical term

$$W_{total} = W_{corr} + W_{mechanical} \tag{29.66}$$

Returning to (29.63), the logarithmic term can be expanded according to

$$\ln(x+1) = x - \frac{x^2}{2} + \frac{x^3}{3} - \cdots$$

where x is equal to $s/(2lf\tau)$. When the latter expression is much smaller than unity, the square and higher terms can be omitted. This condition applies particularly to high loads (small values of s), high-frequency f, and large value of slip l. Empirical values of the constant τ for iron range from 0.06 to 3s. Assuming reasonable values for $\tau = 0.06\,s$, $f = 10\,cps$, $l = 0.01\,cm$, and s (distance between asperities) $= 10^{-4}$ cm, then $s/(2lf\tau) = 0.008$. Therefore, whenever experimental conditions approximate those cited and higher terms of the logarithmic expansion can be neglected, we obtain

$$W_{\text{corr}} = \frac{ncks}{f\tau} \qquad (29.67)$$

This expression is equivalent to assuming, from the very start, a linear rate of oxidation or of gas adsorption on clean iron, where k/τ is the reaction-rate constant. The linear rate reasonably approximates the actual state of affairs for very short times of adsorption or oxidation.

Since the number of asperities along one edge of unit area is equal to $\sqrt{n}$, it follows that $s + c$ is approximated by $1/\sqrt{n}$. Also, recalling that $n\pi(c/2)^2 = L/p_m$, the terms n, c, and s can be eliminated from (29.67), or

$$W_{\text{corr}} = \frac{k_0 L^{1/2}}{f} - \frac{k_1 L}{f} \qquad (29.68)$$

where

$$k_0 = \frac{2}{\sqrt{p_m\pi}} \frac{k}{\tau} \quad \text{and} \quad k_1 = \frac{4}{p_m\pi} \frac{k}{\tau}$$

Combining (29.65), (29.66), and (29.68), we have the final expression for fretting as measured by weight loss corresponding to a total of C cycles:

$$W_{\text{total}} = (k_0 L^{1/2} - k_1 L)\frac{C}{f} + k_2 lLC \qquad (29.69)$$

29.8 CONVERSION FACTORS

Multiply millimeters penetration per year (mm/y) by $2.74 \times$ density (g/cm³) to obtain grams per square meter per day (gmd).
Multiply gmd by 0.365/density to obtain mm/y.

Metal	Density (g/cm³)	2.74 × Density	0.365/Density
Aluminum	2.70	7.40	0.135
Brass (red)	8.75	24.0	0.0417
Brass (yellow)	8.47	23.2	0.0431
Cadmium	8.65	23.7	0.0422
Columbium (niobium)	8.57	23.5	0.0426
Copper	8.96	24.6	0.0407
Copper–Nickel (70–30)	8.95	24.5	0.0408
Iron	7.87	21.6	0.0464
Iron–silicon (Duriron) (84–14.5)	7.0	19.2	0.0521

Metal	Density (g/cm^3)	2.74 × Density	0.365/Density
Lead (chemical)	11.35	31.1	0.0322
Magnesium	1.74	4.77	0.210
Nickel	8.90	24.4	0.0410
Nickel–copper (Monel) (70–30)	8.84	24.2	0.0413
Silver	10.49	28.7	0.0348
Tantalum	16.6	45.5	0.0220
Tin	7.30	20.0	0.0500
Titanium	4.51	12.4	0.0809
Zinc	7.13	19.5	0.0512
Zirconium	6.49	17.8	0.0562

Multiply inches penetration per year (ipy) by 696 × density to obtain milligrams per square decimeter per day (mdd). Multiply mdd by 0.00144/density to obtain ipy.

29.8.1 Additional Conversion Factors

Multiply	by	to obtain
ipy	25.4	mm/y
mdd	0.1	gmd
mA/cm^2	10	A/m^2
A/m^2	0.093	A/ft^2
psi	6.89×10^3	Pa (pascals)
Å	0.1	nm (nanometer)

29.8.2 Current Density Equivalent to a Corrosion Rate of 1 gmd

$$1\,\text{gmd} = 1.117n/W \text{ amperes per square meter}$$

where W is the gram atomic weight[a]

Reaction	A/m^2
$Al \rightarrow Al^{3+} + 3e^-$	0.124
$Cd \rightarrow Cd^{2+} + 2e^-$	0.0199
$Cu \rightarrow Cu^{2+} + 2e^-$	0.0352
$Fe \rightarrow Fe^{2+} + 2e^-$	0.0400
$Fe \rightarrow Fe^{3+} + 3e^-$	0.0600
$Pb \rightarrow Pb^{2+} + 2e^-$	0.0108
$Mg \rightarrow Mg^{2+} + 2e^-$	0.0919
$Ni \rightarrow Ni^{2+} + 2e^-$	0.0381
$Sn \rightarrow Sn^{2+} + 2e^-$	0.0188
$Zn \rightarrow Zn^{2+} + 2e^-$	0.0342

[a]Multiply A/m^2 by 0.1 to obtain mA/cm^2.

29.9 STANDARD POTENTIALS

$25°C$, see also Section 3.8.

	$\phi°$ (volts)
$FeO_4^{2-} + 8H^+ + 3e^- \rightarrow Fe^{3+} + 4H_2O$	1.9
$Co^{3+} + e^- \rightarrow Co^{2+}$	1.82
$PbO_2 + SO_4^{2-} + 4H^+ + 2e^- \rightarrow PbSO_4 + 2H_2O$	1.685
$NiO_2 + 4H^+ + 2e^- \rightarrow Ni^{2+} + 2H_2O$	1.68
$Mn^{3+} + e^- \rightarrow Mn^{2+}$	1.51
$PbO_2 + 4H^+ + 2e^- \rightarrow Pb^{2+} + 2H_2O$	1.455
$Cl_2 + 2e^- \rightarrow 2Cl^-$	1.3595
$Cr_2O_7^{2-} + 14H^+ + 6e^- \rightarrow 2Cr^{3+} + 7H_2O$	1.33
$O_2 + 4H^+ + 4e^- \rightarrow 2H_2O$	1.229^a
$Br_2(l) + 2e^- \rightarrow 2Br^-$	1.0652
$Fe^{3+} + e^- \rightarrow Fe^{2+}$	0.771
$O_2 + 2H^+ + 2e^- \rightarrow H_2O_2$	0.682
$I_2 + 2e^- \rightarrow 2I^-$	0.5355
$O_2 + 2H_2O + 4e^- \rightarrow 4OH^-$	0.401
$Hg_2Cl_2 + 2e^- \rightarrow 2Hg + 2Cl^-$	0.2676
$AgCl + e^- \rightarrow Ag + Cl^-$	0.222
$SO_4^{2-} + 4H^+ + 2e^- \rightarrow H_2SO_3 + H_2O$	0.17
$Cu^{2+} + e^- \rightarrow Cu^+$	0.153
$Sn^{4+} + 2e^- \rightarrow Sn^{2+}$	0.15
$AgBr + e^- \rightarrow Ag + Br^-$	0.095
$Cu(NH_3)_2^+ + e^- \rightarrow Cu + 2NH_3$	−0.12
$Ag(CN)_2^- + e^- \rightarrow Ag + 2CN^-$	−0.31
$PbSO_4 + 2e^- \rightarrow Pb + SO_4^{2-}$	−0.356
$HPbO_2^- + H_2O + e^- \rightarrow Pb + 3OH^-$	−0.54
$2H_2O + 2e^- \rightarrow H_2 + 2OH^-$	−0.828
$Zn(NH_3)_4^{2+} + 2e^- \rightarrow Zn + 4NH_3$	−1.03

^{a}At pH 7, 0.2 atm O_2, $\phi = 0.81$ V.

Source: Data from *Oxidation Potentials*, W. Latimer, Prentice-Hall, Englewood Cliffs, NJ, 1952 (with permission).

29.10 NOTATION AND ABBREVIATIONS

activity	a
activity coefficient	γ
ampere	A
angstrom	Å
atmosphere	atm
atom percent	at.%
calorie	cal

centimeter	cm
corrosion current	I_{corr}
corrosion current density	i_{corr}
corrosion potential	ϕ_{corr}
corrosion potential of a galvanic couple	ϕ_{galv}
coulomb	C
critical crack depth	a_{cr}
current	I
current density	i
decimeter	dm
density	d
electromotive force (emf)	E
equivalent	eq
exchange current density	i_0
faraday	F
foot	ft
Flade potential	ϕ_F
gallon	gal
Gibbs free energy	G
gram	g
gram calorie	cal
grams per square meter per day	gmd
hour	h
inch	in.
inch penetration per year	ipy
joule	J
kilogram	kg
kilometer	km
liter	L
mean ion activity coefficient	$\gamma_\pm$
megapascal	MPa
meter	m
microampere	μA
micrometer (micron)	μm
milliampere	mA
milligram	mg
milligrams per square decimenter per day	mdd
milliliter	mL
millimeter	mm
millimeters penetration per year	mm/y
millivolt	mV
minute	min
molal solution	M
nanometer	nm
normal solution	N

ohm	Ω
ounce	oz
overpotential	η
overvoltage	ε
parts per million	ppm
pascal	Pa
potential	ϕ
potential on the standard hydrogen scale	ϕ_H, ϕ(S.H.E.)
pound	lb
pound per square inch	psi
pressure	p
relative humidity	R.H.
saturated calomel electrode	S.C.E.
second	s
standard hydrogen electrode	S.H.E.
stress-corrosion cracking	S.C.C.
Tafel slope	β
volt	V
weight percent	wt.%
yield strength	Y.S.

REFERENCES

1. F. Mansfeld and K. Oldham, *Corros. Sci.* **11**, 787 (1971); F. Mansfeld, in *Advances in Corrosion Science and Technology*, Vol. 6, M. Fontana and R. Staehle, editors, Plenum Press, New York, 1976, p. 163.
2. C. Wagner and W. Traud, *Z. Elektrochem.* **44**, 391 (1938).
3. W. Langelier, *J. Am. Water Works Assoc.* **28**, 1500 (1936).
4. C. Hoover, *J. Am. Water Works Assoc.* **30**, 1802 (1938).
5. S. Powell, H. Bacon, and J. Lill, *Eng. Chem.* **37**, 842 (1945).
6. F. Stone, in *Chemistry of the Solid State*, W. Garner, editor, Butterworths, London, 1955, p. 385.
7. F. P. Bowden and D. Tabor, *The Friction and Lubrication of Solids*, Oxford University Press, New York, 1950.

INDEX

KEY: *T* refers to tables. *F* refers to figures.

Corrosion and Corrosion Control, by R. Winston Revie and Herbert H. Uhlig
Copyright © 2008 John Wiley & Sons, Inc.